Pesticide Risk in
GROUNDWATER

Edited by
Marco Vighi
Istituto di Entomologia Agraria
Milan, Italy
and
Enzo Funari
Istituto Superiore di Sanità
Rome, Italy

LEWIS PUBLISHERS

Boca Raton   New York   London   Tokyo

Library of Congress Cataloging-in-Publication Data

Pesticide risk in groundwater / edited by Marco Vighi and Enzo Funari.
    p. cm.
    Includes bibliographical references and index.
    ISBN 0-87371-439-3
    1. Pesticides—Environmental aspects. 2. Groundwater—Pollution.
 I. Vighi, M. II. Funari, E.
 TD427.P35P447   1995
 363.73′84—dc20   94-42771
                                                           CIP

© 1995 by CRC Press, Inc.
Lewis Publishers is an imprint of CRC Press

No claim to original U.S. Government works
International Standard Book Number 0-87371-439-3
Library of Congress Card Number 94-42771
Printed in the United States of America 1 2 3 4 5 6 7 8 9 0
Printed on acid-free paper

Preface

The use of synthetic pesticides, starting in the 1940s, resulted in a huge improvement in pest control with enormous consequences not only for quantitative and qualitative agricultural productivity but also for human health through the control of the vectors of several serious diseases. There is no doubt that the recent improvement in welfare in many areas of the world is, at least in part, a result of pesticide usage. On the other hand, since the 1950s, enthusiasm for these successes has been counterbalanced by the growing awareness of the unwanted consequences of pesticide use and of the risks for man and the environment. Thus, in the last few decades, the need for and appropriate management of pesticides has been recognized, either by the scientific community or by the administrative authorities, in order to make the benefits produced by pesticides consistent with the protection of the environment and human health.

Pesticides are substances intentionally designed for their toxicity to living organisms. Therefore, even if their toxic activity should be specifically directed against some target pests and growing efforts are devoted to produce compounds with even more specific action, adverse effects on nontarget organisms can hardly be avoided.

Risk to man from pesticides can occur through various exposure routes. Undoubtedly, the highest risks, and those most easily supported by documentary evidence, occur to persons directly exposed to contact with these compounds, i.e., workers in the chemical industry and, especially, farmers. These kinds of effects can be quantified by means of occupational health studies and can be controlled through the adoption of suitable security measures. Another source of pesticide risk is represented by residues in treated agricultural products or stored foodstuffs.

Acceptable Daily Intakes (ADI) of pesticides for man have been developed and updated annually by the FAO/WHO Joint Meeting on Pesticide Residues (JMPR).[1] The JMPR also establishes the Maximum Residue Limits (MRL) of pesticides in agricultural products, in accordance with Good Agricultural Practice (GAP). On these bases, pre-sale pesticide treatments in most countries are strictly regulated and the observance of treatment criteria help avoid or at least minimize the risk for man due to pesticide residues in food.

Finally, a third component of pesticide risk is due to exposure to low levels of pesticides present in various environmental compartments, not in consequence of direct treatment, but as a result of processes of environmental distribution and fate on a more or less wide scale in space and time. Man can be exposed to this kind of risk through air and water pollution and through consumption of living organisms in which pesticides are bioaccumulated. In this case the quantification of the exposure and the evaluation of the risk are much more difficult. They require a multidisciplinary approach based on toxicology, ecotoxicology, and environmental chemistry.

The quantification of exposure requires extensive and reliable monitoring data, which are seldom available, or can be estimated by means of suitable predictive models. Complex processes regulating the biogeochemical cycle of a molecule must be taken into account, together with partition properties, transport patterns and transformation reactions of the substance, and environmental characteristics (climate, type of soil, water balance, etc.). On the other hand, the study of the effects produced by long-term exposure to low concentrations is more difficult than the determination of acute risk.

Environmental quality objectives should be developed in order to define acceptable loads on the territory. This should be considered along with agronomic needs, and social and economic evaluations should be made to plan a suitable strategy of control and a fitting pesticide management and land use for the protection of man and the environment.

Pesticide pollution of groundwater falls in this framework. It is the result of agricultural practices, the properties of the substance and its behavior in the soil environment, as well as the characteristics of aquifers and their vulnerability.

Groundwaters are extremely precious resources and in many countries they represent the most important drinking water supply. They are generally microbiologically pure and, in most cases, they do not need any treatment. On the contrary, a bacterial community is part of the normal biota of surface waters and microbiological contamination is likely to occur. Thus, surface waters must, in any case, be treated before use for human consumption, as also imposed by most national and international standards and guidelines (see, for example, the EEC Directive on the quality of surface waters[2]).

Therefore, groundwaters are, ideally, the best water source for humans. Yet, groundwaters may be exposed to various kinds of contaminations which deteriorate quality and, if used for drinking purposes, represent a threat to human health. In particular, many pesticides may contaminate vulnerable groundwaters because of their extensive use and high leaching potential. Thus, the control of groundwater quality, in compliance with drinking water objectives derived from toxicological properties of chemicals, is a key point for the protection of human health.

Quality objectives have been proposed for many pesticides by the World Health Organization (Guidelines corresponding to a lifespan exposure[3]) and by the U.S. Environmental Protection Agency (Health Advisories for different periods of exposure[4]). These quality objectives do not necessarily coincide with national standards that are substantially based on toxicological quality objectives but take into account other factors of a technological, economical, social, political, and "philosophical" nature. This latter is the case of the standard for pesticides in the EEC Directive on drinking water.[5] Since pesticides are toxic substances by definition, mainly xenobiotics, and that drinking water is a fundamental resource intended for regular daily consumption, the European Community maintained "the general principle that drinking water should not contain any pesticides and that, given the limitations of current analytical techniques the limit of 0.1 µg/l for individual pesticide represents a practical zero value that can be used for enforcement purposes".

Thus, in EEC member states, control actions for the protection of drinking water must have as a final goal not a true toxicological limit but a more stringent "legal" standard. This kind of standard establishes a very important principle in the protection of human health and, conceptually, it must be considered as a regulatory decision to attain the final goal of a clean environment. Nevertheless, from a practical point of view, compliance with the EEC standard requires a long-term integrated strategy of management of pesticides and land use.

In fact, to assess the potential threat to European drinking water resources, The Scientific Advisory Committee on Toxicity and Ecotoxicity (CSTE) of the EEC stressed the urgent need for more information, including:

- Patterns of pesticide usage
- Studies of pesticide behavior in soils
- Effects of hydrological conditions and soil characteristics on pesticide behavior in the environment
- Behavior and fate of pesticides in groundwater
- Efficiency of water treatment techniques for the removal of pesticide residues

Nevertheless, one serious shortcoming of the EEC Directive is the absence of a legal tool that should be applied to achieve the "practical zero" standard. Only recently a subsequent directive concerning the placing of plant protection products on the market[6] states a series of qualifications for a compound in order to be used without producing an unacceptable threat to the environment and human health. In the framework of this directive, the Annex VI (Uniform Principles for the Evaluation of Plant Protection Products) has been proposed including the principles for the protection of groundwater quality against the use of pesticides.

In contrast to a complex integrated strategy of goundwater protection, the easiest and fastest intervention for controlling pesticide pollution is by banning one or more compounds at risk, as in the case of atrazine pollution in Italy. Such a control measure would produce, in general, a series of uncontrolled negative repercussions in the agronomic, economic, and environmental sectors and the remedy would be worse than the problem. This means that interventions aimed at protecting groundwater from pesticide pollution must be carefully planned on the basis of sound interdisciplinary knowledge.

This book is an attempt to provide an overview of the main aspects related to pesticide pollution of groundwater and is divided into five main sections as follows:

Section I — Occurrence in Groundwater and Use of Pesticides

To indicate the extent of the problem of groundwater contamination by pesticides on a worldwide basis, experimental data of groundwater monitoring has been reviewed and discussed. On the basis of this evaluation, herbicides prove to be the class of pesticides of major concern for groundwater. Therefore, agronomic problems and needs are examined in particular for this class of compounds. The need for chemical weed control is explained and herbicide use patterns are described in relation to various crops and agronomic practices. The agronomic consequences of banning a herbicide in the absence of an adequate strategy are examined as well.

Section II — Environmental Distribution and Fate and Exposure Prediction

To evaluate environmental exposure and, in particular, the possibility of the occurrence in groundwater, the distribution patterns of a substance in the environment and the characteristics of the soil and groundwater compartments should be carefully studied. In this part, predictive approaches to estimate the distribution and fate of pesticides are described and a chapter is devoted to hydrogeological aspects affecting the vulnerability of aquifers.

Section III — Pesticides and Human Health

Pesticides are evaluated in relation to their toxicology. The criteria and procedures by WHO and the U.S. EPA to define quality objectives are critically described. A comparison is made between monitoring data on pesticides in groundwaters and their quality objectives.

Section IV — Possible Solutions

Various strategies to control groundwater pollution problems are evaluated first by examining *a posteriori* interventions to be adopted in emergency situations. Different water treatment options are described from technical and economic points of view.

On the other hand, the most important rules of a strategy of pesticide management for the prevention of groundwater pollution are described. The main lines of action are the following:

- **The chemical approach:** the methods for the selection of environmental friendly active ingredients and the industrial strategy to improve environmental behavior are described;
- **The agronomic approach:** the perspectives of different alternatives to chemical pest management are discussed;
- **The land use approach:** examples of agricultural land planning as a tool for comprehensive, long range environmental protection are given.

Section V — Legislation and Economy

A review of the state-of-the-art of drinking water regulations in the EEC, USA, and other OECD countries is given. A comparative evaluation of different possible strategies for the regulatory management of pesticides at different time and space scales is proposed. Finally, the economic implications of groundwater pollution and its control are described and exemplified with a real case study.

REFERENCES

1. "Pesticides Residues in Food" Reports of the Joint Meeting of the FAO Panel of Experts on Pesticide Residues and the WHO Expert Group on Pesticide Residues, FAO Plant Production and Protection Series, Rome.
2. EEC Council Directive 75/440 of 16 June 1975 Concerning the quality of surface waters intended for the abstraction of drinking water in Member States. O.J.N.L 194, 25/77 (1975) pp. 34-39.
3. WHO *Guidelines for drinking water quality. Vol. 1. Recommendations* (Geneva: World Health Organisation, 1984) pp. 130.
4. "Drinking Water Regulations and Health Advisories by Office of Water", U.S. Environmental Protection Agency, Washington, DC (1992).
5. EEC Council Directive 80/778 of 15 July 1980 Relating to the quality of water intended for human consumption. O.J.N.L 229, 30/87 (1980) pp. 11-29.
6. EEC Council Directive 91/414 Concerning the placing of plant protection products on the market. O.J.N.L 230/1 (1991) pp. 1-32.

The Editors

Marco Vighi is a professor of Agricultural Ecotoxicology at the Faculty of Agriculture of the University of Milan. He received his degree at the University of Milan. From 1969 to 1983 he worked at the Water Research Institute of the Italian National Research Council. He is currently involved in research on environmental risk assessment of organic contaminants, particularly pesticides.

Since 1991 he has been designated a member of the Scientific Advisory Committee on Toxicology and Ecotoxicology (CSTE) by the European Economic Community. He has also taken part in or is presently a member of working groups and consultation meetings promoted by several international organizations (OECD, FAO/ UNEP, WHO, European Science Foundation) in the field of environmental pollution and risk assessment.

Professor Vighi has authored over 100 scientific publications on several aspects of environmental contamination, from eutrophication of surface waters to environmental toxicology.

Enzo Funari is a senior researcher in the Department of Environmental Hygiene at the National Institute of Health in Italy. He received his degree in biological sciences from the University of Rome in 1977. He worked in the Pesticide Unit of the National Institute of Health until 1979 and then in the Biochemical Toxicology area until 1987. Since then he has been with the Department of Environmental Hygiene.

For five years he was responsible for and carried out research programs regarding the risk of groundwater contamination within the framework of collaborations with the Italian Ministry of the Environment. Between 1988 and 1992, he collaborated with the World Health Organization in a review of the guidelines for drinking water quality. The first volume of these guidelines were published in 1993. As a product of his work in this discipline, he has authored and co-authored many scientific papers and research reports.

Contributors

Vincenzo Angileri
Institute of Agricultural Engineering
University of Milano
Milano, Italy

Lars Bergman
Department of Economics
Stockholm School of Economics
Stockholm, Sweden

Antonio Berti
Institute of Agricultural Entomology
University of Milano
Milano, Italy

Antonio Di Guardo
Institute of Agricultural Entomology
University of Milano
Milano, Italy

Loredana Donati
Istituto Superiore di Sanità
Laboratorio di Igiene Ambientale
Roma, Italy

Enzo Funari
Istituto Superiore di Sanità
Laboratorio di Igiene Ambientale
Roma, Italy

Giuseppe Giuliano
Water Research Institute (IRSA)
National Research Council
Rome, Italy

David I. Gustafson
Monsanto
St. Louis, Missouri

Per-Olov Johansson
Department of Economics
Stockholm School of Economics
Stockholm, Sweden

Costantino Nurizzo
Politecnico di Milano
D.I.I.A.R. Sez. Ambientale
Milano, Italy

Guido Premazzi
European Union
Joint Research Centre
Ispra, Italy

Donatello Sandroni
Institute of Agricultural Entomology
University of Milano
Milano, Italy

Maurizio Sattin
C. N. R.
Centro di Studio Sulla Biologia E. Controllo
 Delle Piante Infestanti
Padova, Italy

Tore Söderqvist
Department of Economics
Stockholm School of Economics
Stockholm, Sweden

Alessandro Toccolini
Institute of Agricultural Engineering
University of Milano
Milano, Italy

Marco Vighi
Institute of Agricultural Entomology
University of Milano
Milano, Italy

Giuseppe Zanin
Istituto di Agronomia
Universita di Padova
Padova, Italy

Giuliano Ziglio
Department of Civil and Environmental
 Engineering
University of Trento
Trento, Italy

Contents

Section V
Legislation and Economy

Section I
Occurrence in Groundwater
and
Use of Pesticides

CHAPTER 1

Pesticide Levels in Groundwater: Value and Limitations of Monitoring

Enzo Funari, Loredana Donati, Donatello Sandroni, and Marco Vighi

CONTENTS

INTRODUCTION

Even if the environmental risk from pesticides was recognized relatively early, the occurrence of pesticides in groundwater was detected much later. The high concern for pesticide risk in the 1960s was mainly related to DDT and other organochlorinated compounds. They were banned or severely restricted in all developed countries at the beginning of the 1970s. As for other less persistent pesticides, the general belief was that these compounds could not leach into groundwater under normal conditions. This perception was mainly due to a lack of knowledge about the important features of chemicals movement through the soil.[1]

One of the first references to the discovery of pesticides other than chlorinated hydrocarbons in groundwater was by Richard et al.[2] In the late 1970s the number of detections rapidly increased, as well as the concern of the public authorities regarding the problem of groundwater contamination by these kinds of chemicals.

More or less systematic monitorings have been planned all over the world since the early 1980s, demonstrating the extent of the problem and helping to understand many factors regulating the leaching capability of pesticides and the vulnerability of groundwater resources. In this chapter some examples of the information available on pesticide occurrence in groundwater will be described and discussed. The results of a worldwide literature review on this topic will be provided to give a picture of the present knowledge about groundwater contamination and to evaluate the value and limitations of this approach for the collection of monitoring data. Moreover, the information obtained from a systematic survey carried out on a national basis will be described and compared with the former.

Finally, the role of various pesticide classes in groundwater contamination will be evaluated.

LITERATURE REVIEW OF PESTICIDE OCCURRENCE IN GROUNDWATER

DESCRIPTION AND RESULTS

The review has been made mainly through a survey conducted on specialized scientific journals from 1987 to the first few months of 1993. In addition, some gray documents elaborated by national agencies in different countries and data from other relevant reviews have also been used. In these latter cases, the reviews can contain data obtained before 1987. For each pesticide, in addition to the data on their levels, further information has been gathered whenever possible to include the geographical location of the data (country and region), some description of the pedology and hydrogeology of the area examined as well as the groundwater depth, the period of sampling, the extent of the surveys examined, and/or the ratio of the positive results.

The complete results of the review are reported in the Appendix to this chapter. A synthesis of the results for the different groups of pesticides (herbicides, insecticides, nematicides and fungicides) is shown in Tables 1 to 3 and for some degradation products in Table 4.

All the detected active ingredients are reported in the tables, divided into chemical categories. For each compound, the number of data sets is reported, together with the range of mean and maximum measured concentrations. If available, the percentage of positive frequency, its range among different studies, and the depth of sampled groundwater are also reported. The percentage of positive frequency has been calculated only for the compounds for which a sufficient number of wells were analyzed. In some studies, only the positive percentage was reported, without the total number of analyses performed. In other cases, positive frequency was not indicated. Thus, the information reported in the tables must be taken as partial and approximate.

Herbicides represent the main class of pesticides responsible for groundwater contamination. Indeed, 32 herbicides have been reported to be present in groundwater, of which 29 at maximum levels above 0.1 µg/L (Table 1 and Tables 1 to 16 in the Appendix). Among these substances, triazines cause the most concern. Nine triazines have been reported to be contaminants of groundwater. The data set for these herbicides ranges from only one study for hexazinone, prometon, and terbutryne to 41 for atrazine. A large data set is also available for cyanazine, metribuzin, and simazine, while only a few investigations have been conducted on propazine and terbuthylazine, and these are probably insufficient for defining their occurrence in groundwater. Other categories of concern are amides and phenoxy derivatives, even if the data set for these herbicides is not as large as that available for triazines.

Among single herbicides, atrazine shows the highest degree of frequency in the contamination of groundwater (38% of positive frequency), with the largest number of data sets (41) followed by alachlor (10% of positive frequency out of 33 data sets). High frequency (20%) is also shown by molinate but with a relatively low number of data sets (4).

The range of concentrations for most active ingredients is extremely wide. For some compounds, maximum values detected in the different data sets may range over more than four orders of magnitude (from less than 0.1 to more than 1000 µg/L). Extremely high values, of some hundreds of µg/L, should be taken, in general, as exceptional events, probably due in some cases to pollution of groundwater body wells from point sources instead of leaching from agricultural areas in normal use conditions.[3-5] In other cases, extremely high figures can be typical of low depth groundwater in permeable soils, in relation to treated areas and to treatment periods. In general, for all compounds for which a relatively high number of data is available the most frequent values of

Table 1   Synthesis of the Results of the Review on Herbicide Occurrence in Groundwater

Chemical Category	Herbicide	Number of Data Set[a]	Range of Groundwater Depth (m)	Positive Frequency[b]	Range of Positive Frequencies	Range of Mean Levels (μg/L)	Range of Maximum Levels (μg/L)
Amides	Alachlor	33	1 – <60	10%	<1 – 25%	0.03 – 9.4	>0.02 – 3000
"	Metolachlor	20	0.8 – <60	6%	<1 – 15%	0.4 – 0.92	0.1 – 1800
"	Propachlor	4		<1%	<1 – 4%	0.11 – 0.4	0.28 – 2.98
"	Propanil	1	<15				1.23
Benzoic acid derivatives	Chlorthal-dimethyl	7	3 – <60	≤1%	<1 – 60%	0.02 – 109	0.03 – 1039
"	Dicamba	7			<1 – 10%	0.17 – 0.6	1.1 – 320
"	TBA	1				12.5	79
Benzonitriles	Dichlobenil	1	0.6 – 1.6				0.05
Chloroaliphatic acids	TCA	1	0.85 – 1			670	1270
Diazines	Bentazone	9	2 – >50	2%	1 – 38%	0.12 – 0.2	0.02 – 40
"	Bromacil	8	4 – 40	6%	4 – 19%	0.2 – 30	1.8 – 1250
Dinitroanilines	Pendimethalin	2			1 – 2%	0.06	<1 – 0.12
"	Trifluralin	9		2%	<1 – 12%	0.03 – 5.6	0.02 – 14.9
Phenolic derivatives	Dinoseb	9	1 – 4		3 – 9%	0.07 – 0.09	>0.1 – 2100
Phenoxyderivatives	2,4-D	6		6%	<1 – 85%	0.2 – 1.4	0.2 – 49.5
"	MCPA	6	12 – <15	≤1%	<1 – 4%	0.1 – 0.26	0.02 – 5.5
"	Mecoprop	7	1.6 – 29	0.1%	21%	<0.1	0.03 – 1000
"	2,4,5-T	3		<1%	<1%	0.21	0.21 – 17
Pyridine derivatives	Picloram	3		<1%	<1 – 1%	0.16 – 1.4	0.63 – 49
Tiocarbamates	Molinate	4	<15	20%	13 – 21%		6.3 – 154
"	Tiocarbazil	1	<15				0.09
Triazines	Atrazine	41	0.9 – <60	38%	2 – 98%	0.1 – 27	0.11 – 1500
"	Cyanazine	17	1 – <60	3%	<1 – 10%	0.27 – 0.51	0.09 – 130
"	Hexazinone	1				8	9
"	Metribuzin	17	3 – <60	3%	<1 – 17%	0.11 – 1.6	0.03 – 940
"	Prometon	1				16.6	29.6
"	Propazine	2					0.2 – 0.4
"	Simazine	18	0.8 – 12	2%	<1 – 35%	0.03 – 22	0.08 – 35
"	Terbuthylazine	4	0.8 – 1			0.05 – 0.49	0.07 – 1.92
"	Terbutryne	1	4				2.4
Urea derivatives	Isoproturon	2		3%	3%	<0.05	0.08 – 0.1
	Linuron	1				1.9	2.7

a Usually there is one data set per study, but, in some cases, there can be two or more data sets in one study (for instance, referring to different geographic areas).
b This value has been calculated as the ratio between the total positive wells and the total wells, only for compounds for which sufficient data sets are available (either at least 100 wells examined or 3 data sets) on the frequency of contamination and the positive or total number of wells.

Table 2   Synthesis of the Results of the Review on Insecticide and Fungicide Occurrence in Groundwater

Chemical Category	Compound	Number of Data Set[a]	Range of Groundwater Depth (m)	Positive Frequency[b]	Range of Positive Frequencies	Range of Mean Levels (μg/L)	Range of Maximum Levels (μg/L)
Alicyclic halogenated hydrocarbons	Aldrin	1				0.1	0.1
"	Chlordane[c]	1				1.7	1.8
"	Dieldrin	2				0.01 – 0.02	0.02 – 0.1
"	Endosulfan	1				0.3	0.4
"	Lindane	3				0.04 – 0.1	0.08 – 47
"	Toxaphene[d]	1				3205	4910
Aromatic halogenated hydrocarbons	DDT	1				1.7	402
"	TDE	1				4.8	6.2
Aromatic hydrocarbons	Chlorothalonil[e]	1				0.02	12.6
Carbamates	Aldicarb[f]	15	0.06 – 45	23%	<1 – 40%	9 – 80	>7 – 515
"	Carbofuran[d,f]	9		10%	<1 – 20%	0.02 – 5.3	0.06 – 176
"	Methomyl	1				9	9
"	Oxamyl[f]	5	7 – 23		<1%	<0.1 – 4.3	0.2 – 395
Organophosphate	Diazinon	1				162	478
"	Fonofos	3		2%	<1 – 4%	0.03 – 0.1	0.03 – 0.9
"	Malathion[d]	1				41.5	53
Organophosphate	Methamidophos[d]	1				4.8	105
"	Methyl-parathion	1				88.4	256
"	Parathion[d]	1				0.03	0.04
"	Sulprofos	1				1.4	1.4

a Usually there is one data set per study, but in some cases there can be two or more data sets in one study (for instance, referring to different geographic areas).
b This value has been calculated as the ratio between the total positive wells and the total wells, only for compounds for which sufficient data sets are available (either at least 100 wells examined or 3 data sets) on the frequency of contamination and the positive or total number of wells.
c Also used as fungicide.
d Also used as acaricide.
e Fungicide.
f Also used as nematicide.

Table 3 Synthesis of the Results of the Review on Nematicide Occurrence in Groundwater

Chemical Category	Compound	Number of Data Set[a]	Range of Groundwater Depth (m)	Positive Frequency[b]	Range of Positive Frequencies	Range of Mean Levels (μg/L)	Range of Maximum Levels (μg/L)
Aliphatic halogenated hydrocarbons	DBCP	4		50%		0.01 – 5	0.02 – 1240
"	1,2-Dichloropropane[c]	10	9 – 40		4 – 38%	0.8 – 8.5	4 – 1200
"	1,3-Dichloropropene	8	1 – 24			0.23 – 2530	0.89 – 8620
"	Ethylene dibromide[c]	8	<60	11%	3 – 25%	0.9 – 6.5	0.24 – 140

a Usually there is one data set per study, but in some cases there can be two or more data sets in one study (for instance, referring to different geographic areas).
b This value has been calculated as the ratio between the total positive wells and the total wells, only for compounds for which sufficient data sets are available (either at least 100 wells examined or 3 data sets) on the frequency of contamination and the positive or total number of wells.
c Also used as insecticide.

Table 4   Synthesis of the Results of the Review on the Occurrence of Degradation Products of Pesticides in Groundwater

Products	Parent Compound	Number of Data Set[a]	Range of Groundwater Depth (m)	Positive Frequency[b]	Range of Positive Frequencies	Range of Mean Level (µg/L)	Range of Maximum Level (µg/L)
Aldicarb sulfone	Aldicarb	1	1.1 – 1.6				61
Anthranil acid isopropylamide	Bentazone	2	2 – 5				<0.1 – 0.2
Deethylatrazine	Atrazine	11	0.6 – 5	46%	3.5 – 67%	0.09 – 0.54	0.1 – 7.6
Deethyldeisopropylatrazine	Atrazine, Simazine	2		13%	11 – 67%		0.37 – 1
Deethylterbuthylazine	Terbuthylazine	1	0.8 – 1			0.05	0.07
Deisopropylatrazine	Atrazine, Simazine	7	0.6 – 5	44%	3.4 – 75%	0.01 – 0.68	0.06 – 3.54
3,4-dichloroaniline	Propanil	1	<15				0.3
2,6-dichlorobenzamide	Dichlobenil	2	0.6 – 40			0.3 – 180	
Hydroxyalachlor	Alachlor	1		0.2%		0.91	
3-hydroxycarbofuran	Carbofuran	1		0.4%		0.98	
3-ketocarbofuran	Carbofuran	1		0.4%		0.03	
MITC	Metolachlor	4	0.55 – 23			0.05 – 0.6	

[a] These results refer to wells if not specified differently.
[b] This value has been calculated as the ratio between the total positive wells and the total wells, only for compounds for which sufficient data sets are available (either at least 100 wells examined or 3 data sets) on the frequency of contamination and the positive or total number of wells.

reported maximum concentration range from less than one to few µg/L, with a few exceptions of very high figures (above 100 µg/L).

From the survey, 19 insecticides have been reported as contaminants of groundwater (Table 2 and Tables 17 to 20 in the Appendix), but the data set is not always adequately large. Among the insecticides detected in groundwater, aldicarb and carbofuran are described in a relatively large number of studies and show a high frequency of detection. For these compounds, high figures for the maximum concentrations (some tens or even more than 100 µg/L) are more frequent than for herbicides. The few nematicides and soil disinfectants detected in groundwater (Table 3 and Tables 21 to 24 in the Appendix) show, in general, a high frequency of positive findings and a very wide variability of maximum values, with a frequent occurrence of very high figures.

With the exception of a sporadic study on chlorotalonil (Table 2 and Table 20 in the Appendix), no other data are reported for the occurrence of fungicides in groundwater.

The data on the presence of degradation products of pesticides (Table 4 and Table 25 in the Appendix) generally show they contaminate groundwater at relatively low levels. Some exceptions refer to products such as aldicarb sulfone and 2,6-dichlorobenzamide, for which very few data are available. It is not surprising that the more frequently detected compounds are degradation products of triazines and, in particular, of atrazine.

VALUE AND LIMITATIONS OF THIS REVIEW AND MORE GENERALLY OF LARGE-SCALE REVIEWS

A review like this can give important information about the occurrence of pesticides in water and on the role played by various compounds or by different classes of pesticides in water pollution. Nevertheless, the reliability of such a review of the literature is affected by several limitations depending either on the general procedure of the review itself or on some specific features of the single papers quoted. Some of those limitations can be synthesized in the following questions.

Is the Review Exhaustive and Significantly Representative of the Real Worldwide Situation?

Since the major scientific journals have been reviewed, together with the more relevant sources of gray literature, such as reports of some important national and international agencies and organizations [e.g., U.S. Environmental Protection Agency, U.S. Department of Agriculture, European Economic Communities (EEC) Committees, etc.], it can be hypothesized that the majority of available information has been examined and that the more significant and reliable data have been collected. Nevertheless, a look at the geographic distribution of data collected in the presented review indicates an extreme lack of uniformity. In fact, all studies taken into account have been carried out in developed countries, in particular in North America and West Europe.

Obviously, this does not mean that water pollution is an exclusive problem of North America and Western Europe. Environmental contaminations of Eastern European countries is a problem of extremely high concern. As for developing countries, even if the overall consumption of pesticides is lower than in Europe and North America, their massive use in some intensive agricultural areas is well known.

A more suitable explanation of the dishomogeneous distribution of water pollution data derives from the distribution of frequency of studies, which is affected by several nonscientific reasons (availability of technical and economic resources, concern for environmental problems, political and public pressure, etc.). Moreover, in the case of production of data in developing countries, their availability could sometimes be more problematic. Only sporadically are they published in large diffusion international journals, and the reports of public organizations are of extremely restricted circulation and availability. On the contrary, technical reports produced in North America and EEC countries are generally written in English and are often widely distributed. Thus, a large, worldwide literature review, such as that presented in the previous pages, cannot provide a complete and highly significant picture of the real environmental situation.

Are the Single Papers Suitably Planned in Order to be Representative of a Local Situation?

This aspect is particularly relevant for specific papers published in scientific journals. In general, these kinds of studies are not aimed at producing exhaustive monitoring data. On the contrary, they are planned in order to give precise answers to specific scientific problems. Therefore, often only selected compounds or particular groups of chemical substances are analyzed, in conjunction with the objectives of the research. Moreover, even the study sites could be selected in relation to specific aspects. As a consequence, a literature review based on scientific papers could give a misleading picture of environmental pollution. More suitable to this objective are technical reports produced by public agencies or organizations. Nevertheless, in this case too, it is very important to know the criteria used in planning environmental monitoring. Even these kinds of studies could be planned in consequence of particular problems (e.g., the concern of the public towards some kinds of substances or for some particular site). Often, legal, political, and social aspects could affect the scientific significance of a survey.

Moreover, it is quite impossible, or at least very difficult from a technical and economic point of view, to organize a survey on all pesticides used for agricultural and nonagricultural applications. Thus, a selection must be made, and the criteria for this selection must be carefully evaluated. A significant survey of water pollution should take into account several factors, such as loads and use patterns, environmental characteristics, hydrology and hydrogeology of the site, properties of compounds of interest, etc.

Are the Analytical Methodologies Appropriate and Comparable?

The detection of pesticides in natural waters, and especially in groundwater, requires very sensitive analytical methods, due to the concentration levels, usually very low, of these compounds. In particular, as groundwater could be used as a drinking water supply without any treatment, it must comply with the quality standards set up for human consumption. This means that, for example, in the EEC countries, concentrations as low as 0.1 µg/L or lower must be considered as significant.

For many pesticides, such a low detection limit is not easy to attain or, at least, requires complex methods and sophisticated instruments. In some cases the selection of pesticides to be looked for is conditioned by the availability of suitable analytical methods. Moreover, reporting a pesticide as being "not detected" does not mean it is not present at significant levels and does not have the same value for all compounds. Much attention must be paid to the reporting of detection limits for different chemicals.

It is very difficult to evaluate critically and carefully all these aspects in a wide literature review, since specific observations would be required for each single paper. Thus, as a consequence, some statements deriving from such a review can not be definitely conclusive. For example:

- it is difficult to define if the most frequently detected compounds are really the most commonly occurring in the environment or if they are the most frequently investigated due to other reasons
- it is difficult to evaluate if currently unreported compounds are really not present in water or if they are not looked for due to several reasons (analytical difficulties, low public concern, etc.)
- the evaluation of "not detected" reports is highly problematic because they could have a different meaning for various compounds and could be different, for the same compound, in various studies

Nevertheless, even taking all these limitations into account, useful information can be obtained from a wide literature review, at least regarding general trends. In particular, from the data presented previously, some more significant observations could be made, taking into account only

Table 5   Mean of Maximum Values Measured in Each Data Set and Indicators of Leaching Capability for Pesticides Detected in More than One Single Data Set in the Worldwide Literature Review

	Molecule	Number of Data Sets	Geometric Mean µg/L	Water Distribution %[a,c]	GUS[b,c]
	Herbicides				
1	Alachlor	33	4.5	87	1.1 – 3.3
2	Atrazine	41	6.7	95	3.4 – 4.2
3	Bentazone	9	1.8	99.9	5.0 – 7.5
4	Bromacil	8	34	98	4.8 – 5.6
5	Chlortal Dimethyl (DCPA)[d]	7	106	14	0.7 – 0.8
6	Cyanazine	17	1.9	98	2.4 – 4.9
7	2,4-D[d]	6	2.1	75	0.1 – 3.5
8	Dicamba[d]	7	6.9	99.9	4.2 – 5.1
9	Dinoseb[d]	9	12.4	72	1.9 – 2.5
10	Isoproturon	2	0.09	96	2.2 – 2.9
11	MCPA[d]	6	0.7	92	1.4 – 2.9
12	Mecoprop[d]	7	0.8	99.9	3.7 – 5.9
13	Metolachlor	20	3.9	63	2.2 – 3.4
14	Metribuzin	17	2.3	99.0	2.2 – 8.9
15	Molinate	4	17	83	1.9 – 4.6
16	Pendimethalin	2	0.1	3	<0
17	Picloram[d]	2	9.9	99.9	5.3 – 9.1
18	Propazine	2	0.3	84	2.5 – 2.6
19	Simazine	18	1.0	96	3.4 – 4.1
20	2,4,5-T[d]	3	1.7	99.8	3.3 – 5.5
21	Terbuthylazine	4	0.6	81	1.8 – 3.7
22	Trifluralin	9	0.5	19	<0
	Insecticides				
23	Aldicarb	15	52	99.6	1.8 – 4.3
24	Carbofuran	9	7.2	99.4	3.9 – 6.0
25	Dieldrin	2	0.04	40	1.4 – 1.5
26	Fonofos	3	0.29	21	0.4 – 0.6
27	Lindane	3	5.5	39	1.5 – 2.0
28	Oxamyl	5	4	99.7	2.5 – 7.2
	Nematicides				
29	DBCP	4	12	16	3.5 – 5.5
30	1,2-Dichloropropane	10	39	1	4.4 – 6.6
31	1,3-Dichloropropene	8	36	1	2.5 – 4.4
32	Ethylene Dibromide	8	5.2	4	3.6 – 5.6

[a] Calculated with the standard fugacity model level I (see text).
[b] GUS (Groundwater Ubiquity Score) > 2.8: leacher; 1.8 < GUS < 2.8: transition; GUS < 1.8: improbable leacher (see text).
[c] Values of physicochemical properties needed for the calculation are shown in Table 6.
[d] Anionic compounds.

these compounds detected in more than one sporadic study (Table 5). In Table 5, an attempt has been made to compare the presence of the pesticides detected in groundwater bodies with the intrinsic properties responsible for their environmental behavior. For this purpose all the pesticides for which at least two data sets showing their presence in groundwaters are available have been included in this table.

For every pesticide, the affinity for water, the tendency to leach, and the geometric mean of the maximum levels have been calculated. The affinity for water has been estimated by means of an evaluative model (fugacity model level I) applied in standard form.[6] The model allows a comparative screening evaluation of the partitioning of a chemical among the different environmental compartments (air, water, soil, sediment, biota) of an ideal unit of world of 1 Km^2 with proportions among compartments similar to those of the real world (for more details on this approach see Chapter 3). For a preliminary screening evaluation, it can be assumed that a percent distribution in the water compartment higher than 80% indicates a high affinity for water, and a percent distribution between 70 and 80 indicates a medium-high affinity.

Table 6 Selected Values of Physicochemical Properties and Range of Half-Life in Soil for Pesticides Detected in Groundwater

	Molecule	Water Solubility g/L	Vapor Pressure Pa	Log Kow	Log Koc	DT50 Days
	Herbicides					
1	Alachlor	0.24[8]	0.003[8]	2.8[10]	2.2[11]	4[12] – 70[8]
2	Atrazine	0.03[8]	4e – 5[8]	2.3[8]	2.1[b]	60 – 160[8]
3	Bentazone	0.5[8]	5e – 4[8]	0.2[b]	0.0[13]	19 – 77[13]
4	Bromacil	0.81[8]	3e – 5[8]	2[10]	1.8[b]	150[20] – 350[21]
5	Chlortal dimethyl (DCPA)	5e – 4[8]	3e – 4[14]	4.4[a]	3.6[15]	60 – 100[14]
6	Cyanazine	0.17[8]	2e – 7[8]	1.8[16]	1.6[b]	10[22] – 108[23]
7	2,4-D	0.62[8]	1.4[8]	2.8[9]	1.7[11]	1 – 35[24]
8	Dicamba	6.5[8]	0.004[8]	0.5[10]	0.3[b]	14 – 25[8]
9	Dinoseb	0.1[8]	0.18[8]	3.0[a]	2.8[b]	40 – 120[25]
10	Isoproturon	0.05[8]	3e – 6[8]	2.2[8]	2.0[b]	12 – 29[8]
11	MCPA	0.82[8]	2e – 4[8]	2.5[a]	2.3[b]	7[6] – 50[26]
12	Mecoprop	0.62[8]	3e – 4[8]	0.1[8]	–0.1[b]	8 – 28[14]
13	Metolachor	0.53[8]	0.002[8]	3.4[8]	2.3[17]	15[14] – 107[27]
14	Metribuzin	1.05[8]	6e – 5[8]	1.6[8]	1.4[b]	4[28] – 301[29]
15	Molinate	0.88[8]	0.75[8]	2.8[8]	1.9[18]	8 – 160[8]
16	Pendimethalin	3e – 4[8]	0.004[8]	5.1[8]	4.2[11]	30[8] – 170[13]
17	Picloram	0.43[8]	8e – 5[8]	0.3[a]	1.4[11]	30 – 330[8]
18	Propazine	0.005[8]	4e – 6[8]	2.9[16]	2.7[b]	80 – 100[8]
19	Simazine	0.005[8]	8e – 7[8]	1.9[8]	1.7[b]	30[30] – 60[13]
20	2,4,5-T	0.15[8]	7e – 7[8]	0.8[a]	0.9[15]	12[31] – 59[32]
21	Terbuthylazine	0.008[8]	1e – 4[8]	3.0[8]	2.8[b]	30[8] – 170[33]
22	Trifluralin	9e – 4[8]	0.015[8]	5.1[8]	4.1[19]	30[11] – 255[13]
	Insecticides					
23	Aldicarb	6.0[8]	0.013[8]	1.1[10]	0.9[b]	7[10] – 28[23]
24	Carbofuran	0.32[8]	0.003[8]	1.3[8]	1.1[b]	23[11] – 117[8]
25	Dieldrin	2e – 4[10]	5e – 4[10]	3.7[10]	3.5[b]	868[23] – 1000[11]
26	Fonofos	0.0013[8]	0.028[b]	3.9[8]	3.7[b]	25[10] – 120[11]
27	Lindane	0.007[8]	0.006[8]	3.8[10]	3.4[9]	266[23] – 2000[25]
28	Oxamyl	280[8]	0.03[8]	1.0[9]	0.8[b]	6[23] – 180[11]
	Nematicides					
29	DBCP	1.2[9]	77[9]	2.3[9]	1.6[10]	30 – 200[9]
30	1,2-dichoropropane	2.7[9]	6600[9]	2.0[9]	1.8[b]	100 – 1000[25]
31	1,3-dichloropropene	2.0[8]	6000[8]	1.2[a]	1.0[b]	7 – 30[25]
32	Ethilene dibromide	4.3[8]	1500[8]	1.7[a]	1.5[b]	28 – 180[25]

[a] Calculated according to Hansch and Leo.[34]
[b] Calculated according to Karickhoff.[35]

The Groundwater Ubiquity Score (GUS) index[7] represents another screening approach developed to evaluate the leachability of pesticides by means of organic carbon sorption coefficient (Koc) and disappearance time in soil (DT50) (see Chapter 4).

Molecular properties needed for the calculation of fugacity partition and GUS index are shown in Table 6. For each physicochemical property, a single value is reported, selected on the basis of a critical review of the literature. For soil half-life, a range of values is reported, due to its variability in function of environmental parameters. Thus, also for the GUS index, a range of values is indicated in Table 5.

It must be emphasized that the use of Koc as an index of soil binding capability is suitable for apolar, nonionized compounds, for which partition with the organic matrix of soil is the most important mechanism determining affinity for the soil compartment. For polar and ionized organic chemicals, the interactions with the inorganic matrix of soil play a significant role, increasing soil bindings for the cationic substances and enhancing leachability for the anionic ones.

In the selection of Table 5, herbicides are clearly the most frequently detected compounds. Of 32 active ingredients, 22 are herbicides and only 10 belong to the classes of insecticides and nematicides; no fungicides are present in the selection. With only a few exceptions, all the compounds detected in groundwater are characterized by very high affinity for the water compartment. Among the 22 nonsporadically detected herbicides, 16 show high affinity for the water compartment (Table 5). Among the three cases showing a medium high affinity (2,4 D, dinoseb,

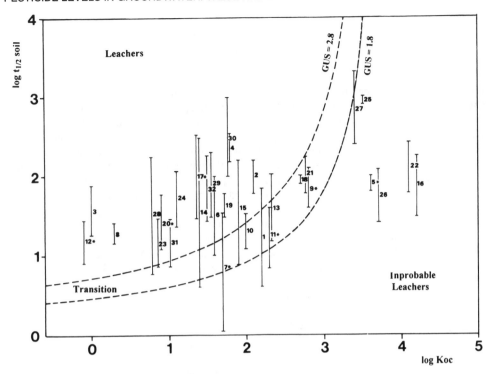

Figure 1   Persistence and mobility properties of pesticides found in groundwater. The dashed lines represent the threshold values of the GUS function, indicating the boundaries of the "leachers" and "improbable leacher" chemicals. Pesticides are numbered according to Table 5.

and metolachlor), the first two are reasonably underestimated due to the anionic properties of these compounds.[8] The same is true for DCPA, showing an apparently low affinity for water. Only two compounds (pendimethalin and trifluralin) have been detected in spite of their very low affinity for water. Yet, pendimethalin has been evidenced only in two cases at very low maximum levels (0.12 and less than 0.1 µg/L). Trifluralin has been found even at relatively high levels. Nevertheless, some doubts on the normal agricultural application as a cause of contamination have been raised for the highest levels.[3]

In addition to the fugacity model, the use of the GUS index demonstrates that the majority of the compounds detected in groundwater falls in the "leacher" or "transition" area, with the exception of some underestimated anionic compounds and of the only two real improbable leachers, pendimethalin and trifluralin, previously mentioned (Figure 1).

The few insecticides detected with relatively high frequency and concentrations show very high affinity for water, with the exception of lindane. Yet, the presence of lindane can be explained by its very high persistence. The same is valid for dieldrin; moreover, dieldrin and fonofos show low frequency of detection and very low concentrations. All the detected insecticides are used mainly (aldicarb), or at least partially, for soil treatment. The mode of application of these compounds may reasonably enhance their leaching capability. A very particular behavior is typical for compounds used as nematicides and fumigants for soil treatment. All those compounds show high water solubility (more than 1 g/L), but, due to their extremely high volatility (vapor pressure in the order of kPa), their affinity for the water compartment appears low, and, in the fugacity model, they are distributed mainly in the air compartment. In this case, the driving force determining environmental fate is the mode of application. Soil fumigants are injected in the root zone, and, after application, the soil is compacted to retain the fumigant longer. Thus, leaching into groundwater is very likely, favored by high solubility and low soil sorption capability, especially if there is rain or the fields are irrigated.

For herbicides, the geometric means of the maximum values range from less than 1 to a few µg/L, with the exceptions of bromacil (34 µg/L) and DCPA (106 µg/L). In the case of bromacil, the highest level detected, 125 µg/L, has been found after an application rate much higher than normal and in very shallow groundwater (4.5 to 6 m).[36] Other compounds (insecticides, nematicides) often show higher levels. In particular, aldicarb, 1,2-dichloropropane and 1,3-dichloropropene are frequently detected at levels of some hundreds of μg/L. The attempt to correlate mean values with indicators of leachability (percent distribution in water, GUS index) has proven unsuccessful. Precise relationships between parameters used for the quantification of the occurrence in groundwater (concentrations, percent detection, etc.) and factors affecting environmental exposure (loads, physicochemical properties, leaching capability, etc.) are not possible with the available set of data. Occurrence in groundwater depends, in general, on a complex combination of different factors, such as quantities applied, modes of application, environmental characteristics, and intrinsic properties of the compounds. As previously mentioned, in different situations, one or another factor can act as the principal driving force in determining the presence in groundwater. Moreover in a wide-ranging review like this, a description of the problem is even further complicated by the nonhomogeneous significance of the data collected. Extremely high concentrations or high frequency of detection could be the consequence of a very specific and particularly oriented monitoring plan (e.g., in known hot spots or areas at risk).Therefore, the above-mentioned observations should be taken only as indications of very general trends. A possible relationship between the pesticides detected in groundwater bodies and their use has also been examined.

In general, at least for herbicides, the detected compounds are the most commonly and extensively used. An evaluation of the ten most important herbicides used in the world in the last 10 years indicate the following compounds: [37]

glyphosate	2, 4 D
alachlor	atrazine
metolachlor	propanil
thiobencarb	MCPA
paraquat	trifluralin

Seven of these compounds appear in the list of herbicides detected in groundwater according to the described literature survey, and six are present in Table 5. The absence of the other compounds can be easily explained, at least for glyphosate and paraquat, with the cationic properties of those molecules, well known for being extremely strongly bound to the inorganic component of soil and commonly defined as practically immobile and non leacher.[7,38,39] This is true only to a minor extent for thiobencarb, which, on the other hand, has a relatively low persistence.[40-42]

Another evaluation in the function of loads can be made on the basis of data on pesticide use in the EEC.[30] Twelve out of the 22 herbicides detected in groundwater in more than one sporadic study belong to the herbicides used in the EEC at a rate of more than 500 tons per year (Table 7). This relation between use and occurrence in groundwater is not as obvious as it might appear. In fact, for other kinds of compounds, such as insecticides, this rule is not respected. Extensively used compounds, such as pirethroids and organophosphates, are not usually detected in groundwater. For these compounds, other factors (such as use patterns, physicochemical properties, persistence, etc.) are more important than loads in determining occurence in groundwater.

EVALUATION OF A NATIONAL-SCALE SURVEY

In order to overcome some general limitations related to a wide-range review, such as those presented in the previous pages, and to compare the results of this survey with monitoring data produced on the basis of a planned activity, a different set of data coming from a national survey has been examined. Of course, this survey is more reduced in terms of geographic extension but is more significant in evaluating the risk of the different compounds as well as the factors playing a major role in groundwater pollution. Data refers to the results of an extensive monitoring carried out in Germany, reported by Fielding et al.[30] The data, produced by the different Federal States (Länder) and collected by the Federal Agency for the Environment (Unweltbundesamt), refers to about 50,000 measurements of more than 170 compounds, including pesticides and a few isomers and metabolites, for a total of 133 active ingredients. This large number of compounds covers almost all the active ingredients used in significant amounts in Germany. Analyses were performed

Table 7 Use of Different Classes of Pesticides in Europe

Molecules	>2000 Tons/Year	Amount Used 1000 – 2000 Tons/Year	500 – 1000 Tons/Year
Herbicides	atrazine glyphosate isoproturon MCPA mecoprop	alachlor chlortoluron chlormequat chlorotalonyl 2,4-D dichlorprop ioxynil metabenzthiazuron TCA trifluralin metamitron	bentazone carbetamide chloridazon carbendazim simazine triallate metolachlor molinate paraquat pendimethalin propanil ethidimuron
Insecticides		dimethoate malathion methyl bromide parathion	azinphos (et.+met.) phorate
Fungicides	mancozeb maneb metam sodium	prochloraz	captan propiconazole propineb tridemorph zineb ziram
Nematicides and soil disinfectants	1,3-dichloropropene	1,2-dichloropropane	

From Fielding, M. et al., Pesticides in Ground and Drinking Water, Commission of the European Communities, Water Pollution Research Reports, No. 27, Brussels, 1991.

either on ground or surface waters. Unfortunately, in many cases, data are reported jointly, and it is not possible to differentiate between the two water resources. Thus, the survey must be taken as representative of the general occurrence of pesticides in water.

A synthesis of the results of the survey is shown in Tables 8 to 11 for the different groups of active ingredients. Only active ingredients (excluding metabolites) have been reported in the tables. Also, banned or regulated persistent chlorinated hydrocarbons (DDT, lindane, aldrin, dieldrin, endrin) have been excluded. As previously mentioned, DDT and other persistent chlorinated hydrocarbons, due to their massive use all over the world for decades and to their extremely high persistence, are at present distributed worldwide in all environmental compartments.[43] Thus, the presence of persistent chlorinated hydrocarbons must be interpreted as the long-term result of an old worldwide pollution and not as the consequence of present agricultural practice. Therefore, in the tables, 127 active ingredients are described, and the following observations can be made.

a. Forty-five out of 76 monitored herbicides (59%), have been detected in water. For insecticides, the corresponding figure is 7 out of 24 (29%), and, for fungicides, it is 7 out of 14 (50%). Only four soil fumigants and nematicides have been monitored, and two have been detected. Considering the frequency of detection, for herbicides, out of a total of more than 36,000 determinations, 12% were positive findings. The percentage of positive findings for insecticides is a little more than 2%, out of more than 2300 determinations. Only 605 determinations were made on fungicides and less than 3% were positive findings. A very high level of detection is typical for the few soil fumigants investigated, in particular for 1,2-dichloropropane.

b. For herbicides and insecticides, about 60% of the compounds detected in water show levels higher than 0.1 µg/L. The percentage of samples with a concentration higher than 0.1 µg/L ranges from 7 to 100% of the total positive findings. This means that, for a large number of compounds, measured levels could be higher than the allowable concentration in drinking water, in agreement with the EEC Directive.[44]

Table 8 Data on Herbicides Found in Water in Germany[30] Listed According to Decreasing Percentage of Positive Findings

	Herbicides	Number of Measurements	Positive Findings Number	Positive Findings %	% >0.1 µg/l	Use (Tons/Year)	Water Distribution %[a]
1	Trichloroacetic acid	2	2	100	0	50 – 1000	d
2	Dikegulak	62	47	75.8	1	<50	99.9
3	Atrazine	4648	1436	45.2	0.4	>1000	95
4	Dichlorprop	200	73	36.5	0.3	>1000	99.7
5	T-2,4,5	126	45	35.7	0	<50	99.8[d]
6	Carbetamide	36	11	30.6	0	200 – 500	99.8
7	Simazine	4345	791	18.2	0.2	100 – 200	96
8	Bromacil	1139	202	17.7	0.8	<50	97
9	MCPP	735	112	15.2	0.3	>1000	99.9
10	Bentazone	283	41	14.5	0.5	200 – 500	99.9
11	Monuron	88	12	13.6	0.1	<50	b,c
12	Hexazinone	168	21	12.5	0.4	<50	99.9
13	Triallate	24	3	12.5	0	200 – 500	60
14	MCPA	457	51	11.2	0.2	500 – 1000	92[d]
15	Dalapon	9	1	11.1	0	<50	99.7
16	Chlortoluron	672	68	10.1	0.5	200 – 500	95
17	Bromoxynil	34	3	8.8	0	<50	98
18	Diuron	6987	60	8.7	0.3	200 – 500	91
19	Terbuthylazine	3204	257	8.0	0.1	<50	81
20	Isoproturon	667	45	6.7	0.4	>1000	96
21	MCPB	65	4	6.1	0	<50	57[d]
22	D-2,4	955	45	4.7	0.7	200 – 500	75[d]
23	Propazine	3847	178	4.6	0.5	<50	84
24	Methabenzthiazuron	652	28	4.2	0.3	200 – 500	91
25	Metolachlor	1520	56	3.6	0.3	<50	90
26	Pendimethalin	146	5	3.4	0.3	<50	3
27	Metazachlor	1524	48	3.1	0.4	200 – 500	b
28	Metribuzin	155	4	2.5	0.4	<50	99
29	Chloridazon	92	2	2.1	0	200 – 500	96
30	DB-2,4	48	1	2.1	1	<50	c,d
31	Trifluralin	154	3	1.9	0	200 – 500	19
32	Fenoprop	62	1	1.6	0.3	<50	d
33	Prometryn	329	5	1.5	0	<50	68
34	Dicamba	77	1	1.2	0	<50	99.9[d]
35	Ametryn	171	2	1.1	0	<50	81
36	Monolinuron	172	2	1.1	0.5	<50	61
37	Dichlobenyl	101	1	1	1	<50	b
38	Cyanazine	735	7	1	1	<50	98
39	Atraton	128	1	<1	0.1	<50	b,c
40	Metobromuron	628	3	<1	0	<50	94
41	Metoxuron	645	4	<1	0.2	<50	99
42	Sebuthylazine	1957	7	<1	0	<50	b,c
43	Linuron	568	2	<1	0.3	<50	81
44	Terbutryn	1563	5	<1	0	<50	57
45	Desmetryn	1500	2	<1	0.4	<50	94
46	Alachlor	62	0	0	0	<50	87
47	Amitrol	16	0	0	0	100 – 200	99.9
48	Asulam	2	0	0	0	<50	b,c
49	Aziprotryn	6	0	0	0	<50	b
50	Bifenox	11	0	0	0	<50	12
51	Buturon	86	0	0	0	<50	b,c
52	Butylate	1	0	0	0	<50	24
53	Chlorbufam	17	0	0	0	<50	91
54	Chloroxuron	123	0	0	0	<50	73
55	Chlorpropham	14	0	0	0	<50	77
56	Clopyralid	7	0	0	0	<50	b
57	Cycloate	1	0	0	0	<50	21
58	Dimefuron	8	0	0	0	<50	93
59	Dinoseb	48	0	0	0	<50	73[d]
60	Dinoterb	37	0	0	0	<50	46

Table 8 (continued) Data on Herbicides Found in Water in Germany[30] Listed According to Decreasing Percentage of Positive Findings

	Herbicides	Number of Measurements	Positive Findings Number	%	% >0.1 µg/l	Use (Tons/Year)	Water Distribution %[a]
61	Dipropetryn	59	0	0	0	<50	40
62	EPTC	1	0	0	0	<50	46
63	Fenuron	87	0	0	0	<50	99.8
64	Fluazifop buthyl	1	0	0	0	<50	12
65	Fluometuron	17	0	0	0	<50	89
66	Ioxynil	36	0	0	0	200 – 500	29
67	Metamitron	154	0	0	0	<50	99.8
68	Methoprotryn	93	0	0	0	<50	b
70	Nitrofen	1	0	0	0	<50	b,c
71	Phenmedipham	29	0	0	0	100 – 200	52
72	Picloram	25	0	0	0	<50	99.9[d]
73	Prometon	27	0	0	0	<50	84
74	Propham	31	0	0	0	<50	70
75	Pyridate	26	0	0	0	<50	b
76	Triclopyr	11	0	0	0	<50	81

[a] Calculated with the standard fugacity model level I (see text).
[b] No data about Kow.
[c] No data about vapor pressure.
[d] Anionic compound.

Table 9 Data on Insecticides Found in Water in Germany[30] Listed According to Decreasing Percentage of Positive Findings

	Insecticides	Number of Measurements	Positive Findings Number	%	% <0.1 µg/l	Use (Tons/Year)	Water Distribution %[a]
1	Dimethoate	133	37	27.8	0.3	<50	99.8
2	Propetamphos	61	5	8.2	0	<50	b
3	Etrinphos	61	2	3.3	0	<50	72
4	Chlorfenvinphos	579	1	1.5	0.1	<50	77
5	Heptachlor	91	1	1.1	0	<50	b,c
6	Disulfoton	106	1	<1	0.1	<50	28
7	Parathion ethyl	256	2	<1	1	100 – 200	40
8	Aldicarb	44	0	0	0	<50	99.6
9	Azinphos ethyl	62	0	0	0	<50	44
10	Azinphos methyl	7	0	0	0	<50	89
11	Bromophosethyl	71	0	0	0	<50	b,c
12	Carbaryl	18	0	0	0	<50	95
13	Carbofuran	38	0	0	0	<50	99
14	Chlordane	56	0	0	0	<50	38
15	Cypermethryn	9	0	0	0	<50	0.1
16	Diazinon	120	0	0	0	<50	68
17	Dichlorvos	21	0	0	0	200 – 500	99
18	Endosulfan	216	0	0	0	<50	0.2
19	Fenitrothion	56	0	0	0	<50	63
20	Malathion	20	0	0	0	<50	91
21	Methoxychlor	73	0	0	0	<50	8
22	Mevinphos	41	0	0	0	<50	b
23	Parathion methyl	150	0	0	0	<50	80
24	Propoxur	50	0	0	0	<50	96
25	Thiometon	53	0	0	0	<50	b

[a] Calculated with the standard fugacity model level I (see text).
[b] No data about Kow.
[c] No data about vapor pressure.

Table 10   Data on Fungicides Found in Water in Germany[30] Listed According to Decreasing Percentage of Positive Findings

	Fungicides	Number of Measurements	Positive Findings		% >0.1 µg/l	Use (Tons/Year)	Water Distribution %[a]
			Number	%			
1	Pentachlorphenol	11	5	45.5	0	<50	c
2	Metalaxyl	55	3	5.5	0	<50	99.4
3	Fenpropimorph	26	1	3.8	0	500 – 100	21
4	Triadimenol	64	2	3.1	0.5	100 – 200	76
5	HCB	185	4	2.2	0	<50	<1
6	Triadimefon	64	1	1.6	0	<50	77
7	Vinclozolin	99	1	1	1	100 – 200	81
8	Captan	1	0	0	0	<50	74
9	Dichlofuanid	1	0	0	0	<50	46
10	Folpet	1	0	0	0	<50	b
11	Oxadixyl	54	0	0	0	<50	99.8
12	Prochloraz	5	0	0	0	500 – 100	15
13	Procymidon	1	0	0	0	<50	65
14	Quintozen	38	0	0	0	<50	6

[a] Calculated with the standard fugacity model level I (see text).
[b] No data about Kow.
[c] Anionic compound.

Table 11   Data on Nematicides and Soil Disinfectants Found in Water in Germany[30] Listed According to Decreasing Percentage of Positive Findings

	Nematicides and Soil Disinfectants	Number of Measurements	Positive Findings		% >0.1 µg/l	Use (Tons/Year)	Water Distribution %[a]
			Number	%			
1	Dichlorpropene	14	6	49.2	0	<50	1.5
2	Dichloropropane-1,2	305	290	95.1	0.48	<50	1.1
3	DBCP	1	0	0	0	<50	10.5
4	Ethylene dibromide	1	0	0	0	<50	b,c

[a] Calculated with the standard fugacity model level I (see text).
[b] No data about Kow.
[c] No data about vapor pressure.

c. Among the 16 herbicides with positive findings totalling more than 10%, 10 are used in large amounts in Germany (more than 50 tons in 1989). Among the 31 undetected compounds, only three were used in significant amounts. Almost all the detected compounds show a high or very high affinity for the water compartment, quantified as percent distribution calculated with the fugacity model. The only exceptions, out of 45 compounds, are pendimethalin and trifluralin. Even if the percentage of positive findings is relatively low for these two compounds (3.4 and 1.9% respectively) their possible occurrence in groundwater confirms the findings of the worldwide review described previously. The frequency of compounds with low water affinity is higher among undetected herbicides (12 over 31). Insecticides are not so massively used in Germany as are herbicides. Only in the case of two compounds (parathion ethyl and dichlorvos) were more than 50 tons per year used. It is difficult to find a precise relationship with uses or intrinsic properties of the compounds, due to the relatively small data set. The same is true for fungicides and nematicides, although among the four highly used fungicides, three were detected in water.

INFORMATION NEEDED TO IMPROVE THE USEFULNESS OF MONITORING DATA

On the basis of the limitations pointed out in this chapter, we would like to give some advice regarding the information that should be included to make monitoring surveys more useful. The results of monitoring surveys should indicate the following:

1. How the monitoring survey was designed (e.g., if it refers mainly to an outdoor experiment, or to a national or regional monitoring plan)
2. An adequate description of the geology, hydrogeology and pedology of the area
3. Reasons for the selection of the pesticides examined (on the basis of their use in the area, their predicted leaching potential, the availability of analytical methodologies, public concern, etc.)
4. Representativity of the wells or boreholes with respect to the groundwater examined (on the basis of historical data, etc.)
5. Whether the groundwater was for drinking purposes or other uses
6. Depth of the wells, number of positive and examined wells, analytical sensitivity for each pesticide examined
7. The presumably temporal relationship between the application of the pesticides and their occurrence in the groundwater in the area examined

REASONS FOR CLASSIFYING HERBICIDES AS COMPOUNDS OF MAJOR CONCERN

The empirical results from the two surveys described indicate that, among the various classes of pesticides, herbicides are the most frequently detected in water, particularly in groundwater. This particular behavior of herbicides can be easily explained by taking into account several aspects related to their uses and properties compared with other classes of widely used pesticides (e.g., insecticides and fungicides).

TOTAL LOADS

Herbicides represent about 44% of the total pesticide market, compared to about 29% for insecticides and 21% for fungicides. In particular, for Western Europe and North America, which cover jointly 57% of the world pesticide market (30 and 27% respectively) the figures for herbicide use are about 38% and 67% respectively (Table 12).

MODE OF USE AND TIME OF APPLICATION

Herbicides are often applied directly on soil, whereas other pesticides, with the exception of nematicides and of insecticides used in soil disinfection, are generally spread on plants. The amount of insecticides and fungicides deposited on soil after a treatment is considered as lost and, therefore, must be minimized. It is difficult to quantify the amount of a pesticide, not directly applied to the soil, that can reach the soil surface as an immediate consequence of treatment or by a subsequent wet or dry deposition. This aspect depends on several factors such as:

- application methods and characteristics of spreading apparatus
- physicochemical properties of the compound and volatilization capability
- atmospheric conditions during and after the treatment
- patterns of wet and dry fallout
- characteristics of the crop (dimension of the plants, leaf surface, etc.)

Very roughly, it can be hypothesized that the amount of pesticide reaching the soil surface could range between 20 and 60% of the total amount applied. This significantly reduces the total load of a pesticide to the soil system whereas, for an herbicide directly applied onto soil, losses by volatilization are definitely smaller, even if not negligible. It is not surprising that the majority of insecticides frequently detected in groundwater are applied onto soil. Moreover, the period and the patterns of pesticide treatment must not be forgotten. In normal agricultural practice, herbicides and nematicides are applied by means of a few treatments (in general one or two), either in pre- or post-emergence, in any case at the beginning of the growing season. An example of the herbicide and nematicide treatment periods for some major crops is shown in Figure 2. On the contrary, insecticide and fungicide treatments are repeated for several months during the growing season. In the

Table 12   World Pesticides Market. Data are Expressed in Million Dollars

Geographic Area	Herbicides	Insecticides	Fungicides	Others	Total
Western Europe	3013	1645	2612	646	7916
North America	4700	1445	466	439	7059
Eastern Asia	1712	2305	1575	210	5802
South America	1020	710	333	80	2143
Eastern Europe	990	570	390	115	2065
Others	190	980	169	85	1424
Total	**11,625**	**7655**	**5545**	**1575**	**26,400**

From Agrofarma, *Il Mercato dei Fitofarmaci: Andamento e Rilevazioni*, Milan, 1992, p. 39.

temperate area, where the majority of developed countries are located, in the normal meteoric cycle, rain is distributed mainly in spring and fall whereas summer is usually the driest season. Thus, the whole herbicide load is spread in the rainy season and can be immediately carried to ground or surface waters through leaching or runoff. On the contrary, for other pesticides a large amount of the load is applied in the dry season and could be subject to several processes of disappearance (degradation, volatilization, etc.) before being transported in surface or groundwaters. This is particularly true for insecticides, whereas, for fungicides, wet conditions could enhance treatment needs.

PHYSICOCHEMICAL PROPERTIES

As better explained in the following chapters (see Chapter 3) the leaching capability of an organic chemical depends on physicochemical properties, such as water solubility, soil sorption coefficient (Koc), persistence, etc. Generally, herbicides have relatively high water solubility and a low Koc, in comparison with insecticides and, to a lesser extent, fungicides, and have higher leaching capability. In order to define and demonstrate this statement better, a comparison has been made of herbicides, insecticides, and fungicides. All active ingredients described in the 9th edition of the *Pesticide Manual*[8] have been taken into account for the application of the fugacity model level I in its standard form (see Chapter 3). The physicochemical data needed for the application of the model (water solubility, vapor pressure, octanol–water partition coefficient) have been taken

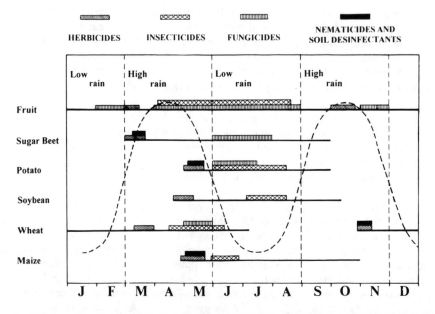

Figure 2   Scheme of the main treatments (herbicides, insecticides, fungicides, and nematicides) on some important crops, in relation to the general trend of precipitations in temperate climates (dashed line).

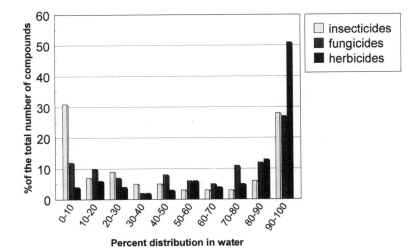

Figure 3   Affinity of pesticides for the water compartment calculated according to the fugacity model level I. The total number of compounds refers to chemicals listed in the *Pesticide Manual*[5] for which the physicochemical properties needed for the calculation are available.

from handbooks or extensive literature reviews.[8–10,16] Complete, reliable information was available for 143 out of 205 herbicides, 152 out of 180 insecticides, and 79 out of 130 fungicides. The fugacity model was applied on these compounds in order to calculate the percent distribution in the water compartment. Figure 3 shows the distribution of water affinity for the various groups of pesticides.

It appears evident that, in general, for all classes of compounds, water is the environmental compartment at risk. This is particularly true for herbicides, with more than 65% of these compounds demonstrating very high water affinity (percent distribution higher than 80%). Lower figures are typical for insecticides (36%). These compounds often show high lipophilicity and are therefore partitioned in large amounts in the organic component of soil. As a consequence, they are low mobile compounds. Many fungicides and soil disinfectants are often highly volatile compounds.

For all the reasons described above, herbicides are typical contaminants of water and, in particular, of groundwater. Therefore, in the following chapters, particular care will be devoted to these compounds, particularly in describing agronomic needs and management strategies.

REFERENCES

1. Hallberg, G.R. "Pesticide Pollution of Groundwater in the Humid United States," *Agric. Ecosystems Environ.* 26: 299-367 (1989).

2. Richard, J.J., G.A. Junk, M.J. Avery, N.L. Nehring, J.S. Fritz, and H.J. Svec. "Analysis of Various Iowa Waters for Selected Pesticides: Atrazine, DDE, and Dieldrin — 1974," *Pestic. Monit. J.* 9:117-123 (1975).

3. Kross, B.C., M.I. Selim, G.R. Hallberg, D.R. Bruner, and K. Cherryholmes. "Pesticide Contamination of Private Well Water, a Growing Rural Health Concern," *Environ. Int.*18(3):231-241 (1992).

4. Kessler, K. "Wisconsin's Groundwater Monitoring Program for Pesticides," in *Pesticides and Groundwater; a Health Concern for the Midwest* (Navarre, MN: The Freshwater Foundation and the U.S. EPA, 1987), pp. 105-113.

5. Holden, P. *Pesticides and Groundwater Quality: Issues and Problems in Four States* (Washington, DC: Academic Press, 1986), 124 pp.

6. Mackay, D. *Multimedia Environmental Models, The Fugacity Approach.* (Chelsea, MI: Lewis Publishers, Inc., 1991) p. 257.

7. Gustafson, D.I. "Groundwater Ubiquity Score: a Simple Method for Assessing Pesticide Leachability," *Environ. Toxicol. Chem.* 8:319-357 (1989).

8. Worthing, C.R. and R.J. Hance, Eds. *The Pesticide Manual. A World Compendium,* 9th ed. (Croydon, UK: British Crop Protection Council, 1991).

9. Howard, P.H., Ed. *Handbook of Environmental Fate and Exposure Data for Organic Chemicals. Vol. I: Large Production and Priority Pollutants,* (Boca Raton, FL: Lewis Publishers, Inc., 1989) p. 574; *Vol. II: Solvents* (Boca Raton, FL: Lewis Publishers, Inc., 1990) p. 546; *Vol.III: Pesticides* (Boca Raton, FL: Lewis Publishers, Inc., 1991) p. 683.

10. Suntio L.R., W.Y. Shiu, D. Mackay, J.N. Seiber, and D. Glotfelty. "Critical Reviews of Henry's Law Constants for Pesticides," *Rev. Environ. Contam. Toxicol.* 103: 1-59 (1988).

11. Johnson, B. "Setting Revised Specific Numerical Values," California Department of Food and Agriculture, Sacramento, CA (October 1989).

12. Cohen, S.Z., S. Creeger, R. Carsel, and C. Enfield. "Potential for Pesticide Contamination of Ground Water Resulting from Agricultural Uses," in *Treatment and Disposal of Pesticide Wastes,* Symposium Series 259, R.F. Krueger and J.N. Seiber, Eds. (Washington, DC: American Chemical Society, 1984), pp. 297-325.

13. Canton, J.H., J.B.H.J. Linders, R. Luttik, B.J.W.G. Mensink, E. Panman, E.J. van de Plassche, P.M. Sparenburg, and J. Tuinstra. "Catch-up Operation on Old Pesticides: an Integration," RIVM, Report no. 678801002, Bilthoven, The Netherlands (1990).

14. Colby, S.R., E.R. Hill, L.M. Kitchen, R.G. Lym, W.J. McAvoy, and R. Prasad. *Herbicide Handbook. VIth edition.* Weed Science Society of America (Champaign, Illinois 1989) p. 301.

15. Jury, W.A., D.D. Focht, and W.J. Farmer. "Evaluation of Pesticide Groundwater Pollution Potential from Standard Indices of Soil-Chemical Adsorption and Biodegradation," *J. Environ. Qual.* 16:422-428 (1987).

16. Finizio, A. "Indagine sui problemi relativi alla determinazione del coefficiente di ripartizione n-ottanolo/acqua (Kow) di antiparassitari" PhD Thesis, University of Milano, Milan, Italy (1993).

17. *Water Quality Workshop: Integrating Water Quality and Quantity into Conservation Planning,* (Ft. Worth, TX: Soil Conservation Service National Technical Center, 1980).

18. Kanazawa, J. "Relationship Between the Soil Sorption Constants for Pesticides and Their Physicochemical Properties," *Environ. Toxicol. Chem.* 8:477-484 (1989).

19. Kenaga, E.E. "Predicted Bioconcentration Factors and Soil Sorption Coefficients of Pesticides and Other Chemicals," *Ecotoxicol. Environ. Safety* 4:26-38 (1980).

20. Gardiner, J.A., R.C. Rhodes, J.B. Adams, and E.J. Soboczenski. "Synthesis and Studies with Labeled Bromacil 2-C14 and Terbacil," *J. Agric. Food Chem.* 17:980-986 (1969).

21. Rao, P.S.C. and J.M. Davidson. "Estimation of Pesticide Retention and Transformation Parameters Required in Nonpoint Source Pollution Model," in *Environmental Impact of Nonpoint Source Pollution,* M.R. Overcash and J.M. Davidson, Eds. (Ann Arbor, MI: Ann Arbor Science Publishers Inc., 1980), pp. 23-67.

22. Smith, C.N., R.A. Leonard, G.W. Langdale, and G.W. Bailey. "Transport of Agricultural Chemicals from Small Upland Piedmont Watersheds," U.S. EPA, Res. Rep. Ser. EPA-600/3-78-056 (1978).

23. Rao, P.S.C., A.G. Hornsby, and R.E. Jessup. "Indices for Ranking the Potential for Pesticide Contamination of Groundwater," *Soil Crop Sci. Soc. Fl. Proc.* 44:1-24 (1985).

24. White, A.M., L.E. Asmussen, E.W. Hauser, and J.W. Turnbull. "Loss of 2,4-D in Runoff from Plots Receiving Simulated Rainfall and from a Small Agricultural Watershed," *J. Environ. Qual.* 5:487-490 (1976).

25. Howard, P.H., R.S. Boethling, W.F. Jarvis, W.M. Meylan, and E.M. Michalenko. *Handbook of Environmental Degradation Rates* (Boca Raton, FL: Lewis Publishers, Inc., 1991) p. 725.

26. Sattar, M.A. and J. Paasivirta. "Fate of the Herbicide MCPA in Soil, Analysis of the Residue of MCPA by an Internal Standard Method," *Chemosphere* 9:365-375 (1980).

27. Walker, A. and P.A. Brown. "The Relative Persistence in Soil of Five Acetanilide Herbicides," *Bull. Environ. Contam. Toxicol.* 34:143-149 (1985).

28. Wauchope, R.D. and R.A. Leonard. "Maximum Pesticide Concentrations in Agricultural Runoff: a Semiempirical Formula," *J. Environ. Qual.* 9:665-672 (1980).

29. Kempson-Jones, G.F. and R.J. Hence. "Kinetics of Linuron and Metribuzin Degradation in Soil," *Pestic. Sci.* 10:449-454 (1979).

30. Fielding M., D. Barcelo, S.A. Helweg, S. Galassi, L. Torstensson, P. Van Zoonen, R. Wolter, and G. Angeletti. *Pesticides in Ground and Drinking Water,* Commission of the European Communities, Water Pollution Research Reports, No. 27, Brussels (1991).

31. Smith, A.E. "Relative Persistence of di- and tri-chlorophenoxyalkanoic Herbicides in Saskatchewan Soils," *Weed Res.* 18:275 (1978).

32. Edwards, W.M. "Proceeding No-Tillage System Symposium 21-2-73," Ohio State University, Columbus, OH (1973).

33. Burkhard, N. and J.A. Guth. "Chemical Hydrolysis of 2-chloro-4,6-bis-(alkylamino)-1,3,5-triazine Herbicides and their Breakdown in Soil under the Influence of Adsorption," *Pestic. Sci.* 12:45-52 (1981).

34. Hansch, C. and A. Leo. *Substituent Constants for Correlation Analysis in Chemistry and Biology* (New York: John Wiley & Sons, 1979) p. 337.

35. Karickhoff, S.W. "Semiempirical Estimation of Sorption of Hydrophobic Pollutants on Natural Sediments and Soils," *Chemosphere*, 10, 833-846, (1981).

36. Hebb, E.A. and W.B. Wheeler. "Bromacil in Lakeland Soil Ground Water," *J. Environ. Qual.* 7:598-601 (1978).

37. Agrofarma and Monsanto — personal communications.

38. Glass, R.L. "Adsorption of Glyphosate by Soils and Clay Minerals," *J. Agric. Food Chem.* 35:497-500 (1987).

39. Calderbank, A. "The Occurrence and Significance of Bound Pesticide Residues in Soil," *Rev. Environ. Contam. Toxicol.* 108:71-103 (1989).

40. Matsunaka, S. and S. Kuwatsuka. "Environmental Problems Related to Herbicides Use in Japan," in *Environmental Quality and Safety. Global Aspects of Chemistry, Toxicology and Technology as Applied to the Environment*, F. Coulston and F. Korte, Eds. (Stuttgart, Germany: Georg Thieme, 1975), pp. 149-159.

41. Horiguchi, H., K. Kawata, and T. Toshida. "Effects of Herbicides on the Soil Environment. I. Degradation of the Herbicide Benthiocarb in the Soil of a Paddy Field," *Kanyo Kagaku Sogo Kenkyusho Nenpo* 1:29-33 (1981).

42. Draper, W.M. and D.G. Crosby. "Photochemistry and Volatility of Drepamon in Water," *J. Agric. Food Chem.* 32:728-733 (1984).

43. Atlas, E.L. and C.S. Giam, "Sea-Air Exchange of High-Molecular Weight Synthetic Organic Compounds," in *The Role of Air-Sea Exchange in Geochemical Cycling,* P. Bouat-Menard, Ed. (Dordrecht, The Netherlands: NATO ASI Series, D. Reidel Publishing Company, 1985), pp. 295-330.

44. EEC Council Directive 80/778 of 15 July 1980 Relating to the Quality of Water Intended for Human Consumption. O.J. N.L 229, 30/87 (1980) pp. 11-29.

45. *Il Mercato dei Fitofarmaci: Andamento e Rilevazioni* (Milan: Agrofarma, 1992), p. 39.

APPENDIX

DATA ON THE PRESENCE OF PESTICIDES IN GROUNDWATER

In the following tables the main results of the survey of pesticides in groundwater are reported. For each compound, all references indicating groundwater detection are reported.

Compounds are divided into different pesticide classes:

- Herbicides: Tables 1 to 16
- Insecticides and fungicides: Tables 17 to 20
- Nematicides: Tables 21 to 24
- Degradation products: Table 25

Table 1 Studies on the Presence of Alachlor in Groundwater

Authors	Country	Region	Pedology and Hydrogeology	Groundwater Depth (m)	Positive:Examined Ratio[a]	Mean Level (µg/L)	Maximum Level (µg/L)
Ontario Min. Env., 1987a	Canada	Ontario			53:351 (15%)		20
Ontario Min. Env., 1987b	Canada	Ontario			2:42 (5%)		2.3
Funari and Sampaolo, 1989	Italy	North			3:459 (<1%)		1.6
Janssen and Puijker, 1988	Netherlands	Vierlingsbeek					0.3
Ritter et al., 1987	U.S.	DE	loamy sand soil	2 – 5			2
Ritter et al., 1987	U.S.	DE	loamy sand soil	3.1 – 4.6			15
Cohen et al., 1986	U.S.	IA, MD, NE, PA		4.6			10 (typical)
McKenna et al., 1988	U.S.	IL	sandy, alluvial aquifer				0.04
Good and Taylor, 1987	U.S.	IL			8:422 (2%)[b]	0.03	6.5
Hallberg, 1985	U.S.	IA	karst area				16.6
Hallberg, 1985	U.S.	IA	alluvial and Pleistocene aquifer				0.7
Hallberg et al., 1987	U.S.	IA			13%		5
Hallberg et al., 1987	U.S.	IA			13%[b]		250
Baker and Austin, 1985	U.S.	IA	karst system		15%[b]		4.2
Kelley, 1985	U.S.	IA			5:70 (7.1%)		0.3
Detroy et al., 1988	U.S.	IA	alluvial and Quaternary and bedrock aquifer	<60			21
Hallberg et al., 1987	U.S.	IA			92:456 (20.3%)	0.75	10
					34:184 (18.5%)		

Reference	Country	State	Description	Depth	Detection		
Kross et al., 1992	U.S.	IA			1.2%	0.67	4.76
Libra et al., 1984	U.S.	IA			9%, 8%[b]	9.4	16.6
Robbins and McCool, 1987	U.S.	KS			5:123 (4%)		12
Snethen and Robbins, 1987a,b	U.S.	KS			1:104 (1%)		6.2
Bushway et al., 1992	U.S.	ME			17:67 (25%)		7.2
Isensee et al., 1988	U.S.	MD	silt loam surface, coarse-textured strata	1 – 1.4	18%[b]		3.6
Mass. Int. Agric. P.T.F., 1986	U.S.	MA			3%		4.4
Klaseus et al., 1988a,b	U.S.	MN			40:725 (5.5%)	0.39	9.76
Spalding et al., 1980	U.S.	NE	silt loam soil	6 – 10	2:14 (14%)		0.71
Junk et al., 1980	U.S.	NE		15	1.35 (3%)[b]		>0.02
Nebraska Dep. Health, 1985	U.S.	NE			5%		0.8
Thompson, 1984	U.S.	PA			10%		3
Kimball and Best, 1987	U.S.	SD	shallow sand and gravel aquifers		12:57 (21%)		3.1
Holden, 1986	U.S.	WI			47:377 (12%)[b]		21 samples >2
Kessler, 1987	U.S.	WI			7%		3000
Williams et al., 1988	U.S.	CT, FL, IA, IL, KS, LA, MA, MD, ME, NE, PA, WI				0.9	113

[a] These results refer to wells if not specified differently.
[b] Samples instead of wells.

Table 2   Studies on the Presence of Atrazine in Groundwater

Authors	Country	Region	Pedology and Hydrogeology	Groundwater Depth (m)	Positive:Examined Ratio[a]	Mean Level (µg/L)	Maximum Level (µg/L)
Ontario Min. Env., 1987b	Canada	Ontario			18:42 (43%)		4.2
Schiavon and Jacquin, 1973	France	Lorraine	clay, clay-loam soil	0.9		16	29
Snegaroff, 1979	France	Charente Maritime	alluvial clay soil			27	46
Roth, 1987	Germany	Baden-Wurttemburg	low-adsorptive soil				0.4
IPS, 1987	Germany						0.26
Hurle et al., 1987	Germany	Baden-Wurttemberg	limestone and sedimentary areas		14:24 (58%)	<0.1	0.76
Stock et al., 1987a,b	Germany	Schleswig-Holstein	humic sand soil	0.5 – 3		0.08	6.75
Schmitz, 1988	Germany					0.74	17
Berri et al., 1983	Italy	Pavia			201:300 (67%)		4
Premazzi et al., 1989	Italy	North			279:318 (88%)		0.8
Funari and Sampaolo, 1989	Italy	North			1154:4064 (28%)		2
Molino et al., 1991	Italy	Lucca		6 – 13	312:437 (71%)		0.37
Janssen and Puijker, 1987	Netherlands	Noord-Brabant		below 3	7:29 (24%)[b]	0.1	0.8
Janssen and Puijker, 1988	Netherlands	Lunteren		12			0.11
Janssen and Puijker, 1988	Netherlands	Vierlingsbeek		2 – 5			0.8
Croll, 1986	United Kingdom	East Anglia	sand soil and shallow water table				0.5
Ritter et al., 1988	U.S.	DE	silt loam soil	6 – 9	11:23 (48%)		45
McKenna et al., 1988	U.S.	IL			21:422 (5%)[b]	1.2	7.3
Good and Taylor, 1987	U.S.	IL	sandy, alluvial aquifer		34.4%		2.5
Hallberg et al., 1987	U.S.	IA			98%[b]		10
Hallberg et al., 1987	U.S.	IA					13
Baker and Austin, 1985	U.S.	IA	Karst system		69.6%[b]		8.2

Reference							
Kelley, 1985	U.S.	IA					3
Detroy et al., 1988	U.S.	IA	alluvial and Quaternary and bedrock aquifers	<60	24:70 (34.3%); 20.3%		21
Kross et al., 1992	U.S.	IA			72:355 (20.3%)	0.9	6.61
Libra et al., 1984	U.S.	IA			4.4%	0.3	1.6
Hallberg et al., 1987	U.S.	IA			48%, 31%[b]	1.14	21.11
Steichen et al., 1986	U.S.	KS			79:184 (43%)		7.4
Robbins and McCool, 1987	U.S.	KS			4:104 (4%)		16
Bushway et al., 1992	U.S.	ME			6:123 (5%)		9.3
Isensee et al., 1988	U.S.	MD	silt loam surface, coarse textured strata	1 – 1.4	27:67 (40%)		95
Klaseus et al., 1988a,b	U.S.	MN			75%[b]	0.12	42.4
Nebraska Dep. Health, 1985	U.S.	NE			266:725 (36.7%)		107
Junk et al., 1980	U.S.	NE			365:451 (81%)		88
Thompson, 1984	U.S.	PA			38%		3
Kimball and Best, 1987	U.S.	SD	shallow sand, gravel aquifer		1:57 (2%)		5.4
Dillaha et al., 1987	U.S.	VA	loamy sand soil		16%	0.46	25.6
Kessler, 1987	U.S.	WI					1500
Ritter et al., 1987	U.S.		loamy sand soil	3.1			54
Isensee et al., 1989	U.S.			1.5 – 3			2
Williams et al., 1988	U.S.	CA, CO, CT, IA, IL, KS, MD, ME, NE, NJ, PA, VT, WI				0.5	40

[a] These results refer to wells if not specified differently.
[b] Samples instead of wells.

Table 3 Studies on the Presence of Bentazone in Groundwater

Authors	Country	Region	Groundwater Depth (m)	Positive:Examined Ratio[a]	Mean Level (µg/L)	Maximum Level (µg/L)
IPS, 1987	Germany		various	11:141 (8%)	0.12	0.85
Schmitz, 1988	Germany					0.85
Funari and Sampaolo, 1989	Italy	North		19:1442 (1%)		40
Cormegna et al., 1992	Italy	Piemonte, Lombardia	<15			24.3
Cormegna et al., 1992	Italy	Piemonte, Lombardia	>50			0.58
Gelderland Province, 1989	Netherlands	Lunteren	12			0.02
Janssen and Puijker, 1987	Netherlands	Noord-Brabant				1
Janssen and Puijker, 1988	Netherlands	Vierlingsbeek	2 – 5	10:26 (38%)[b]	0.2	1
Holden, 1986	U.S.	California				20

a These results refer to wells if not specified differently.
b Samples instead of wells.

Table 4 Studies on the Presence of Bromacil in Groundwater

Authors	Country	Region	Pedology and Hydrogeology	Groundwater Depth (m)	Positive:Examined Ratio[a]	Mean Level (µg/L)	Maximum Level (µg/L)
Milde et al., 1982	Germany	Baden-Wurttemberg				30	147
Stakelbeek, 1989	Netherlands	Bussum		24 – 32			17
Van der Zouwen et al., 1988	Netherlands	Holten		4 – 29			3.2
Vogelaar et al., 1987	Netherlands			15 – 40			1.8
Hebb and Wheeler, 1978	U.S.	FL	sand	4.5 – 6	4:21 (19%)	0.2	1250
Robbins and McCool, 1987	U.S.	KS			5:123 (4%)		16
Cohen et al., 1986	U.S.						300 (typical)
Williams et al., 1988	U.S.	CA, FL				9	22

a These results refer to wells if not specified differently.

Table 5   Studies on the Presence of Chlortal-dimethyl (DCPA) in Groundwater

Authors	Country	Region	Positive:Examined Ratio[a]	Mean Level (µg/L)	Maximum Level (µg/L)
Holden, 1986	U.S.	CA			35
Kross et al., 1992	U.S.	IA	0.4%	0.02	0.03
Spencer et al., 1985	U.S.	NY	6:10 (60%)		600 (monacid-diacid metabolites)
Cohen et al., 1986	U.S.	NY			700 (typical)
Pettit, 1988	U.S.	OR	28%, 65%[b]		431
Kessler, 1987	U.S.	WI	25%		760
Williams et al., 1988	U.S.	NY		109	1039

a These results refer to wells if not specified differently.
b Samples instead of wells.

Table 6   Studies on the Presence of Cyanazine in Groundwater

Authors	Country	Region	Pedology and Hydrogeology	Groundwater Depth (m)	Positive:Examined Ratio[a]	Mean Level (µg/L)	Maximum Level (µg/L)
Ontario Min. Env., 1987a	Canada	Ontario			18:351 (5%)		4
Janssen and Pujiker, 1988	Netherlands	Vierlingsbeek		2 – 5			0.6
Cohen et al., 1986	U.S.	IA, PA					1 (typical)
Hallberg et al., 1987	U.S.	IA			15%[b]		4.6
Hallberg et al., 1987	U.S.	IA					1
Baker and Austin, 1985	U.S.	IA	karst system		16%[b]		7.4
Kelley, 1985	U.S.	IA			7:70 (10%)		1.4
Detroy et al., 1988	U.S.	IA	alluvial and Quaternary and bedrock aquifers	<60	11:355 (3%)		3
Hallberg et al., 1987	U.S.	IA			7:184 (4%)	0.27	0.41
Kross et al., 1992	U.S.	IA			1.2%	0.3	0.84
Libra et al., 1984	U.S.	IA			8%, 9%	0.3	0.48
Isensee et al., 1988	U.S.	MD	silt loam surface, coarse-textured strata	1 – 1.4	13%[b]	0.51	1
Klaseus et al., 1988a,b	U.S.	MN			<7:725 (<1%)		2.9
Nebraska Dep. Health, 1985	U.S.	NE			23:451 (5%)		3.2
Pionke et al., 1988	U.S.	PA	from coarse to fine textured		1:20 (5%)		0.09
Kessler, 1987	U.S.	WI			10%	0.4	130
Williams et al., 1988	U.S.	IA, LA, MD, NE, PA, VT					7

a These results refer to wells if not specified differently.
b Samples instead of wells.

Table 7 Studies on the Presence of 2,4-D in Groundwater

Authors	Country	Region	Pedology and Hydrogeology	Positive:Examined Ratio[a]	Mean Level (µg/L)	Maximum Level (µg/L)
Hallberg et al., 1987	U.S.	IA		<1%[b]		0.2
Kross et al., 1992	U.S.	IA		0.6%	0.2	0.26
Stulken et al., 1987	U.S.	KS		23:27 (85%)		<1
Klaseus et al., 1988a,b	U.S.	MN		15:725	0.22	5.7
Kimball and Best, 1987	U.S.	SD	shallow sand and gravel aquifer	9:57 (16%)		6.7
Williams et al., 1988	U.S.	CT, MS			1.4	49.5

[a] These results refer to wells if not specified differently.
[b] Samples instead of wells.

Table 8 Studies on the Presence of Dicamba in Groundwater

Authors	Country	Region	Pedology and Hydrogeology	Groundwater Depth (m)	Positive:Examined Ratio[a]	Mean Level (µg/L)	Maximum Level (µg/L)
Ontario Min. Env., 1987a	Canada	Ontario			<3:351 (<1%)		2.3
Baker and Austin, 1985	U.S.	IA			2%[b]		6.1
Detroy et al., 1988	U.S.	IA	alluvial and Quaternary and bedrock aquifers	<60	<3:355 (<1%)		2.3
Klaseus et al., 1988a,b	U.S.	MN			7:725 (1%)	0.17	2.1
Kessler, 1987	U.S.	WI			10%		320
Ritter et al., 1987	U.S.		loamy sand soil	3			32
Williams et al., 1988	U.S.	CT, ME				0.6	1.1

[a] These results refer to wells if not specified differently.
[b] Samples instead of wells.

Table 9   Studies on the Presence of Dinoseb in Groundwater

Authors	Country	Region	Pedology and Hydrogeology	Groundwater Depth (m)	Positive:Examined Ratio[a]	Mean Level (µg/L)	Maximum Level (µg/L)
Schmitz, 1988	Germany						>0.1
Vogelaar et al., 1989	Netherlands	Dordrecht		4			0.16
Loch, 1987	Netherlands	Drenthe	low, moderately humic soil	1 – 2		0.9	9.2
Holden, 1986	U.S.	CA					740
Cohen et al., 1984	U.S.	Long Island, NY			6:66 (9%)		4.5
Mass. Int. Agric. P.T.F., 1986	U.S.	MA			7%		36.7
Cohen et al., 1986	U.S.	NY					5
Kessler, 1987	U.S.	WI					2100
Williams et al., 1988	U.S.	MA, ME, NY			3%	0.7	36.7

[a] These results refer to wells if not specified differently.

Table 10   Studies on the Presence of MCPA in Groundwater

Authors	Country	Region	Groundwater Depth (m)	Positive:Examined Ratio[a]	Mean Level (µg/L)	Maximum Level (µg/L)
Schmitz, 1988	Germany					2
Cormegna et al., 1992	Italy	Piemonte, Lombardia	<15	1:23 (4%)		2.01
Funari and Sampaolo, 1989	Italy	Veneto				<0.1
Gelderland Province, 1989	Netherlands	Lunteren	12			0.02
Klaseus et al., 1988a,b	U.S.	MN		<7:725 (<1%)	0.26	2.2
USEPA, 1988	U.S.	MT				5.5

[a] These results refer to wells if not specified differently.

Table 11 Studies on the Presence of Mecoprop in Groundwater

Authors	Country	Region	Groundwater Depth (m)	Positive:Examined Ratio[a]	Mean Level (µg/L)	Maximum Level (µg/L)
Hurle et al., 1987	Germany	Baden-Wurttemberg		8:50 (16%)[b]		0.6
Hurle et al., 1987	Germany	Baden-Wurttemberg		2:5 (40%)[b]		0.03
IPS, 1987	Germany		various	20:95 (21%)	<0.1	0.37
Schmitz, 1988	Germany					1000
Gelderland Province, 1989	Netherlands	Lunteren	12			0.84
Lagas et al., 1988	Netherlands	Ter Apel	1.6 – 2.1			2
Janssen, 1989b	Netherlands	Vierlingsbeek	19 – 29			0.03

[a] These results refer to wells if not specified differently.
[b] Samples instead of wells.

Table 12 Studies on the Presence of Metolachlor in Groundwater

Authors	Country	Region	Pedology and Hydrogeology	Groundwater Depth (m)	Positive:Examined Ratio[a]	Mean Level (µg/L)	Maximum Level (µg/L)
Ontario Min. Env. 1987a	Canada	Ontario			53:351 (15%)		1800
Ontario Min. Env. 1987b	Canada	Ontario			3:42 (7%)		3.7
Schmitz, 1988	Germany						0.45
Funari and Sampaolo, 1989	Italy	North					<0.1
Verdam et al., 1988	Netherlands	Valkenswaard		0.8 – 2.1	1:160 (<1%)		0.21
Cohen et al., 1986	U.S.	IA, PA					0.4 (typical)
Good and Taylor, 1987	U.S.	IL					6
Hallberg et al., 1987	U.S.	IA			4%[b]		4.6
Hallberg et al., 1987	U.S.	IA	karst system				14
Baker and Austin, 1985	U.S.	IA	karst system		7%[b]		5.9
Kelley, 1985	U.S.	IA			3:70 (4%)		0.3
Detroy et al., 1988	U.S.	IA	alluvial and Quaternary and bedrock aquifers	<60	18:355 (5%)		200
Hallberg et al., 1987	U.S.	IA			26:184 (14%)	0.68	12.2
Kross et al., 1992	U.S.	IA			1.5%	0.92	9.9
Libra et al., 1984	U.S.	IA			1%,[b] 4%		0.1
Robbins and McCool, 1987	U.S.	KS			2:123 (2%)		1.6
Klaseus et al., 1988a,b	U.S.	MN			7:725 (1%)	0.42	2.3
Thompson, 1984	U.S.	PA			6%		0.4
Kessler, 1987	U.S.	WI			14%		440
Williams et al., 1988	U.S.	CT, IA, IL, PA, WI				0.4	32.3

[a] These results refer to wells if not specified differently.
[b] Samples instead of wells.

Table 13 Studies on the Presence of Metribuzin in Groundwater

Authors	Country	Region	Pedology and Hydrogeology	Groundwater Depth (m)	Positive:Examined Ratio[a]	Mean Level (µg/L)	Maximum Level (µg/L)
Ontario Min. Env. 1987a	Canada	Ontario			21:351 (6%)		300
Cohen et al., 1986	U.S.	IA					4.3
McKenna et al., 1988	U.S.	IL	sandy, alluvial aquifer	3	21:422 (5%)[b]	0.11	0.4
Kelley et al., 1986	U.S.	IA					8.3
Hallberg et al., 1987	U.S.	IA			<1%[b]		3.6
Baker and Austin, 1985	U.S.	IA	karst system		2.5%[b]		1.9
Kelley, 1985	U.S.	IA			3:70 (4%)		0.3
Detroy et al., 1988	U.S.	IA	alluvial and Quaternary and bedrock aquifers	<60	7:355 (2%)		2
Hallberg et al., 1987	U.S.	IA			19:184 (10.3%)	0.33	6.84
Kross et al., 1992	U.S.	IA			1.9%	0.16	0.72
Libra et al., 1984	U.S.	IA			17%,[b] 17%	1.6	4.4
Snethen and Robbins, 1987a,b	U.S.	KS			8:84 (1%)		0.15
Robbins and McCool, 1987	U.S.	KS			1:123 (<1%)		0.18
Klaseus et al., 1988a,b	U.S.	MN			9:725 (1.2%)	0.32	1.1
Kimball and Best, 1987	U.S.	SD	shallow sand and gravel aquifers		2:57 (4%)		0.03
Kessler, 1987	U.S.	WI			9%		940
Williams et al., 1988	U.S.	IA, IL, KS, WI				0.6	6.8

[a] These results refer to wells if not specified differently.
[b] Samples instead of wells.

Table 14   Studies on the Presence of Simazine in Groundwater

Authors	Country	Region	Pedology and Hydrogeology	Groundwater Depth (m)	Positive:Examined Ratio[a]	Mean Level (µg/L)	Maximum Level (µg/L)
Snegaroff, 1979	France	Charente Maritime	alluvial clay soil			22	35
Hurle et al., 1987	Germany	Baden-Wurttemberg	limestone and sedimentary area			0.03	1.15
Meinert and Hafner, 1987	Germany	Baden-Wurttemberg	sandy loam soil	0.8 – 1		0.4	1.2
Werner, 1987	Germany	Baden-Wurttemberg					0.08
Roth, 1987	Germany	Baden-Wurttemberg					0.2
Hurle et al., 1987	Germany	Rheinland-Pfalz	limestone and sedimentary area		5:20 (25%)	0.06	0.15
IPS, 1987	Germany					<0.1	0.14
Schmitz, 1988	Germany						1
Funari and Sampaolo, 1989	Italy	North		12	31:2553 (1%)		10
Gelderland Province, 1989	Netherlands	Lunteren					0.09
Janssen and Puijker, 1988	Netherlands	Vierlingsbeek		2 – 5			0.5
Cohen et al., 1986	U.S.	CA, PA, MD					3 (typical)
Weaver et al., 1983	U.S.	CA			6:166 (4%)		3.5
Klaseus et al., 1988a,b	U.S.	MN			<7:725 (<1%)		2.6
Pionke et al., 1988	U.S.	PA	from coarse to fine textured		7:20 (35%)	1.4	1.7
Thompson, 1984	U.S.	PA			16%		3.4
Kessler, 1987	U.S.	WI			24%		15
Williams et al., 1988	U.S.	CA, CT, MD, NE, NJ, PA, VT				0.3	9.1

[a] These results refer to wells if not specified differently.

Table 15   Studies on the Presence of Trifluralin in Groundwater

Authors	Country	Region	Pedology and Hydrogeology	Positive:Examined Ratio[a]	Mean Level (µg/L)	Maximum Level (µg/L)
Holden, 1986	U.S.	CA	sandy alluvial aquifers			0.9
McKenna et al., 1988	U.S.	IL		51:422 (12%)[b]	0.03	0.14
Hallberg et al., 1987	U.S.	IA		2:184 (1%)	0.05	0.08
Kelley, 1985	U.S.	IA		1:70 (1%)		0.2
Kross et al., 1992	U.S.	IA		0.4%		14.9
Snethen and Robbins, 1987a,b	U.S.	KS		3:84 (4%)	5.6	5.4
Nebraska Dep. Health, 1985	U.S.	NE		23:451 (5%)		0.4
Kimball and Best, 1987	U.S.	SD	shallow sand and gravel aquifers	1:57 (2%)		0.02
Williams et al., 1988	U.S.	KS, MD, MS, NE			0.4	2.2

[a] These results refer to wells if not specified differently.
[b] Samples instead of wells.

Table 16 Studies on the Presence of Herbicides with a Poor Data Set in Groundwater

Pesticide	Authors	Country	Region	Pedology and Hydrogeology	Groundwater Depth (m)	Positive:Examined Ratio[a]	Mean Level (µg/L)	Maximum Level (µg/L)
Dichlobenil	Lagas et al., 1989	Netherlands			0.6 – 1.6			0.05
Hexazinone	Williams et al., 1989	U.S.	ME				8	9
Isoproturon	IPS, 1987	Germany			various	5:161 (3%)	<0.05	0.08
Isoproturon	Schmitz, 1988	Germany						0.1
Linuron	Williams et al., 1988	U.S.	WY				1.9	2.7
Molinate	Berri et al., 1983	Italy	Pavia			39:300 (13%)		154
Molinate	Funari and Sampaolo, 1989	Italy	North			272:1288 (21%)		9
Molinate	Holden, 1986	U.S.	CA					6.3
Molinate	Cormegna et al., 1992	Italy	Piemonte, Lombardia		<15			9.96
Pendimethalin	Funari and Sampaolo, 1989	Italy	Veneto			1:76 (1%)		<0.1
Pendimethalin	Kross et al., 1992	U.S.	IA			1.6%	0.06	0.12
Picloram	Kross et al., 1992	U.S.	IA			0.6%	0.39	2
Picloram	Williams et al., 1988	U.S.	ME, ND, WI				1.4	49
Picloram	Klaseus et al., 1988a,b	U.S.	MN			7:725 (1%)	0.16	0.63
Prometon	Williams et al., 1988	U.S.	TX			16.6	29.6	
Propachlor	McKenna et al., 1988	U.S.	IL	sandy, alluvial aquifer		4:422 (1%)[b]	0.27	2.98
Propachlor	Hallberg et al., 1987	U.S.	IA			8:184 (4%)	0.4	1.75
Propachlor	Kross et al., 1992	U.S.	IA			0.4%	0.11	0.28
Propachlor	Klaseus et al., 1988a,b	U.S.	MN			7:725 (1%)	0.35	0.52
Propanil	Cormegna et al., 1992	Italy	Piemonte, Lombardia		<15			1.23
Propazine	Funari and Sampaolo, 1989	Italy	North					0.4
Propazine	Williams et al., 1988	U.S.	NE, PA				0.2	0.2
2,4,5-T	Roth, 1987	Germany	Baden-Wurttemberg					17
2,4,5-T	Holden, 1986	U.S.	CA					1.4
2,4,5-T	Klaseus et al., 1988a,b	U.S.	MN			1:725 (0.1%)	0.21	0.21
TBA	Jernals, 1985	Sweden	Malmohus				12.5	79
TCA	Jernals and Klingspor, 1981	Sweden	Malmohus	humic sand soil	0.85 – 1		670	1270
Terbuthylazine	Hurle et al., 1987	Germany	Baden-Wurttemberg		0.8 – 1		0.49	1.92
Terbuthylazine	Hurle et al., 1987	Germany	Baden-Wurttemberg				0.05	0.07
Terbuthylazine	Meinert and Hafner, 1987	Germany	Baden-Wurttemberg	sandy loam soil	0.8 – 1		0.4	1
Terbuthylazine	Vighi et al., 1991	Italy	Piemonte					<1
Terbutryne	Vogelaar et al., 1989	Netherlands	Dordrecht		4			2.4
Tiocarbazil	Cormegna et al., 1992	Italy	Piemonte, Lombardia		<15			0.09

a These results refer to wells if not specified differently.
b Samples instead of wells.

Table 17 Studies on the Presence of Aldicarb in Groundwater

Authors	Country	Region	Pedology and Hydrogeology	Groundwater Depth (m)	Positive:Examined Ratio[a]	Mean Level (µg/L)	Maximum Level (µg/L)
Ahlsdorf et al., 1989	Germany						22.25
Verdam et al., 1988	Netherlands	Ter Apel		0.6 – 2.1			130 (aldicarb and metabolites)
Loch, 1987	Netherlands	Drenthe	low-humic sand soil	1 – 2			130
Smelt et al., 1983	Netherlands	Noord Brabant	sandy soil	2		43	22 (>90% sulfone)
Cohen et al., 1986	U.S.	AR, AZ, CA, FL, MA, ME, NC, NJ, NY, OR					
Holden, 1986	U.S.	CA					50 (typical)
Cohen et al., 1984	U.S.	Long Island, NY		2 – 45	2270:8408 (27%)		47
Soren and Stelz, 1985	U.S.	Long Island, NY					>7
Mass. Int. Agric. P.T.F., 1986	U.S.	MA			59:148 (40%)		515
Klaseus et al., 1988a,b	U.S.	MN			2:725 (0.3%)		44.2
Pettit, 1988	U.S.	OR				9	30.6
Rothschild et al., 1982	U.S.	WI			1.9%	80	10
Jones, 1987	U.S.	WI			104:1400 (7%)		210 (aldicarb and metabolites)
Kessler, 1987	U.S.	WI			309:1234 (25%)		>10
Williams et al., 1988	U.S.	CA, FL, MA, NC, NY, RI, WI				9	111
							315

[a] These results refer to wells if not specified differently.

Table 18 Studies on the Presence of Carbofuran in Groundwater

Authors	Country	Region	Pedology and Hydrogeology	Positive:Examined Ratio[a]	Mean Level (µg/L)	Maximum Level (µg/L)
Hallberg et al., 1987	U.S.	IA		1.3%[b]		0.52
Kelley, 1988	U.S.	IA		2:10 (20%)		1.2
Hallberg et al., 1987	U.S.	IA		10:184 (5.4%)	0.02	0.06
Bushway et al., 1992	U.S.	ME		13:67 (19%)		10
Holden, 1986	U.S.	MD	sandy soil			30
Mass. Int. Agric. P.T.F., 1986	U.S.	MA		16%		36.6
Cohen et al., 1984	U.S.	NY, WI	sandy soils			50 (typical)
Kessler, 1987	U.S.	WI		<1%		15
Williams et al., 1988	U.S.	MA, NY, RI			5.3	176

[a] These results refer to wells if not specified differently.
[b] Samples instead of wells.

Table 19 Studies on the Presence of Oxamyl in Groundwater

Authors	Country	Region	Groundwater Depth (m)	Positive:Examined Ratio[a]	Mean Level (μg/L)	Maximum Level (μg/L)
Mass. Int. Agric. P.T.F., 1986	U.S.	MA		<1%		1
Cohen et al., 1986	U.S.	NY, RI				65 (typical)
Williams et al., 1988	U.S.	MA, NY, RI				395
Janssen and Puijker, 1987	Netherlands	Drenthe	7 – 15	2:8 (25%)[b]	4.3	0.2
Hoogsteen, 1987	Netherlands	Noordbangeres	7 – 23		<0.1	0.2

[a] These results refer to wells if not specified differently.
[b] Samples instead of wells.

Table 20 Studies on the Presence of Insecticides and Fungicides with a Poor Data Set in Groundwater

Pesticide	Authors	Country	Region	Pedology	Positive:Examined Ratio[a]	Mean Level (μg/L)	Maximum Level (μg/L)
Aldrin	Williams et al., 1988	U.S.	MS, SC			0.1	0.1
Chlordane	Williams et al., 1988	U.S.	MS			1.7	1.8
Chlorothalonil	Williams et al., 1988	U.S.	ME, NY			0.02	12.6
DDT	Williams et al., 1988	U.S.	MS, NJ, SC			1.7	402
Diazinon	Williams et al., 1988	U.S.	MS			162	478
Dieldrin	Williams et al., 1988	U.S.	NE, NJ			0.02	0.02
Dieldrin	McKenna et al., 1988	U.S.	IL	sandy, alluvial aquifer	40:422 (9%)[b]	0.01	0.1
Endosulfan	Williams et al., 1988	U.S.	ME			0.3	0.4
Fonofos	Williams et al., 1988	U.S.	IA, NE			0.1	0.9
Fonofos	Kelley, 1985	U.S.	IA				0.9
Fonofos	Hallberg et al., 1987	U.S.	IA		3:70 (4%)		0.03
Lindane	Hurle et al., 1987	Germany	Baden-Wurttemberg		1:184 (0.5%)	0.03	0.08
Lindane	Williams et al., 1988	U.S.	MS, NJ, SC			0.04	47
Lindane	Holden, 1986	U.S.	CA			0.1	46
Malathion	Williams et al., 1988	U.S.	MS				53
Methamidophos	Williams et al., 1988	U.S.	ME			41.5	105
Methomyl	Williams et al., 1988	U.S.	NY			4.8	9
Methyl parathion	Williams et al., 1988	U.S.	MS			88.4	256
Parathion	Williams et al., 1988	U.S.	ND			0.03	0.04
Sulprofos	Williams et al., 1988	U.S.	IA			1.4	1.4
TDE	Williams et al., 1988	U.S.	MS			4.8	6.2
Toxaphene	Williams et al., 1988	U.S.	MS			3205	4910

[a] These results refer to wells if not specified differently.
[b] Samples instead of wells.

Table 21   Studies on the Presence of DBCP in Groundwater

Pesticide	Authors	Country	Region	Positive:Examined Ratio[a]	Mean Level (µg/L)	Maximum Level (µg/L)
DBCP	Cohen et al., 1986	U.S.	AZ, CA, HI, MD, SC			20 (typical)
DBCP	Peoples et al., 1980	U.S.	CA	59:119 (50%)	5	39.2
DBCP	Holden, 1986	U.S.	CA			1240
DBCP	Williams et al., 1988	U.S.	AZ, CA		0.01	0.02

[a] These results refer to wells if not specified differently.

Table 22   Studies on the Presence of 1,2-Dichloropropane (DCP) in Groundwater

Authors	Country	Region	Groundwater Depth (m)	Positive:Examined Ratio[a]	Mean Level (µg/L)	Maximum Level (µg/L)
IPS, 1987	Germany	Schleswig-Holstein	various	5:13 (38%)	0.8	5.1
Hoogsteen, 1986	Netherlands	Drenthe	25 – 40		2.2	9.3
Janssen and Puijker, 1987	Netherlands	Drenthe	35	24:52 (46%)[b]	8.5	165
Harmelink, 1989	Netherlands	Gasselte	15			5.2
Janssen, 1989a	Netherlands	Valtherbos	9 – 15			33
Cohen et al., 1986	U.S.	CA, MD, NY, WA				50
Holden, 1986	U.S.	CA		82:241 (34%)[b]		1200
Mass. Int. Agric. P.T.F., 1986	U.S.	MA		4%		51
Pettit, 1988	U.S.	OR				4
Williams et al., 1988	U.S.	CA, CT, MA, NY			4.5	550

[a] These results refer to wells if not specified differently.
[b] Samples instead of wells.

Table 23 Studies on the Presence of 1,3-Dichloropropene in Groundwater

Authors	Country	Region	Pedology and Hydrogeology	Groundwater Depth (m)	Mean Level (µg/L)	Maximum Level (µg/L)
Stock et al., 1987a,b	Germany	Niedersachsen	humic and peaty sand soil	1 – 2	2530	8620
Stock et al., 1987b	Germany	Niedersachsen	humic and peaty sand soil	1 – 5		803
Rexilius and Schmidt, 1982	Germany	Schleswig-Holstein	sandy soil	1 – 4	0.6	3.2
Rexilius and Schmidt, 1982	Germany	Schleswig-Holstein	sandy soil	11 – 24	0.23	0.89
Loch, 1987	Netherlands	Drenthe	low-moderately humic sand soil	1 – 2	5.7	80
Hoogsteen, 1987	Netherlands	Noordbargeres		7 – 23		3
Van der Pas and Leistra, 1987	Netherlands	Zuid-Holland	sandy soil	1 – 3	<0.6	2.5
Williams et al., 1988	U.S.	NY			123	270

Table 24 Studies on the Presence of Ethylene Dibromide (EDB) in Groundwater

Authors	Country	Region	Groundwater Depth (m)	Positive:Examined Ratio[a]	Mean Level (µg/L)	Maximum Level (µg/L)
Cohen et al., 1986	U.S.	AZ, CA, CT, FL, GA, MA, SC, WA				20 (typical)
Holden, 1986	U.S.	CA				140
Holden, 1986	U.S.	FL	fino a 60	828:7609 (11%)	6.5	
Mass. Int. Agric. P.T.F., 1986	U.S.	MA		15%		6.9
Pettit, 1988	U.S.	OR		3%		0.72
Zalkin et al., 1983	U.S.	SC		3:19 (16%)		0.24
Kessler, 1987	U.S.	WI		25%		12
Williams et al., 1988	U.S.	CA, CT, GA, MA, NY, WA			0.9	14

[a] These results refer to wells if not specified differently.

Table 25   Studies on the Presence of Degradation Products of Pesticides in Groundwater

Products	Parent Compound	Authors	Country	Region	Groundwater Depth (m)	Positive:Examined Ratio[a]	Mean Level (μg/L)	Maximum Level (μg/L)
Aldicarb sulfone	Aldicarb	Lagas et al., 1989	Netherlands	Ter Apel	1.1 – 1.6			61
Anthranil acid isopropylamide	Bentazone	Janssen and Puijker, 1987	Netherlands					<0.1
Anthranil acid isopropylamide	Bentazone	Janssen and Puijker, 1988	Netherlands	Vierlingsbeek	2 – 5			0.2
Deethylatrazine	Atrazine	Hurle et al., 1987	Germany	Rheinland Pfalz			0.09	0.44
Deethylatrazine	Atrazine	Schmitz, 1988	Germany					2.2
Deethylatrazine	Atrazine	Janssen and Puijker, 1987	Netherlands	Noord-Brabant	<3			0.2
Deethylatrazine	Atrazine	Lagas et al., 1989	Netherlands	Achterhock	0.6 – 2.1		<0.1	1.4
Deethylatrazine	Atrazine	Lagas et al., 1989	Netherlands	Ter Apel	1.1 – 1.6			0.1
Deethylatrazine	Atrazine	Janssen and Puijker, 1988	Netherlands	Vierlingsbeek	2 – 5			0.3
Deethylatrazine	Atrazine	Verdam et al., 1988	Netherlands	Valkenswaard	0.8 – 2.1			0.3
Deethylatrazine	Atrazine	Vogelaar et al., 1988	Netherlands	Genderen	1 – 2			0.85
Deethylatrazine	Atrazine	Lagas et al., 1989	Netherlands			43:81 (53%)		1.5
Deethylatrazine	Atrazine	Ontario Min. Env., 1987b	Canada	Ontario		13:42 (31%)		7.6
Deethylatrazine	Atrazine	Kross et al., 1992	U.S.	IA		3.5%	0.54	2.86
Deethyldeisopropylatrazine	Atrazine	Lagas et al., 1989	Netherlands			6:57 (11%)		0.37
Deethyldeisopropylatrazine	Simazine	Lagas et al., 1989	Netherlands			2:3 (67%)		1
Deethylterbuthylazine	Terbuthylazine	Hurle et al., 1987	Germany	Baden-Wurttemberg	0.8 – 1		0.05	0.07
Deisopropylatrazine	Atrazine	Janssen and Puijker, 1988	Netherlands	Vierlingsbeek	2 – 5			0.1
Deisopropylatrazine	Atrazine	Lagas et al., 1989	Netherlands	Valkenswaard	0.8 – 2.1			0.8
Deisopropylatrazine	Atrazine	Lagas et al., 1989	Netherlands	Veluwe/Utrecht	0.6 – 1.6			1.1
Deisopropylatrazine	Atrazine	Lagas et al., 1989	Netherlands			35:82 (43%)		0.98
Deisopropylatrazine	Simazine	Lagas et al., 1989	Netherlands			3:4 (75%)		1.1
Deisopropylatrazine	b	Hurle et al., 1987	Germany	Rheinland Pfalz			0.01	0.06
Deisopropylatrazine	Atrazine	Kross et al., 1992	U.S.	IA		3.4%	0.68	3.54
3,4-dichloroaniline	Propanil	Cormegna et al., 1992	Italy	Piemonte	<15			0.3
2,6-dichlorobenzamide	Dichlobenil	Lagas et al., 1989	Netherlands		0.6 – 1.6			180
2,6-dichlorobenzamide	Dichlobenil	Vogelaar et al., 1987	Netherlands		15 – 40			0.3
Hydroxyalachlor	Alachlor	Kross et al., 1992	U.S.	IA		0.2%	0.91	0.91
3-hydroxycarbofuran	Carbofuran	Kross et al., 1992	U.S.	IA		0.4%	0.38	0.98
3-ketocarbofuran	Carbofuran	Kross et al., 1992	U.S.	IA		0.4%	0.03	0.03
MITC	Metolachlor	Hoogsteen, 1987	Netherlands	Noordbargeres	7 – 23			0.6
MITC	Metolachlor	Janssen, 1989a	Netherlands	Schipborg	2 – 4			0.05
MITC	Metolachlor	Lagas et al., 1988	Netherlands	Hillegom	0.55 – 1.25			0.15
MITC	Metolachlor	Lagas et al., 1989	Netherlands	Schipborg	1			0.1

a These results refer to wells if not specified differently.
b It is not specified in the study if the compound derives from atrazine or simazine, two of the possible parent compounds.

REFERENCES FOR APPENDIX

Ahlsdorf, B., R. Stock, U. Muller-Wegener, and G. Milde. *Zum Verhalten Ausgewahlter Pflanzenschutzmittel in Oberflachennahen Grundwassern Heterogener Lockersedimente in Pflanzenschutzmittel und Grundwasser.* Schriftenreihe Des Vereins für Wasser-Boden-und Lufthydiene 79 (Stuttgart, Germany: Gustav Fischer, 1989), pp. 375–386.

Baker, J.L. and T.A. Austin. "Impact of Agricultural Drainage Wells on Groundwater Quality," U.S. EPA Report No. G007228010 (1985), 126 pp.

Berri, A., C. Allievi, and F. Burzi. "Residui di Atrazina e Molinate nelle Acque di Falda di Zone Agricole Sottoposte a Diserbo," *Acqua-Aria* 3:259–264 (1983).

Bushway, R.J., H.L. Hurst, L.B. Perkins, L. Tian, C. Guiberteau Cabanillas, B.E.S. Young, B.S. Ferguson, and H.S. Jennings. "Atrazine, Alachlor and Carbofuran Contamination of Well Water in Central Maine," *Bull. Environ. Contam. Toxicol.* 49:1–9 (1992).

Cohen, S.Z., S. Creeger, R. Carsel, and C. Enfield. "Potential for Pesticide Contamination of Ground Water Resulting from Agricultural Uses," in *Treatment and Disposal of Pesticides Wastes,* Symposium Series 259, R.F. Krueger and J.N. Seiber, Eds. (Washington, DC: American Chemical Society, 1984), pp. 297–325.

Cohen, S.Z., C. Eiden, and M.N. Lorber. "Monitoring Ground Water for Pesticides," in *Evaluation of Pesticides in Ground Water,* Symposium Series 315, W.Y. Garner, R.C. Honeycutt, and H.N. Nigg, Eds. (Washington, DC: American Chemical Society, 1984), pp. 170–196.

Cormegna, M., F. Mazzini, and G. Campus. "Determinazione in Acque di Falda di Erbicidi Impiegati in Risicoltura," *Inquinamento* 7/8:56–61 (1992).

Croll, B.T. "The Effect of the Agricultural Use of Herbicides on Fresh Waters," in *Effects of Land Use on Fresh Waters,* J.F. De L.G. Solbè, Ed. (Chichester, UK: Ellis Horwood, 1986), pp. 201–209.

Detroy, M.G., P.K.B. Hunt, and M.A. Holub. "Groundwater Quality Monitoring in Iowa: Pesticides and Nitrate in Shallow Aquifers," in *Proc. Agric. Impacts on Ground Water Conf.* (Dublin, OH: Association Ground Water Science and Engineering, National Water Well Association, 1988), pp. 225–278.

Dillaha, T., S. Mostaghimi, R. Renau, P. McClellan, and V. Shanholtz. "Subsurface Transport of Agricultural Chemicals in the Nomine Creek Watershed," American Society of Agricultural Engineering, Report 87–2629 St. Joseph, MI (1987).

Funari, E. and A. Sampaolo. "Erbicidi nelle Acque Potabili," *Ann. Ist. Super. Sanità* 25:353–362 (1989).

Gelderland Province. "Report of the Study into Pesticides at Private Water-Collection Points in Gelderland," Arnhem, The Netherlands (1989).

Good, G. and A.G. Taylor. "A Review of Agrochemical Programs and Related Water Quality Issues," Illinois Environmental Protection Agency Report, Springfield, IL (1987), 43 pp.

Hallberg, G.R. "Groundwater Quality and Agricultural Chemicals: a Perspective from Iowa," *Proc. North Central Weed Control Conf.* 40:130–147 (1985).

Hallberg, G.R., R.D. Libra, K.R. Long, and R.C. Splinter. "Pesticides Groundwater and Rural Drinking Water Quality in Iowa," in *Pesticides and Groundwater: a Health Concern for the Midwest* (Navarre, MN: The Freshwater Foundation and the U.S. EPA, 1987), pp. 83–104.

Harmelink, T.A.M. "Examination of the Leaching of Nutrients on the Hondsrug: a Study of Groundwater Quality," Drenthe Province, Assen, The Netherlands (1989).

Hebb, E.A. and W.B. Wheeler. "Bromacil in Lakeland Soil Ground Water," *J. Environ. Qual.* 7:598–601 (1978).

Hoogsteen, K.J. "Microverontreiniging in Drie Waterwingebieden," *Water* 19:48–52 (1986).

Hoogsteen, K.J. "Analysis Results of Source and Observation Wells in the Noordbargeres Pumping Station Water Extraction Areas," in *Dichloropropane and Groundwater* (Nieuwegein, The Netherlands: KIWA, 1987).

Holden, P. *Pesticides and Groundwater Quality: Issues and Problems in Four States* (Washington, DC: Academic Press, 1986), 124 pp.

Hurle, K., H. Giessl, and J. Kirchoff. *Über Das Vorkommen Einiger Ausgewählter Pflanzenschutzmittel Im Grunwasser.* Schriftenreihe Des Vereins Für Wasser-, Boden- und Lufthydiene 68 (Stuttgart, Germany: Gustav Fischer, 1987), pp. 169–190.

IPS. "Pflanzenschutzwirkstoffe Und Trinkwasser. Ergebnisse Einer Untersuchungsreihe Der Pflanzenschutzindustrie," Industrieverband Pflanzenschutz, Frankfurt, Germany (1987), 36 pp.

Isensee, A.R., C.S. Helling, T.J. Gish, P.C. Kearney, C.B. Coffman, and W. Zhuang. "Groundwater Residues of Atrazine, Alachlor and Cyanazine under No-Tillage Practices," *Chemosphere* 17:165–174 (1988).

Isensee, A.R., R. Nash, and C.S. Helling. "Leaching of Pesticides into Shallow Ground Water," Abstract of 197th National Meeting of the American Chemical Society, Dallas, TX (1989).

Janssen, H.M.J. "Pesticides in the Groundwater at Noordbargeres and Valtherbos," (Nieuwegein, The Netherlands: KIWA, 1989a).

Janssen, H.M.J. "Pesticides in Untreated Water at Vierlingsbeek. An Overview of the Analysis Results," (Nieuwegein, The Netherlands: KIWA, 1989b).

Janssen, H.M.J. and L.M. Puijker. "Bestrijdingsmiddelen in Groundwater," in *Landbouw en Drinkwatervoorziening,* C.G.E.M. Van Beek, Ed. (Nieuwegein, The Netherlands: KIWA, 1987), Mededeling No. 99, Chapter 9, 19 pp.

Janssen, H.M.J. and L.M. Puijker. "Weedkillers Applied on Maize in the Groundwater at Vierlingsbeek," KIWA Report SWE 88.009, Nieuwegein, The Netherlands (1988).

Jernåls, R. "Leaching of 2,3,6-TBA from a Sandy Soil," *Weeds Weed Control* 26:197–208 (1985).

Jernåls, R. and P. Klingspor. "Leaching of TCA from Arable Fields," *Weeds Weed Control* 22:120–128 (1981).

Jones, R. "The Aldicarb Experience 2. Results of Monitoring and Research Programs," Technical Report of the Union Carbide Agricultural Products Company, Research Triangle Park, NC (1987).

Junk, G.A., R.F. Spalding, and J.J. Richard. "Areal, Vertical and Temporal Differences in Groundwater Chemistry: II. Organic Constituents," *J. Environ. Qual.* 9:479–483 (1980).

Kelley, R.D. "Synthetic Organic Compound Sampling Survey of Public Water Supplies," Iowa Department of Water, Air, Waste Manage Report (April 1985), 32 pp.

Kelley, R.D. "Little Souix River Pesticide Monitoring Report," Iowa Department of Natural Resources Report (March 1988), 20 pp.

Kelley, R., G. Hallberg, L. Johnson, R. Libra, C. Thompson, R. Splinter, and M. Detroy. "Pesticides in Groundwater in Iowa," in *Agriculture Impacts on Groundwater* (Worthington, OH: National Water Well Association, 1986), pp. 229–351.

Kessler, K. "Wisconsin's Groundwater Monitoring Program for Pesticides," in *Pesticides and Groundwater; a Health Concern for the Midwest* (Navarre, MN: The Freshwater Foundation and the U.S. EPA, 1987), pp. 105–113.

Kimball, C.G. and W.A. Best. "Analysis and Evaluation of Field Site Groundwater Monitoring," 1987 Annual Progress Report RCWP, project 20, South Dakota State University, SD (1987).

Klaseus, T.G., G.C. Buzicky, and E.C. Schneider. "Pesticides and Groundwater: Surveys of Selected Minnesota Wells," Report Prepared for Legislative Committee Minn. Resour., Minn. Dep. of Health and Minn. Dep. Agric., Minneapolis, MN (1988a), 95 pp.

Klaseus, T.G., G.C. Buzicky, E.C. Schneider, and J.W. Hines. "Pesticides and Groundwater: Surveys of Selected Minnesota Wells," in *Proc. Agric. Impacts on Ground Water Conf.* (Dublin, OH: Association Ground Water Science and Engineering, National Water Well Association, 1988b).

Kross, B.C., M.I. Selim, G.R. Hallberg, D.R. Bruner, and K. Cherryholmes. "Pesticide Contamination of Private Well Water, a Growing Rural Health Concern," *Environ. Int.* 18(3):231–241 (1992).

Lagas, P. et al. "Pesticide Field Study: Reports from the 1st, 2nd and 3rd Inspections in 1988," RIVM Report N. 728473002, Bilthoven, The Netherlands (1988).

Lagas, P. et al. "The Threat Posed by Pesticides to Groundwater Quality," *H₂O* 14:424–427 (1989).

Libra, R.D., G.R. Hallberg, G.R. Ressmeyer, and B.E. Hoyer. "Groundwater Quality and Hydrogeology of Denovian-Carbonate Aquifers in Floyd and Mitchell Counties, Iowa," Iowa Geol. Surv., Open-file Report 84–2 (1984), 106 pp.

Loch, J.P.G. "Vulnerability of Groundwater to Pesticide Leaching," in *Vulnerability of Soil and Groundwater to Pollutants,* W. Van Duijvenbooden, and H.G. Van Waegeningh, Eds. (The Hague, The Netherlands: TNO Committee on Hydrological Research, 1987), pp. 797–807.

Massachusetts Interagency Pesticide Task Force. "1985 Summary Report: Interagency Pesticide Monitoring Program," Department of Health, Commonwealth of Massachusetts, Boston, MA (1986), 14 pp.

McKenna, D.P., S.F.J. Chou, R.A. Griffin, J. Valkenberg, L. Leseur-Spencer, and J.L. Gilkeson. "Assessment of the Occurrence of Agricultural Chemicals in Groundwater in a Part of Mason County, Illinois," in *Proc. Agric. Impacts on Ground Water Conf.* (Dublin, OH: Association Ground Water Science and Engineering, National Water Well Association, 1988), pp. 389–406.

Meinert, G. and M. Häfner. *Möglichlkeiten und Probleme Bei Der Anwendung Von Pflanzenbehandlungsmitteln in Wasserschutgebieten.* Schriftenreihe Des Vereins für Wasser-, Boden-, und Lufthygiene 68 (Stuttgart, Germany: Gustav Fischer, 1987), pp. 51–63.

Milde, G., J. Pibyl, M. Kiper, and P. Friesel. "Problems of Pesticide Use and the Impact on Groundwater," in *Memoires Prague Congress, Intern. Assoc. Hydrogeol. 1.* (Prague, Czechoslovakia: Novinar Publ. House, 1982), pp. 249–260.

Molino, C., M. Franchi, F. Giampaoli, and G. Disperati. "Esperienze nella Ricerca e Monitoraggio dei Pesticidi nelle Acque Potabili," *Acqua-Aria* 7:663–666 (1991).

Nebraska Department of Health. "Domestic Well Water Sampling in Central Nebraska: Laboratory Findings and Their Implications," Lincoln, NE (1985), 16 pp.

Ontario Ministry of Environment. "Pesticides in Ontario Drinking Water, 1985," Water Resources Branch, Toronto, Canada (1987a), 31 pp.

Ontario Ministry of Environment. "Pesticides in Ontario Drinking Water, 1986," Water Resources Branch, Toronto, Canada (1987b), 56 pp.

Peoples, S.A., K.T. Maddy, W. Cusik, T. Jackson, C. Cooper, and A.S. Frederickson. "A Study of Samples of Well Water Collected from Selected Areas in California to Determine the Presence of DBCP and Certain Other Pesticide Residues," *Bull. Environ. Contam. Toxicol.* 24:611–618 (1980).

Pettit, G. "Assessment of Oregon's Groundwater for Agricultural Chemicals," in *Proc. Agric. Impacts on Ground Water Conf.* (Dublin, OH: Association Ground Water Science and Engineering, National Water Well Association, 1988), pp. 542–563.

Pionke, H.B., D.E. Glotfelty, A.D. Lucas, and J.B. Urban. "Pesticide Contamination of Groundwaters in the Mahatango Creek Watershed," *J. Environ. Qual.* 17:76–84 (1988).

Premazzi, G., G. Chiaudani, and G. Ziglio. "Scientific Assessment of EC Standards for Drinking Water Quality," Commission of the European Communities, Report EUR 12427 EN (1989), 200 pp.

Rexilius, L. and H. Schmidt. "Untersuchungen Zum Versickerungshalten Von 1,3-Dichlorpropen und Methylisothiocyanat Auf Baumschulflächen," *Nachrichtenbl. Dtsch. Pflanzenschutzdienst* 34:161–165 (1982).

Ritter, W., A. Chirnside, and R. Scarborough. "Pesticide Leaching in a Coastal Plain Soil," American Society of Agricultural Engineering, Report 87–2630, St. Joseph, MI (1987).

Ritter, W., A Chirnside, and R. Lake. "Best Management Practices Impacts on Water Quality in the Appoquinimink Watershed," American Society of Agricultural Engineering, Report 88–2034, St. Joseph, MI (1988).

Robbins, V. and P. McCool. "Pesticide Screening of Public Water Supply Wells," Kansas State Department Health and Environment, Bur. of Water Protection, Topeka, KS, Mimeo Rep., (1987).

Roth, M. *Groundwasserbelasting Durch Pflanzenschutzmittel in Baden-württmber. Konzeptionergebnisse-ausblik.* Schriftenreihe Des Vereins für Wasser-, Boden- und Lufthygiene 68 (Stuttgart, Germany: Gustav Fischer, 1987), pp. 143–148.

Rothschild, E.R., R.J. Manser, and M.P. Andersen. "Investigation of Aldicarb Groundwater in Selected Areas of the Central Sand Plain of Wisconsin," *Ground Water* 20:437–445 (1982).

Schiavon, M. and F. Jacquin. "Etude de la Présence d'Atrazine dans les Eaux de Drainage," in *Comptes Rendus Des Journées D'études Sur Les Herbicides* (Versailles, France: Columa, 1973), pp. 35–43.

Schmitz, M. "Vorkommen Von Planzenschutzmittel-wirkstoffen in Brunnen, Uferfiltrat, Quellen- und Grundwasser," BGW, Germany (April 1988).

Smelt, J.H., A. Dekker, M. Leistra, and N.W.H. Houx. "Conversion of Four Carbamoyl Oximes in Soil Samples from Above and Below the Soil Water Table," *Pestic. Sci.* 14:173–181 (1983).

Snegaroff, J. "La Pollution des Eaux par les Triazines Herbicides dans le Secteur du Marais Charentais," *Phytiatr.-phytopharm.* 28:249–261 (1979).

Snethen, D. and V. Robbins. "Farmstead Well Contamination Study," Kansas State Department Health and Environment, Bur. Water Protection, Topeka, KS, Mimeo Rep., (1987a).

Snethen, D. and V. Robbins. "Farmstead Well Contamination Factor Study," Kansas State Department Health and Environment, Bur. Water Protection, Topeka, KS, Mimeo Rep., (1987b).

Soren, J. and W. Stelz. "Perspectives on Nonpoint Source Pollution," U.S. EPA Report 440/15–85–001 (1985), pp. 101–108.

Spalding, R.F., G.A. Junk, and J.J. Richard. "Pesticides in the Groundwater Beneath Irrigated Farmland in Nebraska — August 1978, " *Pestic. Monit. J.* 14:70–73 (1980).

Spencer, W.F., M. Cliath, J. Blair, and R. Lemert. "Transport of Pesticides from Irrigated Fields in Surface Runoff and Tile Drainage," ARS, USDA, Agricultural Conservation Research Report N. 31, Riverside, CA (1985).

Stakelbeek, A. "Bromacil at the Grindweg Pumping Station at Bussum," H_2O 1:26–29 (1989).

Steichen, J., J. Koelliker, and D. Grosh. "Kansas Farmstead Well Survey for Contamination by Pesticides and Volatile Organics," in *Proc. Agric. Impacts on Ground Water Conf.* (Dublin, OH: National Water Well Association, 1986), pp. 530–541.

Stock, R., P. Friesel, and G. Milde. *Grundwasserkontamination Durch Pflanzenbehandlungsmittel in der Niederen Geest Schleswig-holsteins und Im Emsland.* Schriftenreihe Des Vereins für Wasser-, Boden- und Lufthygiene 68 (Stuttgart, Germany: Gustav Fischer, 1987a), pp. 209–223.

Stock, R., P. Friesel, G. Milde, and B. Ahlsdorf. "Grundwasserqualitäs-beeinflüssung Durch Planzenschutzmittel," in *Impact of Agriculture on Water Resources* (Eschborn, Germany: Deutscher Verein des Gas- und Wasserfaches, 1987b), pp. 239–265.

Stulken, L.E., J.K. Stamer, and J.E. Carr. "Reconaissance of Water Quality in the High Plains Aquifer Beneath Agricultural Lands, South-Central Kansas," U.S. Geol. Surv., Water Res. Inv. Report 87–4002 (1987), 25 pp.

Thompson, E.F. "The Canestoga Headwaters," 1984 Annual Progress Report RCWP, project 19, (1984), 137 pp.

U.S. Environmental Protection Agency. "MCPA, Health Advisor," U.S. EPA, Office of Drinking Water (1988).

Van der Pas, L.J.T. and M. Leistra. "Movement and Transformation of 1,3-Dichloropropene in the Soil of Flower-Bulb Fields," *Arch. Environ. Contam. Toxicol.* 16:417–422 (1987).

Van der Zouwen, H. et al., "A Railway Through a Catchment Area: an Unguarded Crossing?," H_2O 24:700–770 (1988).

Verdam et al., "Pesticides in Groundwater Beneath Vulnerable Soils," RIVM Report 72843001, Bilthoven, The Netherlands (1988).

Vighi, M., G.P. Beretta, V. Francani, E. Funari, C. Nurizzo, F. Previtali, and G. Zanin. "Il Problema della Contaminazione da Atrazina della Acque Sotterranee: un Approccio Multidisciplinare," *Inq. Amb.* 20(9):494–510 (1991).

Vogelaar, A.J., C.G.E.M. van Beek, L.M. Puijker, and H.M.J. Jansen. "De Aanwezigheid van de Bestrijdingsmiddelen Bromacil en Amitrol in Het Gronwater Onttrokken Op Enkele Geselekteerde Winningen," KIWA Report SWO-87.329, Nieuwegein, The Netherlands (1987), 27 pp.

Vogelaar, A.J. et al. "The Influence of Agricultural Activities on the Quality of Shallow Groundwater in the Genderen Water Extraction Area," KIWA Report SWO-88.204, Nieuwegein, The Netherlands (1988).

Vogelaar A.J. et al. "Evaluation of the Threat of Pesticides to the Planned Extraction Site at De Biesbosch Polder at Dordrecht," KIWA Report SWO-89.282, Nieuwegein, The Netherlands (1989).

Weaver, D., R. Sava, F. Zalkin, and R. Oshims. "Pesticide Movement to Ground Water. Volume I: Survey on Ground Water Basins for DBCP, EDB, Simazine and Carbofuran," California Department of Food and Agriculture, Sacramento, CA (1983).

Werner, G. *Strategien und Ergebnisse der Überwachung der Rohwasserqualität von Grundwasseförderungsanlagen auf Kontamination Durch Pflanzenbehandlungsmittel.* Schriftenreihe Des Vereins für Wasser-, Boden- und Lufthygiene 68 (Stuttgart, Germany: Gustav Fischer, 1987), pp. 149–167.

Williams, W.M., P.W. Holden, D.W. Parsons, and M.N. Lorber. "Pesticides in Groundwater Data Base Interim Report," U.S. EPA, Office of Pesticide Programs (December 1988).

Zalkin, F., M. Wilkerson, and R. Oshima. "Pesticide Movement to Ground Water. Volume II: Pesticide Contamination in the Soil Profile at DBCP, EDB, Simazine and Carbofuran," California Department of Food and Agriculture, Sacramento, CA (1983).

CHAPTER 2

Agronomic Aspects of Herbicide Use

Maurizio Sattin, Antonio Berti, and Giuseppe Zanin

CONTENTS

WEEDS IN THE AGROECOSYSTEM

INTRODUCTION

As shown in the previous chapter, among the various classes of pesticides, herbicides are the compounds of major concern for groundwater. The frequency of herbicide findings and the relevance of groundwater contamination due to these products, at least in developed countries, impose a fundamental question. Are herbicides really needed, and is the large amount of herbicide used for agricultural and nonagricultural purposes really justified?

0-87371-439-3/95/$0.00+$.50
© 1995 by CRC Press, Inc.

To answer this question it is important to emphasize that weeds are an extremely complex problem, with highly different implications in relation to various agronomic, climatic, biological, and ecological situations. On the other hand, up to now, in spite of some alternative methods (see Chapter 8), chemical treatment is still the most effective strategy for weed control. This is true for weeds more than for other kinds of pests. In particular, for insects, even if chemical control represents the main practice in intensive agriculture, alternative strategies and integrated pest management can substantially reduce, if not completely eliminate, insecticide use.

Therefore, in order to understand the complexity of the problem, a detailed description of the agronomic impact of weeds and of the possibilities of weed control in different situations is needed and will be given below.

WHAT IS A WEED?

For most practitioners the answer is intuitive and almost invariably human-focused and, although the debate has been going on for a long time,[1] there is a lack of a clear-cut and generally accepted definition. This is because "weed" is a concept of humans, not of nature, and, consequently, it depends who says that a plant, at a certain time and place is objectionable or interferes with our activities.[2] There are two contrasting viewpoints: human-focused and ecological; among ecological definitions, Aldrich's[3] has been widely cited ("A weed is a plant that originated in a natural environment and, in response to imposed or natural environments, evolved, and continues to do so, as an interfering associate with our crops and activities") because it provides both an origin and continuing change of perspective. The most common and straight-forward definition still puts humans at the center and considers the weed as a plant growing in the wrong place at the wrong moment.

Weeds can be classified in various ways, Grime,[4] considering some biological traits, indicated three main plant life "strategies" (sets of traits): competitive, ruderal, and stress-tolerant. Other widely used classifications are Raunkiaer's (based on how plants survive unfavorable conditions and, in the case of perennials, on the position of the buds relative to the soil surface) and the one based on periodicity of germination.[5]

Weeds are a successful and biologicalally important component of the agroecosystem and more generally of anthropic environments and both their relative and absolute abundance have increased steadily in the last century.[6,7] This prosperity is particularly noticeable considering the enormous effort that has been devoted to their control. Such extraordinary resilience is most likely ascribable to their high phenotypic and genotypic plasticity[8] that makes them a competitive, dynamic, and persistent component of anthropic environments. Some important characteristics that contribute to make weeds successful are summarized in Table 1.

WEED COMPETITIVITY

The commonly observed proportionality between resource availability and growth and the strong influence that the latter exerts on competitive ability (i.e., the ability to suppress or reduce the growth of competitors) of a species, provide a conceptual framework for understanding the impact of neighboring plants.

Since Darwin[11] put forward his theory of evolution, competition has been considered one of the major driving forces of plant evolution and of the structure and dynamics of plant communities. Although Harper[12] has strongly and justifiably argued against the term "competition" and for replacing it with "interference", the former is still widely used and is not likely to be abandoned in the near future. There are several definitions given to the term competition,[13-16] and the debate is still ongoing.[17,18] For the purpose of this chapter, the definition given by Keddy[19] will be adopted: "the negative effects which one organism (in weed science a plant) has upon another (another plant) by consuming, or controlling access to, a resource that is limited in availability"; defining a resource as "any substance or factor which is consumed by an organism and which can lead to increased growth rates as its availability increased."[15]

Water, light, and mineral nutrients are the major resources for which weeds and crops compete,[10,20] and their availability strongly influences the competitive ability of plants[21] and consequently

Table 1   Ideal Characteristics of Weeds

Related to physiology and growth:

- high relative growth rates at seedling stage
- high rate of photosynthesis
- rapid development of exploitative root systems
- rapid partitioning of photosynthates into new leaf area production
- rapid vegetative growth to reproductive phase
- ability to compete interspecifically by special means (rosette, allelochemicals, etc.)
- freedom from environmental constraints — "general purpose genotype" with high capacity for acclimation to changing environment, tolerance and plasticity

Related to reproductive phase:

- breeding systems that provide some outcrossing but also allow self-fertilization
- copious seed production under favorable conditions with some seed production occurring over a range of favorable and stressful conditions
- continuous seed production for as long as growing conditions permit
- cross-pollination, when it occurs, by unspecialized visitors or wind
- if a perennial, vigorous vegetative reproduction or regeneration from fragments

Related to crop practices:

- morphologic and physiologic similarity to crop (crop mimics)
- timing of seed maturity to coincide with crop harvest
- resistance or tolerance to chemical herbicides
- resistance to mechanical control — regeneration from rhizomes or other vegetative propagules
- seed dormancy, longevity in soil; discontinuous germination over long periods of time

Table adapted from References 9 and 10.

weed flora composition. The mechanisms that regulate the diffusion of a species in a certain habitat can be defined on the basis of different requirements for resources that exist between this species and the others that are already established.[22] When two individuals interfere (i.e., compete), most likely their competitive abilities differ and one will become dominant (at least in the absence of particular environmental effects such as fire or grazing). The process starts with asymmetric competition between individuals and thereafter the effects of the dominant upon the other individual (subordinant) are steadily enhanced through two positive feedback loops.[19] First, the dominant sequesters more resources, so reducing availability for the subordinant, and at the same time is better able to forage for additional resources itself by reinvesting newly captured resources in new growth. This further reduces resource availability for the subordinant. Secondly, the more successful the dominant is at interfering with neighboring plants, the greater is the availability of resources for supporting its further growth. This again increases its ability to sequester resources enhancing both rates of resources acquisition and damage to the competitor. If competitors belong to the same species, the competition is called intraspecific; if not, the competition is interspecific (e.g., weeds and crops).

The variation in the effects of weeds on crop production is mainly due to the relative competitive status of the weeds and crop, which, in turn, depends on the morphologic and physiologic characteristics of the competing species. Gaudet and Keddy[23] showed that simple traits such as biomass, height, and canopy diameter explained most (74%) of the measured competitive ability. It is worth mentioning that most major weeds are not considered strong competitors in the botanical sense because they generally lack the ability to invade and survive in established vegetation.[10] In cultivated fields, weeds benefit from environmental stress-reducing practices that are used to favor the crop.

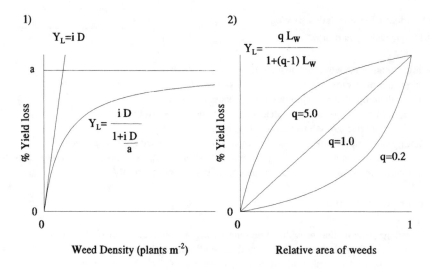

Figure 1 (1) Rectangular hyperbolic model for relating yield loss to weed density. Y_L = Yield loss; D = weed density; a = asymptote of the hyperbola; i = slope of the tangent of the hyperbola for a weed density equal to 0. (2) Hyperbolic model for relating yield loss to relative cover of weeds. L_W = relative cover of weeds ($L_W = LAI_W/(LAI_W + LAI_C)$, where LAI_W = Leaf Area Index of weeds and LAI_C = Leaf Area Index of the crop); q = relative damage coefficient of the weed. The hyperbola is plotted for three different values of q.

The level of competition is, at least for annual species, density dependent. Crop yield loss as a function of weed density[24] or weed cover[25] is generally described by a rectangular hyperbola or similar functions (Figure 1). Some competitive species (e.g., *Xanthium strumarium*) cause heavy yield loss even at very low densities showing a steep initial increase of the curve, while others (e.g., *Panicum miliaceum*) cause an almost linear increase of yield loss (Figure 2).

WEED FLORA DYNAMICS

Weed communities are continuously shifting under the selective pressure of ever-changing environmental conditions (in a broad sense) and human activities. The decline, in the more advanced agricultures (i.e., Western Europe, North America, and Japan), of "mixed" farming systems based on crop rotation in favor of intensive production has reduced weed flora diversity,

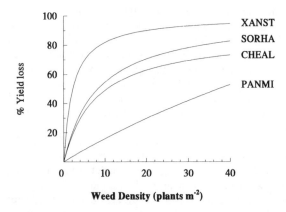

Figure 2 Examples of the hyperbolic relationship between weed density and yield loss of soybean. XANST = *Xanthium strumarium;* SORHA = *Sorghum halepense;* CHEAL = *Chenopodium album;* PANMI = *Panicum miliaceum.*

but not necessarily density.[26] The amount and organization of genetic variation within weed populations will largely determine their capacity to respond to the imposed selection pressure. If this continues for a certain length of time, some species, and within species some races (agroecotypes), will increase their frequency.[27]

From a botanical point of view, most weed species are considered colonizing plants that dominate early successional stages of vegetation in disturbed environments. A disturbance has been defined as "the mechanisms which limit the plant biomass by causing its partial or total destruction."[4] Different forms of disturbance have different selectivity and, generally, there is an inverse correlation between intensity of disturbance and selectivity. Among different forms of disturbance, a distinction can be drawn between those involving the immediate removal or killing of the plants and those that simply reduce or inhibit growth rates. The effects of different disturbances (caused by cropping practices) will be discussed more fully later in this chapter.

Anthropic habitats, and particularly arable land, are characterized by frequent and often highly predictable disturbances. This leads to a formation of ephemeral plant communities where weeds have a short life cycle with abundant and prolonged flowering and seed production. Furthermore, seeds mature quickly and have a high level of viability and longevity. The dynamics of weed communities are strictly related to the soil seed bank characteristics, of which the most important is seed dormancy. Dormant seeds confer a certain inertia to the weed flora.[28]

Weed flora can also evolve following the introduction of new weed seeds (or vegetative propagules) through intentional or, more frequently, unintentional transport by humans or animals, mechanical tools, wind, contamination of crop seed, etc. This phenomenon, called biological invasion, occurs when weed species disperse to and became reproductively established at a new site. Turner[29] distinguishes, on the basis of a geographical scale, an interregional long-distance invasion and a more localized one within a region. The former is primarily human mediated and biologicalally and agronomically more dangerous because it involves species that are less well known in the invaded site.

WEED PERSISTENCE

Most seeds undergo a period of dormancy after dispersal. Seeds that are viable but do not germinate under environmental conditions known to be suitable for the species are dormant. Dormancy may arise in a variety of ways in either the seed coat or the embryo. There are also different types of dormancy although it is frequently difficult to distinguish between them. Table 2 summarizes types, causes and characteristics of seed dormancy.

Because of dormancy, the seeds of each species present in the soil exist in a range of age classes and states of germinability and most go through annual cycles of dormant and nondormant periods until they germinate or die. Normally, dormancy is released during the season preceding the period with favorable conditions for seedling development, whereas dormancy is induced in the season preceding the period with harmful conditions for plant survival.[31]

Dormancy is the main way of seed survival in the soil.[32] Seeds on the surface and in the soil make up the soil seed bank (or simply seed bank), seeds are added to it intermittently, but lost gradually by germination, decay, and predation. In cultivated fields, the seed bank may be as high as 70,000 to 90,000 seeds per m^2 in the upper 15 to 25 cm of soil.[33] Thompson and Grime[34] draw a distinction between "transient" and "persistent" seed banks. A transient seed bank is defined as one in which none of the seed input remains in the soil for more than 1 year. All of the seeds produced in 1 year will be removed from the seed bank before they can be replaced by newly produced seeds. Instead, persistent seed banks have some seeds surviving in the soil for 2 or more years before they germinate.

The highly effective persistence mechanisms of certain weed seeds are undesirable from an agricultural point of view because only a fraction of them may germinate in any year, and conventional weed control practices do not kill ungerminated seeds. The lack of germination of buried dormant seeds, and, more generally, the existence of an effective seed bank, is a key factor that makes weeds difficult to control. What Wyse[35] says is very true: "...it may be time to declare that the weed seed pool is the cancer of weed science and deserves to become a high priority research area for state and national funding."

Table 2   Types, Causes, and Characteristics of Seed Dormancy

Type	Cause(s) of dormancy	Characteristics of the embryo
Physiologic	Physiologic inhibiting mechanisms of germination in the embryo	Fully developed; dormant
Physical	Seed coat impermeable to water	Fully developed; nondormant
Combinational	Impermeable seed coat; physiologic inhibiting mechanisms of germination in the embryo	Fully developed; dormant
Morphologic	Underdeveloped embryo	Underdeveloped; nondormant
Morphophysiologic	Underdeveloped embryo; physiologic inhibiting mechanisms of germination in the embryo	Underdeveloped; dormant

From Baskin, J. M. and C. C. Baskin, Physiology of dormancy and germination in relation to seed bank ecology, in *Ecology of Soil Seed Banks,* San Diego, CA, Academic Press, 1989. With permission.

ECONOMIC IMPACT OF WEEDS

The economic impact of weeds in agriculture is difficult to assess. Losses mainly fall into two categories: direct losses, including qualitative and quantitative reduction of agricultural output, and indirect losses due to the higher input required when weeds are present. The first category encompasses reduction in yield due to competition and lowered grain quality through contamination, while the second includes the cost of weed control (herbicides plus application), increased cost of harvesting and land cultivation, and loss in land value. Weed control and economics are inseparable in an agriculture production context.[36] Knowledge of the cost of weeds is therefore a fundamental step towards assessing the efficiency of current control methods and provides a rationale for allocation of resources for research and development of new herbicides and technologies.

When weeds are left uncontrolled, yield losses can reach 100% depending upon the duration and intensity of competition between weeds and crop, the crop itself, and the environment. In Italy, for example, the mean yield loss in absence of control measures is about 50% in sugarbeet, 33% in soybean, 24% in winter wheat, and 35% in corn.[37] Similar yield losses have been reported for winter wheat and barley by Bourdôt and Saville.[38] Some tropical crops like groundnut, cassava, and upland rice appear to be less competitive with yield losses ranging from 60 to 100%.[39] Worldwide, production losses due to weed competition, despite weed control, have been estimated at 5, 10, and 15% in highly developed, intermediate and least developed countries respectively, giving an overall average loss of 11.5%[40] During the period from 1975 to 1979, in the U.S., annual average losses caused by weeds in crops and pastures were estimated at about $7500 million (U.S.).[32] Table 3 shows the losses caused by the 35 most important weeds during the same period. It is

Table 3   Losses Caused to Agricultural Crops in the U.S. by the 35 Most Important Weeds

Weed species	% Loss[a]	Weed species	% Loss[a]
Amaranthus spp.	11.9	*Bromus* spp.	1.5
Setaria spp.	9.8	*Polygonum convolvulus*	1.4
Digitaria spp.	6.5	*Kochia scoparia*	1.2
Xanthium spp.	6.4	*Allium vineale*	1.0
Sorghum halepense	5.8	*Cassia obtusifolia*	0.8
Abutilon theophrasti	4.0	*Datura stramonium*	0.8
Echinochloa crus-galli	3.9	*Cynodon dactylon*	0.7
Ipomea spp.	3.8	*Stellaria — Cerastium* spp.	0.6
Cyperus spp.	3.8	*Lamium amplexicaule*	0.6
Chenopodium spp.	3.6	*Sida* spp.	0.6
Polygonum spp.	3.2	*Portulaca oleracea*	0.5
Panicum spp.	2.7	*Solanum* spp.	0.2
Brassica — Chorispora — Sisymbrium spp.	2.4	*Rumex* spp.	0.2
Agropyron repens	2.3	*Toxicodendron* spp.	0.2
Convolvulus — Calystegia spp.	2.1	*Eruca — Barbarea* spp.	0.1
Ambrosia spp.	1.9	*Taraxacum officinale*	0.1
Cirsium spp.	1.9	*Dactyloctenium aegytium* spp.	0.1
Avena fatua	1.6	Other weeds	11.8

[a] % of loss of the total losses for U.S. agricultural crops. Value of total losses = $7,468 billion (U.S.).

Table adapted from Reference 41.

Table 4   Direct, Indirect (% of Total) and Total Losses ($ ha⁻¹) Caused by Weeds for the Main Agricultural Crops in the Southern U.S.

	Corn	Wheat	Cotton	Soybean
(1) Loss in Yield	34.34	32.23	19.31	31.13
(2) Loss in quality of the product	0.77	9.59	5.63	1.55
(3) Cost of herbicides	40.57	41.72	51.66	51.91
(4) Loss in extra land preparation	10.24	7.44	20.94	11.95
(5) Loss in land value	0.58	0.70	0.53	0.33
(6) Loss due to increased cost of harvest	13.50	8.32	1.93	3.13
Direct losses (1+2)	35.11	41.82	24.94	32.68
Indirect losses (3+4+5+6)	64.89	58.18	75.06	67.32
Total losses ($ ha⁻¹)	33.63	10.24	39.06	37.38

Table adapted from References 42 and 43.

important to point out that the most competitive and widespread weeds cause losses greater than 5%. It has been calculated that the annual cost of wild oats to the Australian wheat industry is around $43 million (Aus.).[36] This consists of almost $30 million (Aus.) expenditure for herbicides and their application and over $13 million (Aus.) due to the loss in wheat yield caused by the competitive effects of the wild oats plants that survived the herbicide treatment. Utilizing data published by Bridges,[42,43] it is possible to estimate the losses for the main crops in the U.S. (Table 4). The repartition between reduction of agricultural output (direct losses) and additional costs (indirect losses) is roughly 34 and 66%, respectively.

The economic impact of a weed over a certain area can be estimated as follows:

$$Ct = [(Ca + Ch) + Cc] \cdot N \tag{1}$$

Ct = total cost (U.S. $); Ca = cost of application ($ ha⁻¹); Ch = cost of herbicide ($ ha⁻¹); Cc = cost of competition of the surviving weeds (value of the yield loss caused by the weeds that survive after the weed control treatment) ($ ha⁻¹); N = number of hectares (ha) infested.

The cost of competition is as follows:

$$Cc = (Y1 \cdot Ywf) \cdot P \tag{2}$$

where $Y1$ = % yield loss; Ywf = weed free yield (t ha⁻¹); P = grain price ($ t⁻¹).

The yield loss can be expressed with the hyperbolic relationship proposed by Cousens.[24]

$$Y1 = \frac{i \cdot D}{1 + \dfrac{i \cdot D}{a}} \tag{3}$$

where D = weed density; i = tangent of the hyperbola for $D = 0$ (i.e., potential competitivity of the weed at very low densities); a = asymptote of the hyperbola (i.e., maximum yield loss for D approaching infinity).

The yield loss must be calculated on the basis of weed density after the weed control treatment (Da):

$$Da = Db\ H$$

where Da = density before the treatment; H = kill rate of the herbicide (fraction of weeds controlled by the treatment).

As an example, for *Abutilon theophrasti* infesting corn in the Po Valley (Italy), $Ca = 15$ $ ha⁻¹; $N = 60,000$ ha; $Ywf = 9$ t ha⁻¹; $P = 156$ $ t⁻¹. For this species, the parameters of the weed density-yield loss relationship have been estimated at $i = 3.48\%$ and $a = 48.9\%$.[43] Considering a postemergence weed control treatment with dicamba ($Ch = 28$ $ ha⁻¹; $H = 0.9$) and a Da equal to 10 plants m⁻²,

Dp = 1 plant m^{-2}
Y1 = 3.25%
Cc = 45.6 $ ha^{-1}
Ct = $5,317,800

This calculation shows that velvetleaf, which has recently spread in the Po Valley, has a big economic impact. In reality, the impact can be greater than that calculated, because the weed control must be continued in the future in the face of the impossibility of eradicating this species when it has formed a persistent seed bank.[8]

CHEMICAL WEED CONTROL

WORLD MARKET SIZE

Economically, herbicides are the most important category of pesticides (44%), and they have shown the highest average annual increment between 1960 and 1990 (16.6%) (Table 5). In 1990, the world herbicide market reached $11,625 million (U.S.), of which 81% was concentrated in the three geographic areas where agriculture is more developed: North America, Western Europe, and the Far East (Table 6), where chemical weed control is almost invariably the main weed control technique, with a tendency to carry out more than one treatment per crop (e.g., sugarbeet).

Table 5   World Pesticide Market

Agro-Chemicals	Millions of dollars		Real growth per annum (%)[a]
	1960	1990	
Fungicides	340	5545	11.5
Insecticides	310	7655	12.4
Herbicides	170	11,625	16.6
Other chemicals	30	1575	15.7
Total	850	26,400	12.1

[a] based on dollar trend.

Table adapted from Reference 45. With permission.

Table 6   World Herbicide Market per Region (1990)

Region	Millions of dollars	%
North America	4700	40.4
Western Europe	3013	25.9
Far East	1712	14.7
Latin America	1020	8.8
Eastern Europe	990	8.5
Other countries	190	1.7
Total	11,625	100.0

Table adapted from Reference 45. With permission.

Table 7   Herbicide Use (Million Dollars) per Crop Sector (1990)

	Millions of dollars	% on total world market
Small grain cereals	2380	20.5
Corn	1830	15.7
Orchard and vegetables	1680	14.5
Soybean	1520	13.1
Rice	950	8.2
Cotton	407	3.5

Table adapted from Reference 45. With permission.

Table 8 World Market Size (Millions of Dollars) of Different Herbicide Groups and Expected Growth Rate

Herbicide Group	1972	1990	Real growth per annum[a] (%) 1972–1990	1995	Real growth per annum[a] (%) 1990–1995
Triazines	1235	1730	+1.9	1485	−3.0
Amides	495	1250	+5.3	1100	−2.5
Carbamates	590	1060	+3.3	1030	−0.5
Ureas	770	840	+0.5	730	−2.8
Dinitroanilines	420	830	+3.9	730	−2.5
Phenoxiacetic acids and related compounds	610	660	+0.4	540	−3.9
Diazines	65	750	+14.6	710	−1.1
Diphenyl-ethers	50	590	+14.7	680	+2.9
Sulphonylureas	—	530	—	1085	+15.4
Imidazolinones	—	405	—	730	+11.0
Others	345	2980	+12.7	4260	+7.6
Total	4580	11625	+5.3	13080	+2.4

[a] based on dollar trend.

Table adapted from Reference 45. With permission.

Herbicide use on cereals (including rice), soybean, and cotton makes up 60% of the market (Table 7). They are also used in nonagricultural areas; in the U.S. this sector makes up 32% of the market [43] and in Italy, about 10%.

The most widely sold herbicide families are the triazines, amides, and carbamates, which comprise some of the herbicides most commonly used on maize (atrazine, alachlor, metolachlor, EPTC), on rice (propanyl), and on soybean (metribuzin, alachlor, metolachlor). However, the herbicide families with the highest average annual growth since 1972 are the diazines and diphenyl-ethers (14 to 15%) (Table 8). The projections for 1995 show a significant decline in use for all families (1 to 4%) with the exception of two new ones: sulphonylureas (+15%) and imidazolinones (+11%).[45] Because of slow market growth in developing countries and the policy of rationalization introduced in the more developed nations, an average world annual increment of 2.4% has been forecast between 1990 and 1995, when overall spending should reach $13,080 million (U.S.). U.S. consumption of herbicides in 1987 was estimated at 293,000 t.[47] Since this market represents one-third of the world total, with a 5.3% average annual increase in herbicide use (in real terms) (Table 8) and the forecast of a 2.4% increase between 1990 and 1993, it can be assumed that annual world consumption will reach over 1 million tons in 1993.

Arable land and permanent crops cover around 1.4 billion hectares worldwide (Table 9). Assuming that between 80 and 85% of herbicides are for agricultural use and that they are distributed uniformly on cultivated land, it means that an average of about 0.5 kg ha^{-1} is used every year. Obviously, in more advanced agriculture, chemical weed control is widespread (although this varies with country and crop) and is almost invariably the main weed control technique. Table 10 gives some figures relating to the U.S. (Illinois), France, and Italy. Analyzing the global utilization of arable land, cereals (Table 11) account for 51% of the total, of which wheat, rice, and corn are the most widespread. Among other crops (Tables 12 and 13), soybean is the most widely cultivated, with 56 million ha. This gives an insight into why more than 50% of herbicides are used on a few crops, where the choice among different active ingredients (a.i.) is also wider.

Table 9 Year 1989 — Land Use Throughout the World

Region	Arable land	Hectares (× 1000) Permanent crops	Total
Africa	168,162	18,833	186,995
North-Central America	266,981	6853	273,834
South America	116,102	26,032	142,134
Asia	420,334	32,300	452,634
Europe	126,014	13,851	139,865
Oceania	49,618	999	50,617
Ex-USSR	226,100	4530	230,630
Total	1,373,311	103,398	1,476,709

Table adapted from Reference 48.

Table 10   Average Herbicide Doses (g ha⁻¹) Applied
            in Different Crops (1988)

	Corn	Soybean	Rice	Wheat	Sugarbeet
Illinois (USA)	3125	1400	—	—	—
France	2000	1600–2480	—	1140	3300–3800
Italy	3090	—	6500	1500–4100	7210

Table adapted from References 49, 50, and 51.

Table 11   Harvested Area for Cereals (1990)

	Principal cereals (ha × 1000)					Cereal total
Region	Wheat	Rice	Corn	Barley	Sorghum	(ha × 1000)
Africa	9031	5750	20,914	5323	17,799	74,832
North-Central America	43,084	1778	37,607	7858	5970	101,838
South America	9773	5583	15,962	642	1353	34,437
Asia	84,355	131,470	39,895	11,870	18,451	310,370
Europe	27,190	456	10,254	17,083	168	66,002
Oceania	9901	129	71	2601	407	14,466
Ex-USSR	48,214	610	4414	26,116	204	105,411
Total	231,548	145,776	129,117	71,493	44,352	707,356

Table adapted from Reference 48.

Table 12   Harvested Area for the Principal Oil Seed Crops (1990)

	ha × 1000			
Region	Soybean	Groundnut	Sunflower	Rapeseed
Africa	446	5877	990	95
North-Central America	23,643	87	835	2663
South America	17,757	346	2926	46
Asia	12,593	12,826	2925	10,877
Europe	1009	19	3913	3172
Oceania	56	23	77	76
Ex-USSR	835	3	4623	460
Total	56,339	19,968	16,280	17,389

Table adapted from Reference 48.

Table 13   Harvested Area for the Principal Root and Tuber Crops
            and Sugar Crops (1990)

	Root and tuber crops (ha × 1000)			Sugar crops (ha × 1000)	
Region	Potato	Sweet potato	Cassava	Sugarcane	Sugarbeet
Africa	796	1208	8932	1202	93
North-Central America	765	191	204	2701	581
South America	843	150	2517	5327	48
Asia	4868	10,236	3965	7229	1211
Europe	4718	6	—	4	3491
Oceania	48	118	16	414	—
Ex-USSR	5816	—	—	—	3267
Total	17,854	11,909	15,634	16,877	8691

Table adapted from Reference 48.

AVAILABLE HERBICIDES

The 1991 edition of the *Pesticide Manual*[52] lists 238 commonly used herbicides plus 8 safeners. At the moment, it is believed that there are more than 250 herbicides available on the world market. This enormous number of a.i. is grouped into more than 20 families, each with different chemical structures and, as a consequence, different characteristics and weed control properties. Table 14 reports, for each family, the year of introduction of the first and most recent herbicides, plus the

Table 14   Herbicide Groups

Herbicide families	no. of a.i.	Year of introduction of herbicides		
		first	median	last
Amides	23	1960	1972	1985
Aryloxyphenoxypropionate	10	1975	1985	1989
Benzoic acid derivatives	4	1952	1960	1961
Benzonitriles derivatives	4	1960	1963	1969
Carbamates	21	1951	1966	1988
Cyclohexanediones	5	1976	1985	1987
Diazines	7	1962	1963	1976
Diphenyl-ethers	11	1966	1975	1983
Dinitroanilines	11	1960	1971	1974
Imidazolinones	4	1982	1983	1984
Inorganic compounds	4	1901	—	1910
Organophosphates compounds	9	1962	1974	1984
Phenoles	3	1932	1945	1967
Phenoxiacetic acid and related compounds	11	1942	1952	1989
Pyridines	9	1972	—	1985
Quaternary ammonium salts	3	1958	—	1973
Sulfonylureas	16	1980	1985	1989
Triazines and triazinones	23	1956	1965	1974
Ureas	25	1951	1966	1979
Other compounds	27	1947	1969	1987
Safeners	8	1972	1985	1989
Total	287	—	—	—

Table adapted from Reference 52.

median year. It is interesting to note that the innovation rate of some so-called "mature" families is almost zero (i.e., triazine, ureas, and carbamates), while for others, distinguished by a median year post–1965, it is quite high (sulphonylureas and aryloxyphenoxypropanoates). Nonetheless, "mature" herbicides, such as the phenoxy-carboxylics (auxin types) (e.g., 2,4 D), the triazines (e.g., atrazine), and some amides still retain a good market share and are among the more widely used.

The continuous synthesis of new herbicidal molecules is justified both by those withdrawn from the market because of their low efficacy or unfavorable environmental and toxicologic characteristics and, by the need to deal with new or previously unsolved specific weed problems. Very often these problems are created by the continuous use of the herbicides themselves (e.g., resistant biotypes), and, as a consequence, chemical weed control becomes a short-term technique that solves a certain problem but, at the same time, creates the conditions for the appearance of others that will, in turn, be solved by other products. Thereafter, the continuous introduction of new a.i. becomes a necessity to guarantee the reliability of control.

The most recent families of herbicides (sulphonylureas, aryloxyphenoxypropanoates, cyclohexanediones, and imidazolinones) have a common set of characteristics:

1. Use at low dosage due to a strong biologicalal action. Some are also extremely effective in killing a wide range of weeds;
2. Specificity for a single target site, which often is determined by a single or very few genes. They are enzyme inhibitors with a very specific target (i.e., acetolactate synthase for the sulphonylureas and imidazolinones, acetyl CoA carboxylase for the aryloxyphenoxypropanoates and cyclohexanediones);
3. Because of the above characteristics, the probability of developing resistance phenomena is high;[53]
4. Prevalent and sometimes exclusive postemergence activity. Some are active only against grass weeds, while others are specific against certain dicotyledonous weeds (e.g., some of the sulphonylureas);
5. Wide variety of choice within each family, sometimes with similar characteristics (i.e., aryloxyphenoxypropanoates) and others with very different ones (i.e., sulphonylureas and imidazolinones).

The basic chemical structure of these new herbicide families permits the "build up" of many derivatives characterized by very different herbicidal actions, with the potential of solving many

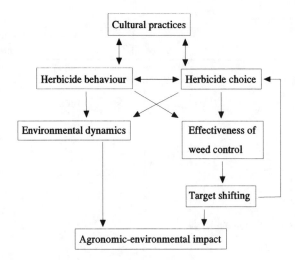

Figure 3   Relationships between cultural practices and agronomic and environmental impacts of herbicides.

of the weed control problems of major crops.[54] This set of characteristics will inevitably bring about new changes to chemical weed control. A higher number of better-targeted, postemergence treatments and a more frequent use of mixtures of different herbicides is foreseeable. Such mixtures are already used in Western Europe,[55] and by means of their wider action spectrum, better control of mixed infestations and reduction of resistance phenomena and costs can be achieved.

CHEMICAL WEED CONTROL IN RELATION TO AGRONOMIC PRACTICES

The biological and environmental effects of herbicides and agronomic techniques are strictly interdependent (Figure 3). The availability of specific herbicides allows new cultivation methods, and, on the other hand, cultural practices can modify the action and fate of the herbicides.[56] Agronomic techniques can vary the impact of chemical weed control modifying: (1) the target (i.e., the weed community), (2) the efficacy of the a.i., (3) the environmental dynamics of the herbicides (persistence, runoff, leaching etc.).

The most important cropping practices that determine the environmental impact and agronomic performances of herbicides are tillage system, crop rotation, and fertilization.

TILLAGE SYSTEMS

Traditionally, land was prepared for cropping by means of large-scale tillage operations that inverted the top soil layer, followed by secondary tillages involving additional seed-bed preparation (Conventional Tillage — CT). The need for better soil conservation management (i.e., erosion and soil fertility) and the advent of herbicides played a major role in the success of reduced tillage systems. Chisel and disk plowing are forms of Minimum Tillage (MT), in which the soil is not completely inverted as it is in moldboard plowing, and some crop residues are left on the soil surface, while No-Tillage (NT) is generally considered as the sowing of a crop onto a previous crop residue or cover crop, where the only soil disturbance is caused by the planting equipment.

In MT and NT systems, the crop residues act as a mulch, affecting the top soil layers by lowering evaporation and gaseous exchanges and raising moisture content. The mulch also exerts a moderating effect on temperature fluctuations by influencing thermal conduction, thermal diffusion, and heat storage,[57] therefore creating a slow warming of the soil during spring. For this reason, in temperate regions, NT causes sowing to be delayed and slow initial crop growth.[58] Soil under MT and NT also shows less compaction, which gives higher microporosity and greater moisture retention[59] and consequently a lower temperature, which influences the flora composition. Moisture loving species, such as *Equisetum* spp.[60] are favored in comparison to others with better water use efficiency (C_4 plants), especially during initial growth stages. Increased earthworm activity and permanent soil cracking phenomena also determine an increase in the water infiltration rate.[61] With repeated MT and NT the soil pH decreases (even by more than 1 unit)[62] due to the accumulation of organic matter and nitrogen in the top soil layers.[63]

Under MT and NT, rhizomes or root systems of perennial weeds are not fragmented, giving rise to denser but more localized infestations and weed seeds tend to accumulate in the surface soil layers,[64-66] where they are in a better position to germinate (especially for the species producing seeds with low dormancy), but are also subject to predation. In this situation, if weed control is effective, the seed bank in the surface layer declines rapidly, but, if it is not, a substantial and rapid increase is observed.[67] Generally, with MT and NT, some annual grasses characterized by low seed longevity and difficulty to control (e.g., *Bromus* spp., *Poa* spp. *Avena*. spp.)[65,68-71] and biennial and perennial weeds are likely to increase.[72,73] Additionally, some species that produce small wind disseminated seeds and that are normally unable to establish themselves in disturbed fields may become increasingly important; examples include *Conyza canadensis, Lactuca serriola,* and *Aster squamatus.* MT and NT also favor the establishment of other weeds such as *Sambucus ebulus, Rubus* spp. and *Phytolacca americana,* which have bird dispersed seeds.[74] This is due to the more frequent presence of birds on uncultivated land where earthworms and seeds are more abundant.

Herbicide activity is strongly influenced by the tillage system. With MT and NT, the higher content of organic matter in the upper horizons acts in two ways: first, the mulch acts as a "shield", intercepting part of the herbicide and so less chemical reaches the soil and weed seeds. This capture by plant residues may or may not influence herbicide efficacy depending on the timing of rainfall: if it rains immediately after the treatment, herbicide performance is similar to that under CT. Second, the higher content of decomposed organic matter in the upper soil layer may bind the herbicides, thereby decreasing their effectiveness[75] and affecting their degradation. The lower pH increases the hydrolysis of most herbicides, reducing their persistence and increasing adsorption, particularly of s-triazine[76] and those that are weak bases, and favoring the activity of fungi, which are very effective in the degradation of adsorbed herbicides.[62] Hence, simplifying the tillage system reduces the efficacy (bioavailability) of preemergence root-absorbed herbicides. This, coupled with the difficulty of carrying out mechanical weed control, forces farmers who practice MT or NT into a substantial change in chemical weed control, varying doses, application time, herbicide choice, and number of treatments. The balance among the different effects of reduced cultivation favors a lessening of herbicide persistence.

Water-borne removal of herbicides from a field can occur by surface runoff, laterally by diffusion, and vertically by leaching.[77] MT and NT are valid tools against soil erosion and therefore against the loss of herbicide molecules adsorbed on the soil. On the contrary, herbicide losses in runoff water following intense rains (herbicide molecules adsorbed on the soil plus solute molecules) can be higher due to the "wash off" of herbicide molecules retained by the mulch. Losses in runoff water are related to the chemical characteristics of the molecules and particularly to the partition coefficient between soil organic carbon and water (Koc): the lower the Koc, the greater the runoff losses.[78] Agronomic practices that reduce erosion are therefore effective in reducing herbicide losses only for molecules with a high Koc. The critical period for herbicide runoff is between 0 and 15 days after the application date.[79] Leaching of herbicides is caused by water movement through the soil, which, in turn, depends on the amount and pattern of rainfall and on soil type and the partitioning of the herbicide between the soil solution and soil matrix. Herbicide leaching in MT and NT fields is normally greater than in CT because of the higher infiltration rate due to the greater porosity.[9]

CROP ROTATION

Rotation, designed to maintain soil fertility, plays a major role in weed control. When different crops are grown in rotation, the chances of any weed species gaining predominance are minimal since the cultural practices for each crop and the competitive ability of the crops will vary, alternately favoring some species and discouraging others. The result is a mixed weed flora, with both annuals and perennials. The intensification of crop production, characterized by monocultures and the "heavy" use of inorganic fertilizers and herbicides, coupled with a reduced labor force, has led to the virtual disappearance of crop rotation from more advanced agriculture in temperate regions. In these situations, the weed flora is very simplified with a few predominant species, ecologicalaly well adapted to the soil and crop management system. Several authors have reported weed flora changes when monoculture has replaced crop rotation.[80-82] In monoculture, the continuous use of some a.i. (i.e., EPTC, 2,4D, and other herbicides that are degraded by direct metabolism)[83] also causes an adjustment to the microbial population of the soil, selecting the microorganisms capable of rapidly degrading the herbicides used. This can greatly affect the persistence and therefore the agronomic efficacy of some molecules. Crop rotation minimizes this effect, preserving the efficacy of herbicides.

Crop rotation, therefore, permits a simple weed control, allowing the use of different herbicides at lower doses and makes the "disturbances" more flexible and less dependent on weather conditions and farm management.

FERTILIZATION

Like crop rotation, fertilization influences chemical weed control through its effect on the weed flora: any variation in soil fertility alters the competitive balance among species in the agroecosystems.[10] Long-term studies[84-86] have demonstrated substantial effects of soil fertilization, especially nitrogen, on weed flora. Grasses and nitrophilous dicotyledons are favored,[87] and the weed community diversity is reduced. Usually the crop benefits from high soil fertility, but the opposite may occur. For example, *Lolium multiflorum* responds better to increased nitrogen fertility than winter wheat.[88] If the crop is favored, its higher growth rates, both in terms of weight and leaf area, enhance its competitivity causing a higher seedling mortality of the less competitive weed species and late emerging plants and leads to a reduction in density and number of species.

Nitrate and nitrite ions in the soil can also break dormancy and stimulate seed germination in many species. Popay and Roberts[89] observed that peaks of germination in *Capsella bursa-pastoris* and *Senecio vulgaris* were correlated to high soil nitrate levels.

The effects of fertilization are modified by weather conditions, weed density, and timing of distribution, and increased fertility generally requires a more intensive and efficient weed control.

CHEMICAL WEED CONTROL IN RELATION TO WEED COMMUNITIES

As early as the mid-1950s, Harper[90] predicted shifts in weed flora composition caused by repeated annual herbicide application. Today, interspecific selection caused by herbicides is considered by far the most important factor contributing to changes in weed flora composition.[91] Chemical weed control influences weed communities by means of the differential kill-rates of the herbicides on different species and, within the species, on individuals. These effects are largely due to plant diversity in such characteristics as age, growth rate, morphology, physiology, and biochemical processes. All weed species are collections of biotypes, and, therefore, there is wide genetic variability within populations, which strongly determines their response to selection pressure imposed by the local physical and biotic environment and herbicide stress.[2] Any plant characteristics with a genetic basis that carries resistance to a herbicide may be subjected to selection.

Often the result of an intensive and less than careful chemical weed control is an unbalanced weed flora, which is difficult to control and requires the use of new and expensive herbicides, often creating an endless loop. The two main phenomena involved in the alteration of weed communities as a consequence of herbicide use are compensation and resistant biotypes.

COMPENSATION PHENOMENA

Frequent use of the same selective herbicide (or even herbicides with similar modes of action) on the same fields has imposed new selection factors, either by radical changes in interspecies competition or by increasing tolerance or even resistance of previously susceptible species.[92]

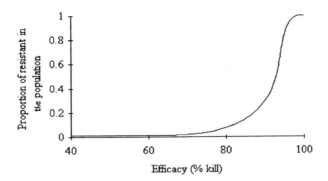

Figure 4   Simulated influence of herbicide efficacy on the maximum level of resistance achieved in a weed population that had 5 continuous years of herbicide application. (From Maxwell, B. D., M. L. Roush, and S. R. Radosevich, *Weed Technol.* 4, 2–13, 1990. With permission.)

Selective pressure will be much greater and, consequently, flora composition will change more substantially and rapidly, the wider the action spectrum and efficacy of the herbicide used (Figure 4). This occurs because ample ecological niches are freed and subsequently filled by the selected weeds. Wide-spectrum herbicides are normally used for presowing (applied before sowing a crop) and preemergence (applied after a crop has been sown but before it appears above the soil) treatments. Better targeted herbicides are used for postemergence treatments (applied after the crop and weeds have appeared above the soil) when the a.i. can be chosen on the basis of the real weed flora.

The selected species are usually phylogenetically similar to the crop, very competitive and difficult to control, and they are favored by the agronomic practices used to increase crop yield. In these conditions, the weed communities become very unbalanced with a dense population of few species and within a few years (4 to 5) the target can change to the point of forcing the weed control program to be completely reset. This kind of weed flora evolution is called compensation.[26] The spread of the graminaceous panicoid weeds (*Sorghum halepense, Digitaria sanguinalis,* etc.) in corn crops treated with atrazine or festucoid grass weeds (e.g., *Alopecurus myosuroides* and *Avena* spp.) in wheat treated with auxin-type herbicides are examples of compensation.

RESISTANT BIOTYPES

The continuous use of certain herbicides may, in the medium-long term (5 to 20 years depending on the situation), determine the appearance of resistant biotypes. Despite the fact that resistance to herbicides was the most recent pesticide resistance to appear, it now causes most concern.

It is worth mentioning that the terms "tolerance" and "resistance" are often used interchangeably; Holt and Le Baron[94] refer to tolerance as "the normal variability of response to herbicides present among plant species." Therefore, one species may be more tolerant than others to a given herbicide, the use of which can cause its increase in the weed population even though all may be controlled to some degree. Resistance implies an altered response to a herbicide by a previously susceptible weed species. Herbicide resistance may therefore be seen as the extreme level of herbicide tolerance. Herbicide resistance can be due to: (1) modification of the biochemical/metabolic target site of a herbicide, and (2) enhanced metabolic detoxification of the herbicide.

Target Site Resistance

Many herbicides act as antimetabolites, blocking the activity of an enzyme by binding to it. Single gene mutations that modify an amino acid can be enough to change the structure of the protein (target site), thus precluding herbicide binding.

The most extensively studied example is the development of weed biotypes resistant to atrazine. This resistance, due to a point mutation in the chloroplast gene that encodes the herbicide binding protein of photosystem II, reduces the susceptibility of the photosynthetic electron transport to symmetrical triazines by roughly 1000 fold.[95] The first report of triazine resistant weeds occurred in 1968 with common groundsel (*Senecio vulgaris*) in western Washington State.[96] It is currently known that at least 57 species of weeds (40 broad-leaved and 17 grasses) have evolved resistance to triazine herbicides somewhere in the world,[46] and it is estimated that over 3 million ha are infested with these genotypes.[55] Resistant weeds are present in all the sectors where triazines are used (corn, vineyards, orchards, railways, etc.).

Some cases of target-site resistance have also been found with dinitroanilines, sulfonylureas, and aryloxyphenoxypropanoates. For dinitroanilines, which are mitotic disrupters that bind to tubulin protein and prevent the formation of microtubules, the resistance is due to an alteration in a tubulin protein. Resistant biotypes of goosegrass (*Eleusine indica*) and green foxtail (*Setaria viridis*) were found in cotton monocultures in Maryland (U.S.)[97] and in barley crops in Canada, respectively, after continuous use of trifluralin.[98]

Another case of target-site resistance involves several populations of grass weeds.[99] These have evolved resistance to diclofop-methyl, an herbicide of the aryloxyphenoxypropanoate family, which is an inhibitor of acetyl CoA carboxylase enzyme.

Resistance to the sulfonylureas seems to be much more dangerous, considering the importance that this family will have in the future and the promptness of the appearance of the first cases of resistance: biotypes of *Lactuca serriola* and *Kochia scoparia*. Schrad were selected after only 4 to 5 years use of a persistent sulphonylurea herbicide.[53] So far, six weed species have shown resistance to this herbicide. Sulphonylureas kill plants by inhibiting the enzyme acetolactate synthase (ALS), that catalyzes the first step in branched chain aminoacids biosynthesis. Resistant weeds have a mutant ALS enzyme that is no longer sensitive to the herbicide. The same action mechanism is shown by the imidazolinones, but the weeds resistant to one family of herbicides are not always resistant to the other because of multiple mutations occurring within the ALS gene, giving rise to different resistance levels.[100]

Enhanced Metabolic Detoxification

The principal mechanisms of detoxification are based on the mono-oxygenase enzyme and on glutathione-S-transferase enzyme. Among crops, the former is present especially in wheat and soybean, which can detoxify several herbicides even if these have very different action mechanisms, while maize has the latter and also other mechanisms that enable it to degrade selective herbicides.[101]

In weeds the appearance of this type of resistance is due to an abrupt rise in detoxification level. Resistance of populations of *Alopecurus myosuroides* in England are ascribed to greater mono-oxygenase activity: these weeds have evolved a biochemical mimicry, that is, a similar mechanism to that developed in wheat,[102] and, consequently, they are also resistant to most herbicides used for wheat. This phenomenon is termed cross-resistance and was already well known for insects.[103] Cross-resistance is defined as the resistance of a pest (e.g., a weed) to a range of pesticides, and it is used in weed science to describe increased resistance to an herbicide as a result of the selection pressure from others.[104] These include a.i. with the same or different action mechanisms. This may indicate that cross-resistance could have several metabolic bases and recent evidence suggests three different possibilities:[105]

- an overproduction of the target-site enzyme in resistant plants;
- an altered enzyme that is less sensitive to the herbicide;
- the development of an enhanced ability for herbicide detoxification, so that less herbicide reaches the target-site.

Resistant biotypes are often less fit than susceptible ones.[106,107] Therefore, resistant plants will be replaced by susceptible individuals over time after a certain herbicide is abandoned or even changed (Figure 5). Alternatively, if the fitness of the resistant biotype is not, or only slightly, inferior to the susceptible one,[108] resistance will decline slowly, if at all.

The development of resistance is one of the most worrying aspects of weed biology: the situation is evolving and extremely complex. Most herbicide families are implicated, and single weed species have developed different resistance mechanisms The establishment of herbicide resistance and tolerance must be prevented, especially for herbicides with a single target site, which

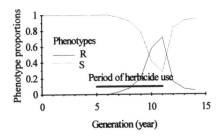

Figure 5 Model simulations describing the evolution of resistance after the introduction and continuous use in the system of an herbicide and subsequent dynamics of resistance in the weed population after the herbicide has been removed. (Maxwell, B. D., M. L. Roush, and S. R. Radosevich, *Weed Technol.* 4, 2–13, 1990. With permission.)

is often determined by a single or very few genes. Newer herbicides (i.e., imidazolinones and sulfonylureas) belong to this category, and resistance is therefore very likely to develop. The key point is to lower herbicide selection pressure: this can be achieved with a set of expedients that include mixing and rotating the a.i. and a wider recourse to mechanical cultivation and crop rotation. In addition to this, herbicides with different chemical characteristics and different action mechanisms should be used to avoid or reduce cross-resistance. It is encouraging that herbicide resistance has not appeared where herbicides have been rotated and mixtures used. No single strategy will be effective by itself.[109]

If resistance phenomena appear, it is essential to understand the type of resistance because the control methods must be formulated on this basis. For instance, dose reduction is a valid tool to prevent target-site resistance: low rates allow greater "dilution" of resistance by susceptible individuals, delaying the appearance of resistant biotypes, while high rates allow only resistant individuals to remain. On the other hand, this tool is not a valid one against the enhanced metabolic detoxification, which is a polygenetically inherited resistance.[55]

CHEMICAL WEED CONTROL IN RELATION TO THE CROP

Herbicide choice is based on selectivity towards the crop, on efficacy in controlling weeds (kill rate), and, sometimes, on its persistence in the soil and on the type of crop rotation.

In general, selectivity is the expression of the differential response of plant species to the applied herbicide. It is therefore the ability, intrinsic or induced, of herbicides to kill certain plants either without any or with limited and acceptable harm to the desired plant (crop). The latter is not a very desirable situation, as even very slight alterations in herbicide activity (e.g., induced by different environmental conditions) may cause poor weed control and/or crop injury. Selectivity depends on numerous interrelated factors. It is a relative rather than an absolute characteristic, influenced by environmental factors and by the kind and amount of herbicide applied, as well as when and how it is applied. Even the most tolerant plant species will be killed if the dosage of herbicide is large enough.

It is possible to distinguish three different types of selectivity:

- selectivity for lack of or minimum interaction between plant and herbicide;
- physiologic selectivity;
- selectivity by manipulation of crop tolerance.

The first type of selectivity is characterized by a separation (complete or almost so) in time and/or space between herbicide and plant. There are several factors that can induce this type of selectivity: different method and time of application, root distribution in soil, and morphologic and ecological characteristics of the plants. Different methods and time of application allow the use of nonselective herbicides if required by specific weed problems. For example, contact herbicides such as glyphosate or paraquat can be used prior to crop emergence or on dormant perennial plants. Treatments carried out under the crop canopy or with shields (in such a way that the herbicide is applied in bands between crop rows) allow the control of weeds that escaped preemergence or early postemergence treatments, by using more efficient but less selective herbicides (e.g., auxin-type herbicides against field bindweed — *Convolvulus arvensis* in corn).

Selectivity based on the different positions through the soil profile of absorbing organs (i.e., roots, coleoptilar node) in relation to herbicides (also defined as positional or placement selectivity), is given by the stratification in the upper layer of the soil of some herbicides. This is a consequence of their low solubility and the strong adsorption by the colloids in the soil. For example, velvetleaf escapes most of the preemergence herbicides applied on a corn crop because of its early and high root elongation rate.[110]

By choosing the proper herbicide and the right crop and weed growth stages, selectivity may be achieved through differential uptake of the herbicide. This must be in contact with plant surfaces long enough to be absorbed in sufficiently large amounts to cause phytotoxicity. Retention, and consequently uptake, is largely influenced by the morphology of the plants (arrangement, shape, number, angle of leaves, position and type of protection of shoot apex and buds, characteristics of the waxy coatings). The selectivity of auxin-type a.i. towards cereal crops is due, at least in part, to these reasons. This selectivity, like many others, does not by itself confer an optimum weed control without damaging the crop.

Selectivity may also be achieved by physiologic differences between weeds and crop. These differences are related to the behavior of the herbicide after it has entered the plants. With this type of selectivity, herbicide and plant therefore interact, but the latter is not damaged either because the target-site is not sensitive (for example in dicotyledonous weeds, the target site of the aryloxyphenoxypropanoates and of the cyclohexanediones — the enzyme acetyl CoA carboxylase — is insensitive to these herbicide families) or, when it is, because the molecule does not reach it. This can be achieved in various ways: (1) some plants are able to divert the herbicides toward "sinks" that are metabolically insensitive to them; (2) translocating the herbicides very slowly (e.g., prometryn in cotton[111]) or very quickly (e.g., diphenamid in tomato[112]) in relation to the site of activity and the site of entry; (3) by detoxification before it reaches the target site. Physiologic selectivity is by far the safest.

Finally, selectivity can be induced by chemical manipulation of the tolerance of crop plants to the herbicides. This is obtained by adding chemical safening agents,[113] commonly referred to as "herbicide safeners," to the commercial formulation. Safeners improve the usefulness of a herbicide by increasing the ratio of crop safety to weed control and by obtaining a wider weed control spectrum covering weeds that are closely related to the crop. Among the mechanisms proposed to explain the action mechanism of safeners, two have received most attention and suggest either a safener-induced enhancement of herbicide detoxification or a competitive antagonism of herbicide and safeners at a common target site.[114]

As mentioned above, herbicide selectivity is not the only factor taken into account in the process of choosing the herbicide to be used, sometimes persistence and type of crop rotation are considered. Persistence becomes important when the following crop is sensitive to the a.i. chosen for the actual crop.[115,116] The key factor is the time between the treatment to the actual crop and the sowing time of the following crop in relation to the degradation of the a.i. This aspect is particularly important in horticultural production, where crops closely follow one another (in 1 year, three or even four crops can be grown on the same field). The availability of very selective herbicides for certain crops, which are also very effective against certain weeds, is also an important point in relation to the crop rotation adopted. With a corn–soybean rotation, it is more convenient to control bindweed in corn rather than to control it in soybean where the herbicides controlling bindweed are more expensive and less efficient. Ideally, chemical weed control should therefore be determined by looking at the entire crop rotation rather than at the single crop.

WEED CONTROL IN NONAGRICULTURAL AREAS

Herbicides are also widely and ever-increasingly used in nonagricultural areas (NAA). In the U.S., this sector represents 23% in value of the market,[117] while in Italy it is about 10%. NAA include very diverse areas: industrial sites (parking and storage areas, power lines, petroleum refineries, etc.), roadsides, railway tracks, archaeological areas, and amenity and sport areas.

In these situations, the concept of a weed is more hazy, and the damage can be of a different nature. Often it is an aesthetic damage, at other times unwanted vegetation can create serious hazards (fire, metal corrosion, etc.) or damage paved surfaces, reduce visibility, obstruct drainage, etc. In any event, unwanted vegetation is a source of injurious weeds for neighboring field crops.

The products and criteria used for weed control can change according to different needs. Sterilization to maintain bare soil is a requirement in many areas such as railways (particularly switches), parking areas, etc. This can be achieved with very persistent herbicides such as the residuals (triazines, ureas, uracils, etc.), but also contact or systemic herbicides (paraquat, glyphosate, auxin-type herbicides) are used to kill existing vegetation or particularly resistant weeds.

In the nonselective weed control, doses can be relatively high and residual herbicides can remain for a number of years before being totally degraded. The choice of a.i. is usually based on cost and persistence, but low solubility and strong soil retention are preferred attributes and should also be taken into account to prevent leaching. The possible effects of herbicide application on nontarget areas and their movement by leaching, runoff, and volatility should always be considered when treating areas in close proximity to desirable vegetation or water caption areas.

Despite the limited amounts of herbicides used in NAA, environmental impact is usually quite high in comparison with agricultural lands. Working in Britanny (France), Gillet[118] showed that losses of simazine, which is mainly used in NAA, reach 40 to 50%, while atrazine losses, used specially in agricultural land, are much lower, with values around 2 to 3%. The higher losses observed in NAA are mainly due to the intense runoff observed for surfaced areas. Gillet also showed that, after an intense rainfall, the simazine concentration in canals and rivers after they had passed through a town was 3- to 20-fold higher than before.

In turfs for sport, uniformity of sward and color are the main aesthetic factors for an ideal playing surface. In this context, even very few weed plants can be a problem (e.g., golf courses) and the herbicide-use criteria are similar to those for selective weed control in a crop. Quite often the weeding is completed by hand.

In other cases (e.g., archaeological areas, roadsides), vegetation management is widely used for controlling undesirable species (mainly woody plants and some dicotyledons) and favoring grasses (e.g., roadsides where a mown strip ensures visibility and prevents encroachment).

AGRONOMIC CONSEQUENCES OF THE BANNING OF A HERBICIDE

Prohibiting the use of a herbicide can be for either toxicologic or environmental reasons. As an example, Table 15 reports the prohibition measures involving herbicides in Italy in recent years.

The agronomic consequences of banning a herbicide are obviously tied to the importance of the a.i., an importance that depends on the one hand on its diffusion in agricultural practice and, on the other, on the presence of alternative herbicides and their cost. The use of a herbicide is more widespread when it is more selective, efficacious, economical, and flexible in use.

The Italian banning of nitrofen in 1981 did not have particular consequences on wheat production, as there were numerous alternatives available with comparable efficacy and costs. The only advantage of this a.i. over the alternatives was its ability to control *Phalaris* spp.,[119] and the increasing spread of these weeds is in part attributable to the banning of nitrofen.

Instead, the consequences of the ban on atrazine have been considerable as this herbicide was used, alone or in a mixture, on almost every corn field, it cost little and was very efficacious and extremely selective. Moreover, it was very flexible, being used both in pre- or post-emergence treatments. It was also reliable, providing good weed control regardless of the pedoclimatic conditions. Atrazine was banned in March 1990, forcing farmers to use a large number of herbicides to achieve a sufficiently wide control spectrum. This had some limited advantages, in that it was necessary to resort to a rotation of a.i., which led to less selection pressure on the weed flora and less probability of the appearance of compensation phenomena. The disadvantages, however, were much greater, because it was necessary to use products that were much less understood toxicologically, were more expensive (e.g., cyanazine), less selective (e.g., linuron), less flexible in use (e.g., terbuthylazine), and less reliable (Table 16). The ban on atrazine increased production costs, since the alternatives are more expensive and caused a reduction in production potential because of lower weed control and less selectivity (Figure 6). Vighi et al.[120] estimated that

Table 15   Banning Measures in the Use of Herbicides in Italy

Herbicide	Motive	Year
2,4,5 T	Toxicology	1970[a]
2,4,5 TP	Toxicology	1970[a]
amitrol	Toxicology	1974[a]
diallate	Toxicology	1977[a]
triallate	Toxicology	1977[a]
sulfallate	Toxicology	1977[a]
nitrofen	Toxicology	1981[a]
atrazine	Environmental contamination	1986,[b] 1987,[b] 1988,[b] 1990[a]
molinate	Environmental contamination	1987,[b] 1988[b]
bentazon	Environmental contamination	1987,[b] 1988[b]

[a] Banned.
[b] Restriction of use.

Table adapted from References 121–127. With permission.

Table 16   Characteristics of Atrazine and of Alternative Preemergence Herbicides

Herbicides	Persistence (months)	Action reliability	Selectivity
atrazine	6–9	very good	very good
terbuthylazine	6–10	good	very good
cyanazine	2–3	very good	good
linuron	3–5	average	average
pendimethalin	4–6	average	average

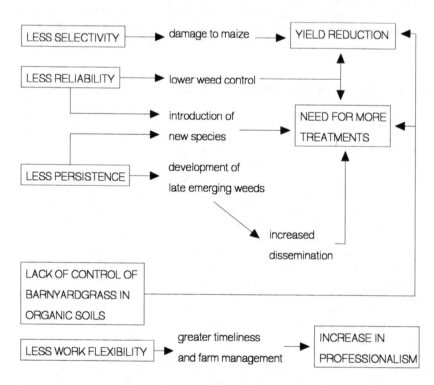

Figure 6   Negative agronomic and weed control consequences of the substitution of atrazine by preemergence herbicides.

Table 17   Herbicides Banned or with Restricted Use in Italy, their Sector of Use and Alternatives Introduced Following the Ban

Banned herbicide	Crop	Target	Alternative Herbicide
2,4,5 TP	Rice	Cyperaceae *Alisma* spp. *Butomus* spp.	bentazon
triallate	Wheat	*Avena* spp.	diclofop-methyl
molinate	Rice	*Echinochloa* spp.	dimepiperate
atrazine	Corn	Broad-leaved weeds	thifensulfuron rimsulfuron
bentazon	Rice	Cyperaceae *Butomus* spp.	bensulfuron

the cost of the atrazine ban in Italy was 0.2 t ha^{-1} of grain, which, in monetary terms, is \$47 to 48 (U.S.) ha^{-1}.

This cost tends, however, to diminish with time, as farmers improve their expertise, and, normally, new alternative a.i. appear (Table 17). The most difficult phase after the banning of an important herbicide lasts between 2 and 4 years. During this period, the farmers alter the weed control technique, exposing themselves, however, to possible errors, and alternative herbicides become available.

Experience from the atrazine ban in Italy has shown that a government ban cannot be an isolated and unexpected act but must be accompanied by instructions and advice to farmers and planned so that there is a brief transition phase involving little cost. This is particularly important when the ban is for environmental reasons, as the decision on alternative techniques is made, in the final analysis, by the farmer. It is this phase that must be managed and directed with care, to avoid inducing even worse environmental effects than those that caused the ban.

REFERENCES

1. Zimdhal, R.L. "Weed science. A plea for thought," U.S.D.A. Cooperative State Research Service, A Symposium Preprint (1991).
2. Norris, R.F. "Have ecological and biological studies improved weed control strategies?," *Proceedings of the First International Weed Control Congress,* Monash University, Melbourne, Australia, J.H. Combellack, K.J. Levick, J. Parsons, and R.G. Richardson, Eds. (Melbourne, Australia: Weed Science Society of Victoria, Inc., 1992), pp. 7–33.
3. Aldrich, R.J. *Weed-Crop Ecology: Principles in Weed Management,* (N. Scituate, MA: Breton Publisher, 1984).
4. Grime, J.P. "The C-S-R- model of primary plant strategies. Origin, implications and tests," in *Plant Evolutionary Biology,* L.D. Gottlieb and S.K. Jain, Eds. (London: Chapman and Hall, 1988).
5. Roberts, H.A., R.J. Chancellor, and T.A. Hill. "The biology of weeds," in *Weed Control Handbook: Principles,* H.A. Roberts, Ed. (Oxford, England: Blackwell Scientific Publications, 1982) pp. 1–36.
6. Ghersa, C.M. and M.L. Rush. "Searching for solutions to weed problems. Do we study competition or dispersion?," *Bioscience* 43:104–109 (1993).
7. Forcella, F. and S.J. Harvey. "Relative abundance in an alien weed flora," *Oecologia* 59:292–295 (1983).
8. Sattin, M., G. Zanin, and A. Berti. "Case history for weed competition/population ecology: velvetleaf (*Abutilon theophrasti*) in corn (*Zea mays*)," *Weed Technol.* 6:213–219 (1992).
9. Baker, J.L. "Effects of tillage and crop residues on field losses of soil-applied pesticides," in *Fate of Pesticides and Chemicals in the Environment,* J.L. Schnoor, Ed. (Chichester, England: John Wiley and Sons, 1992), pp. 175–187.
10. Patterson, D.T. "Comparative ecophysiology of weeds and crops," in *Weed Physiology,* S.O. Duke, Ed., (Boca Raton, FL: CRC Press, 1985), vol. 1, pp. 101–129.

11. Darwin, C. *The Origin of Species,* (London: John Murray, 1859).
12. Harper, J.L. "A Darwinian plant ecology," in *Evolution from Molecules to Man* (Cambridge, England: Cambridge University Press, 1983), pp. 323–345.
13. Donald, C.M. "Competition among crop pasture plants," *Adv. Agron.* 15:1–114 (1963).
14. Harper, J.L. "Approaches to the study of plant competition," *Symposia of the Society for Experimental Biology* 15:1–39 (1961).
15. Tilman, D. *Resource Competition and Community Structure* (Princeton, NJ: Princeton University Press, 1982).
16. Fowler, N. "Competition and coexistence in a North Carolina grassland. II. The effects of the experimental removal of species," *J. Ecol.* 69:843–845 (1981).
17. Tilman, D. "On the meaning of competition and the mechanisms of competitive superiority," *Funct. Ecol.* 1:304–315(1987).
18. Thompson, K. and J.P. Grime. "Competition reconsidered: a replay to Tilman," *Funct. Ecol.,* 2:114–116 (1988).
19. Keddy, P.A. *Competition* (New York: Chapman and Hall, 1989).
20. Zanin, G., A. Berti, and M. Sattin. "Croissance du mais (*Zea mays* L.) en competition avec *Abutilon theophrasti* Medicus," *Proceedings 8th International Symposium on Weed Biology, Ecology and Systematics,* (Dijon, France: COLUMA-EWRS, 1988), pp. 609–618.
21. Sattin, M. and I. Sartorato. "Influenza di due livelli di intensità luminosa su alcune caratteristiche fisiologiche e morfologiche di *Solanum nigrum* L. ed *Amaranthus cruentus* L.," *Riv. Agron.* 25:519–526.
22. Grime, J.P. *Plant Strategies and Vegetation Processes,* (Chicester, U.K.: John Wiley and Sons, 1979).
23. Gaudet, C.L. and P.A. Keddy. "Predicting competitive ability from plant traits: a comparative approach," *Nature* 334:242–243 (1988).
24. Cousens, R. "A simple model relating yield loss to weed density," *Ann. Appl. Biol.* 107:239–252.
25. Kropff, M.J. and C.J.T. Spitters. "A simple model of crop loss by weed competition from early observations on relative leaf area of weeds," *Weed Res.* 31:97–106 (1992).
26. Holzner, W. "Weed species and weed communities," *Vegetatio* 38:13–20 (1978).
27. Barrett, S.C.H. "Genetics and evolution of agricultural weeds," in *Weed Management in Agroecosystems: Ecological Approaches,* M.A. Altieri and M. Liebman, Eds. (Boca Raton, FL: CRC Press, 1988), pp. 57–75.
28. Harper, J.L. "The ecologicalal significance of dormancy," *Proceedings IV International Congress on Crop Protection,* Hamburg, Germany (1957), pp. 415–420.
29. Turner, C.E. "Ecology of invasion by weeds," in *Weed Management in Agroecosystems: Ecological Approaches,* M.A. Altieri and M. Liebman, Eds. (Boca Raton, FL: CRC Press, 1988), pp 41–55.
30. Baskin, J.M. and C.C. Baskin. "Physiology of dormancy and germination in relation to seed bank ecology," in *Ecology of Soil Seed Banks,* M.A. Leck, V.T. Parker, and R.C. Simpson, Eds., (San Diego, CA: Academic Press, 1989), pp. 53–87.
31. Karsen, C.M. "Seasonal patterns of dormancy in weed seeds," in *The Physiology and Biochemistry of Seed Development, Dormancy and Germination,* A.A. Khan, Ed. (Amsterdam, The Netherlands: Elsevier Biomedical Press, 1982), pp. 243–270.
32. Chadoeuf-Hannel, R. "La dormance chez les semences de mauvaises herbes," *Agronomie* 5:761–772 (1985).
33. Roberts, H.A. "Seed banks in soils," *Adv. App. Biol.* 6:1–55 (1981).
34. Thompson, K. and J.P. Grime. "Seasonal variation in the seed banks of herbaceous species in ten contrasting habitats," *J. Ecol.* 67:893–921 (1979).
35. Wyse, D.L. "Future of weed science research," WSSA Abstract, 263 (1991).
36. Auld, B.A., K.M. Menz, and C.A. Tisdell. *Weed Control Economics,* (London: Academic Press, 1987), pp. 177.
37. Zanin, G., A. Berti, and M. Giannini. "Economics of herbicide use on arable crops in north-central Italy," *Crop Prot.* 11:174–180 (1992).
38. Bourdôt, G.W. and D.J. Saville, "The economics of herbicide use in cereal crops in New Zealand," *N.Z. J. Exp. Agric.* 16:201–207 (1988).

39. Kropff, M.J. and K. Moody. "Weed impact on rice and other tropical crops," *Proceedings of the First International Weed Control Congress,* Monash University, Melbourne, Australia, J.H. Combellack, K.J. Levick, J. Parsons, and R.G. Richardson, Eds. (Melbourne, Australia: Weed Science Society of Victoria Inc., 1992), pp. 123–126.

40. Parker, C. and J.D. Fryer. "Weed control problems causing major reductions in world food supply," *FAO Plant Protection Bulletin* 23 (3/4):83–95 (1975).

41. Charadattan, R. and C.J. DeLoach, Jr. "Management of pathogen and insects for weed control in agroecosystems," in *Weed Management in Agroecosystems: Ecological Approaches,* M.A. Altieri and M. Liebman, Eds., (Boca Raton, FL: CRC Press, 1988), pp. 245-264.

42. Bridges, D.C. "Economic losses due to weeds in southern states," *Proc. Southern Weed Science Society,* vol. 44:417–425 (1991).

43. Bridges, D.C. "Economic losses due to weeds in southern states," *Proc. Southern Weed Science Society,* vol. 45:381–391 (1992).

44. Zanin, G. and M. Sattin. "Threshold level and seed production of velvetleaf (*Abutilon theophrasti* Medicus) in maize," *Weed Res.* 28:347–352 (1988).

45. "Agriculture and the Chemical Industry — A Corporate Analysis," (Edinburgh, U.K.: Wood MacKenzie and Co. Ltd., 1991).

46. LeBaron, H.M. "Distribution and seriousness of herbicide-resistant weed infestations worldwide," in *Herbicide Resistance in Weeds and Crops,* J.C. Caseley, G.W. Cussans, and R.K. Atkin, Eds. (Oxford, U.K.: Butterworth Heinemann, 1991), pp. 27–43.

47. Goldburg, R.J. "Environmental concerns with the development of herbicide-tolerant plants," *Weed Technol.* 6:647–652 (1992).

48. *Production Yearbook,* Vol. 44. (Rome: FAO, 1990).

49. Pike, D.R., M.D. McGlamery, and E.L. Knake. "A case study of herbicide use," *Weed Technol.* 5:639–646 (1991).

50. Fougeroux, A. and M. Bourdet. "Les produits phytosanitaires. Evaluation des surfaces et des tonnages par type de traitement en 1988," *La Défense des Végetaux* 259:3–8 (1989).

51. Catizone, P. "Diserbo," in *Agricoltura e Ambiente.* (Bologna, Italy: Edagricole, 1991), pp. 481–540.

52. Worthing, C.R. and R.J. Hance. *The Pesticide Manual.* 9th edition. (Croydon, U.K.: BCPC Publications, 1991).

53. Holt, J.S. "History of identification of herbicide resistant weeds," *Weed Technol.* 6:615–620 (1992).

54. Beyer, E.M., Jr., M.J. Duffy, J.V. Hay, and D.D. Schlueter. "Sulphonylureas," in *Herbicides, Chemistry, Degradation and Mode of Action,* P.C. Kearney and D.D. Kaufman, Eds., (New York: Marcel Dekker, 1988), pp. 117–189.

55. Gressel, J. "Addressing real weed science needs with innovations," *Weed Technol.* 6, 509–525 (1992).

56. Witt, W.W. "Response of weeds and herbicides under no-tillage conditions," in *No Tillage Agriculture: Principles and Practices,* R.E. Phillips and S.H. Phillips, Eds. (New York: Van Nostrand Reinhold Co., 1984), pp. 152–170.

57. Burrows, W.C. and W.E. Larson. "Effect of amount of mulch on soil temperature and early growth of corn," *Agron. J.* 54:19–23 (1962).

58. Lal, R. "Conservation tillage for sustainable agriculture: tropics versus temperate environments," *Adv. Agron.* 42:85–197 (1989).

59. Phillips, R.E. "Soil moisture," in *No Tillage Research. Report. Revue,* R.E. Phillips, G.W. Thomas, and R.L. Blevins, Eds. (Lexington, KY: University of Kentucky, 1981), pp. 23–42.

60. Montegut, J. *Perennes et Vivaces Nuisibles en Agriculture* (Aubervilliers, France: J. Manuel, 1983).

61. Phillips, R.E., R.L. Blevins, G.W. Thomas, W.W. Frye, and S.H. Phillips. "No tillage agriculture," *Science* 208:1108–1113 (1980).

62. Barriuso, E., R. Calvet, and B. Cure. "Simplification du travail du sol et pollution par les pesticides," *Persp. Agric.* 162:31–39 (1991).

63. Moeschler, W.W., G.M. Shear, D.C. Martens, G.D. Jones, and R.R. Wilmouth. "Comparative yield and fertilizer efficiency of no tillage and conventionally tilled corn." *Agron. J.* 64:229–231 (1972).

64. Roberts, H.A. and F.G. Stokes. "Studies on the weeds of vegetable crops. V. Final observations on an experiment with different primary cultivations," *J. Appl. Ecol.* 2:307–315 (1965).

65. Froud-Williams, R.J., R.J. Chancellor, and A.S.H. Drennan. "Potential changes in weed floras associated with reduced cultivation systems for cereal production in temperate regions," *Weed Res.* 21:99–109 (1981).

66. Moss, S.R. "Influence of cultivation on the vertical distribution of weed seeds in the soil," *Proc. 8th Coll. Int. on the Biology, Ecology and Systematics of Weeds* (Paris: Annales ANPP, 1988), pp. 71–80.

67. Catizone, P., M. Tedeschi, and G. Baldoni. "Influence of crop management on weed populations and wheat yield," *Proc. EWRS Symp. on Integrated Weed Control in Cereals,* (Helsinki: European Weed Research Society, 1990), pp. 119–126.

68. Bachtaler, G. "Changes in arable weed infestation with modern crop husbandry techniques," Abstract 6th International Congress on Plant Protection, Vienna, Austria (1967), pp. 167–168.

69. Hammerton, J.L. "Past and future changes in weed species and weed floras," *Proceedings. 9th Brighton Weed Control Conference,* (Croydon, U.K.: BCPC Publications, 1968), pp. 1136–1146.

70. Wicks, G.A. "The role of weed control in reduced tillage systems for drilled crops," 29th North Central Weed Control Conference (1974), pp. 25-29.

71. Cussans, G.W., S.R. Moss, F. Pollard, and B.J. Wilson. "Studies of the effect of tillage on annual weed populations," *Proceedings European Weed Research Society Symposium "The Influence of Different Factors on the Development and Control of Weeds,"* Mainz, Germany, (1979), pp. 115–122.

72. Jan, P., A. Fontaine, and R. Dumont. "Incidence de la simplification du travail du sol sur la flore adventice," in *Simplification de Travail du sol en Production Céréalière,* (Paris: I.T.C.F., 1976), p. 205.

73. Froud-Williams, R.J., R.J. Chancellor, and A.S.H. Drennan. "Influence of cultivation regime upon buried seeds in arable cropping systems," *J. Appl. Ecol.* 20:199-208 (1983).

74. Pollard, F., S.R. Moss, G.W. Cussans, and R.J. Froud-Williams. "The influence of tillage on the weed flora in a succession of winter wheat crops on a clay loam soil and silt loam soil," *Weed Res.* 22:129–136 (1982).

75. Weber, J.B. and S.W. Lowder, "Soil factors affecting herbicide behaviour in reduced-tillage systems," in *Weed Control in Limited-Tillage Systems,* A.F. Wiese, Ed. (Champaign, IL, W.S.S.A., 1985), pp. 227–241.

76. Armstrong, D.E., G. Chesters, and R.F. Harris. "Atrazine hydrolysis in soil," *Soil Sci. Soc. Am. Proc.* 31:61–66 (1967).

77. Devine, M.D., S.O. Duke, and C. Fedtke, *Physiology of herbicide action* (Englewood Cliffs, NJ: PTR Prentice Hall, 1993).

78. McWoy, C.W. "Factors affecting groundwater vulnerability," North Eastern Weed Science Society, Suppl., vol. 42, pp. 16–20 (1988).

79. Wauchope, R.D. "The pesticide content of surface water draining from agricultural fields. A review," *J. Environ. Quality* 7:459–472 (1978).

80. Haas, H. and J.C. Streibig. "Changing patterns of weed distribution as a result of herbicide use and other agronomic factors," in *Herbicide resistance in plants,* H.M. LeBaron and J. Gressel, Eds., (Chicester, U.K.: John Wiley and Sons, 1982), pp. 57–79.

81. Cantele, A., G. Zanin, and M.C. Zuin. "Evolution de la flore adventice du mais en Frioul (Italie Nord-Orientale) et role de la monoculture," *7th International Symposium on Weed Biology, Ecology and Systematics,* (Paris: COLUMA-EWRS, 1984), pp. 437–447.

82. Wilson, B.J. and P.A. Phipps. "A long term experiment on tillage, rotation and herbicide use for the control of *A. fatua* in cereals," *Proceedings of the British Crop Protection Conference, Weeds,* (Croydon, U.K.: BCPC Publications, 1985), pp. 693-700.

83. Roeth, F.W. "Enhanced herbicide degradation in soil with repeated application." *Rev. Weed Sci.* 2:45–65 (1986).

84. Ferrari, C., M. Speranza and P. Catizone. "Weed and crop management of wheat in northern Italy," *7th International Symposium on Weed Biology, Ecology and Systematics,* (Paris: COLUMA-EWRS, 1984), pp. 411–420.

85. Mahn, E.G. "The influence of different nitrogen levels on the productivity and structural changes of weed communities in agro-ecosystems," *7th International Symposium on Weed Biology, Ecology and Systematics,* (Paris: COLUMA-EWRS, 1984), pp. 421–429.

86. Hume, L. "The long-term effects of fertilizer application and three rotations on weed communities in wheat (after 21–22 years at Indian Head, Saskatchewan)," *Can. J. Plant. Sci.* 62:741–750 (1982).

87. Ziliotto, U., G. Zanin, F. Basso, F. Carone, D. De Giorgio, M. Mazzoncini, P. Montemurro, L. Postiglione, G. Stefanelli, and G. Toderi, "Influenza delle lavorazioni del terreno sulla vegetazione infestante: presentazione del problema ed analisi di prove collegiali Italiane," *Riv. Agron.* 26:241–252 (1992).

88. Appleby, A.P., P.D. Olson, and D.R. Colbert. "Winter wheat yield reduction from interference by Italian ryegrass," *Agron. J.* 68: 463–466 (1976).

89. Popay, A.I. and E.H. Roberts, "Ecology of *Capsella bursa-pastoris* (L.) Medik and *Senecio vulgaris* L. in relation to germination behaviour," *J. Ecol.* 58:123–139 (1970).

90. Harper, J.L. "Ecological aspects of weed control," *Outlook on Agriculture* 2:197–205 (1957).

91. Beuret, E. "Influence des pratiques culturales sur l'evolution de la flore adventice: étude du potentiel semencier des sols," *Rev. Suisse Agric.* 21:75–82 (1989).

92. Froud-Williams. R.J. "Changes in weed flora with different tillage and agronomic management systems," in *Weed management in agroecosystems: ecologicalal approaches,* M.A. Altieri and M. Liebman, Eds. (Boca Raton, FL: CRC Press, 1988), pp. 213–236.

93. Maxwell, B.D., M.L. Roush, and S.R. Radosevich. "Predicting the evolution and dynamics of herbicide resistance in weed populations," *Weed Technol.* 4:2–13 (1990).

94. Holt, J.S. and H.M. Le Baron. "Significance and distribution of herbicide resistance," *Weed Technol.* 4:141–149 (1990).

95. Hirshberg, J. and L. McIntosh. "Molecular basis of herbicide resistance in *Amaranthus hybridus,*" *Science* 22:1346–1349 (1983).

96. Ryan, G.F. "Resistance of common groundsel to simazine and atrazine," *Weed Sci.* 18:614–616 (1970).

97. Mudge, L.C., B.J. Gosset, and T.R. Murphy. "Resistance of goosegrass (*Eleusine indica*) to dinitroaniline herbicides," *Weed Sci.* 32:591–594 (1984).

98. Morrison, I.N., B.G. Todd, and K.M. Nawolsky. "Confirmation of trifluralin-resistant green foxtail (*Setaria viridis*) in Manitoba," *Weed Technol.* 3:544–551 (1989).

99. Tardif, F.J. and S.B. Powles. "Target site-based resistance to herbicides inhibiting Acetyl-CoA carboxylase," Proceedings. Brighton Crop Protection conference — Weeds, Vol. 2, (Farnham, Surrey, U.K., 1993), pp. 533–540.

100. Shaner, D.L. "Mechanisms of resistance to acetolactate synthase/acetohydroxyacid synthase inhibitors," in *Herbicide resistance in weeds and crops,* J.C. Caseley, G.W. Cussans, and R.K. Atkin Eds., (Oxford, U.K.: Butterworth Heinemann, 1991), pp. 187–198.

101. Hatzios, K.K. "Biotechnology applications in weed management: now and in the future," *Adv. Agron.* 41:325–375 (1987).

102. Gressel, J. and L.A. Segel. "Modelling the effectiveness of herbicide rotations and mixtures as strategies to delay or preclude resistance," *Weed Technol.* 4, 186–198 (1990).

103. Gressel, J. and Y. Shaaltiel. "Biorational herbicide synergists," in *Biotechnology for crop protection,* P.A. Hedin, J.J. Menn, and R.M. Hollingworth, Eds., ACS Symposium Series 379 (Washington, DC: American Chemical Society, 1988), pp. 4–24.

104. Heap, I. and R. Knight. "The occurrence of herbicide cross-resistance in a population of annual ryegrass, *Lolium rigidum,* resistant to diclofop-methyl," *Aust. J. Agric. Res.* 37:149–156(1986).

105. Cobb, A. *Herbicides and Plant Physiology* (London, U.K.: Chapman and Hall, 1992).

106. Radosevich, S., B.D. Maxwell, and M.L. Roush. "Managing herbicide resistance through fitness and gene flow," in *Herbicide resistance in weeds and crops,* J.C. Caseley, G.W. Cussans, and R.K. Atkin, Eds. (Oxford, U.K.: Butterworth-Heinemann, 1991), pp. 129–143.

107. Holt, J.S. "Interbiotype competition reduces relative fitness and yield of triazine resistant common groundsel (*Senecio vulgaris*)," Weed Science Society of America Abstracts, vol. 33, no. 182 (1993).

108. Valverde, B.E. S.R. Radosevich, and A.P. Appleby. "Growth and competitive ability of dinitroaniline-herbicide resistant and susceptible goosegrass (*Eleusine indica*) biotypes," Proceedings Western Society of Weed Science, p. 81 (1988).

109. Holmberg, M. "10 ways to avoid herbicide-resistant weeds," *Successful Farming,* 11:44–47 (1992).

110. Berti, A., A. Cantele, and G. Zanin. "Fattori biologicali che influenzano la risposta dell'*Abutilon theophrasti* Medicus all'altrazina," *Riv. Agron.* 24:299–307 (1990).

111. Anderson, W.P. *Weed science: principles* (New York: West Publishing Company, 1977).

112. Bingham, S.W. and R. Shaver, "Uptake, translocation and degradation of diphenamid in plants," *Weed Sci.* 6:639–643 (1971).

113. Hoffmann, O.L. "Herbicide antidotes: from concept to practice," in *Chemistry and action of herbicides antidotes,* F.M. Pallos and J.E. Casida, Eds. (San Diego, CA: Academic Press, 1978), pp. 1–13.

114. Hatzios, K.K. "Mechanisms of action of herbicide safeners: an overview," in *Crop safeners for herbicides. Development, uses and mechanisms of action,* K.K. Hatzios and R.E. Hoagland, Eds. (San Diego, CA: Academic Press Inc., 1989), pp. 65–101.

115. Hiltbold, A.E. "Persistence of pesticides in soil," in *Pesticides in soil and water,* W.D. Guenzi, J.L. Ahlrichs, G. Chesters, M.E. Bloodworth, and R.G. Nash, Eds. (Madison, WI: Soil Science Society of America, 1974), pp. 203–222.

116. Walker, A. "Herbicide persistence in soil," *Rev. Weed Sci.* 3:1–17 (1987).

117. Le Baron, H.M. and J. Mc Farland. "Herbicide resistance in weeds and crops. An overview and prognosis," in *Managing resistance to agrochemicals. From fundamental research to practical strategies,* M.B. Green, H.M. Le Baron, and W.K. Moberg, Eds., ACS Symposium Series (Washington, DC: American Chemical Society, 1990), pp. 336–352.

118. Gillet, H. "Les triazines dans l'eau: état des lieux en Bretagne," *Quinzième Conference du COLUMA* (Paris: ANPP, 1992), pp. 55–65.

119. Zanin, G., A. Cantele, S. Della Pietà, G.G. Lorenzoni, F. Tei, and C. Vazzana. "Le erbe infestanti graminacee nella moderna agricoltura: dinamica, problemi e possibili soluzioni," in *Atti S.I.L.M.* (Perugia, Italy: Guerra Guru, 1985), pp. 13–310.

120. Vighi, M., G.P. Beretta, V. Francani, E. Funari, C. Nurizzo, F. Previtali, and G. Zanin. "Il problema della contaminazione da atrazina nelle acque sotterranee: un approccio multidisciplinare," *Ingegneria Ambientale,* 9:494–509 (1992).

121. Foschi, S. and I. Camoni. "Fitofarmaci e ambiente," *La Difesa delle piante,* 9:175–194 (1986).

122. Italian Ministry of Health Regulation. 25/6/1986. Divieto cautelativo nel territorio nazionale dell'impiego di presidi sanitari contenenti il principio attivo Atrazina. G.U. n. 146 del 26/6/1986.

123. Italian Ministry of Health Regulation. 22/12/1986. Divieto cautelativo nel territorio nazionale dell'impiego di presidi sanitari contenenti il principio attivo Atrazina. G.U. n. 298 del 24/12/1986.

124. Italian Ministry of Health Regulation. 3/4/1987, n. 135. Divieto cautelativo nel territorio nazionale dell'impiego di presidi sanitari contenenti i principi attivi Atrazina e Molinate. G.U. n. 80 del 6/4/1987.

125. Italian Ministry of Health Regulation. 30/5/1987, n. 217. Divieto cautelativo nel territorio nazionale dell'impiego di presidi sanitari contenenti il principio attivo Bentazone. G.U. n. 127 del 3/6/1987.

126. Italian Ministry of Health Regulation. 31/3/1988, n. 101. Divieto cautelativo nel territorio nazionale dell'impiego di presidi sanitari contenenti i principi attivi Atrazina, Molinate e Bentazone. G.U. n. 77 del 1/4/1988 (modificata dalla O.M. 13/9/1988, n. 472).

127. Italian Ministry of Health Regulation. 21/3/1990 n. 705/627. Divieti e nuove prescrizioni concernenti l'impiego di alcune sostanze attive diserbanti. G.U. n. 70 del 24/3/1990.

Section II
Environmental Distribution
and
Fate and Exposure
Prediction

CHAPTER 3

Predictive Approaches for the Evaluation of Pesticide Exposure

Marco Vighi and Antonio Di Guardo

CONTENTS

ENVIRONMENTAL MONITORING AND THE NEED FOR PREDICTIVE APPROACHES

Mounting concern in the 1960s regarding environmental pollution resulted in the first major attempts to monitor the environment. National and international organizations devoted large amounts of resources to extensive programs to identify and quantify the presence of potential contaminants in different environmental compartments, particularly water and air. The result of this activity was a huge quantity of data, nevertheless the amount of energy devoted to monitoring was in general out of proportion to the practical usefulness of the information produced.

Examples of monitoring approaches are provided in Chapter 1. They give a picture of the extent of the phenomenon and indicate the need for intervention. However, they are hardly sufficient for planning a proper strategy of groundwater protection without more general approaches capable of interpreting, describing, and predicting environmental distribution and fate patterns.

In fact, traditional environmental monitoring, in the absence of suitable preliminary knowledge, cannot provide the answers to several basic questions:

WHAT ARE THE REALLY IMPORTANT SUBSTANCES TO BE ON THE LOOKOUT FOR

The number of potentially harmful chemical substances is extremely high and for most of them information is very scarce. Although, in some cases, the sensitivity of the scientific community was able to identify emerging problems, blind monitoring activity could not detect serious problems. Thus, a fruitful environmental monitoring should be planned in function of a preliminary hazard assessment based on loadings and use patterns, an estimate of environmental exposure and the effects on living organisms. Even if roughly evaluated, this information is necessary for priority setting.

WHERE IT SHOULD BE SEARCHED

An extensive monitoring of all environmental compartments could be extremely expensive and, in some cases useless. On the other hand, data referring to a single compartment could be completely erroneous. A chemical can move from the compartments where it was introduced to other environmental phases. For instance, water monitoring of a very volatile compound could produce no positive result even if this does not mean that the chemical is of no concern.

WHAT IS THE MEANING OF AN EXPERIMENTAL MEASUREMENT

The measurement of a given environmental concentration gives a picture of an instantaneous situation but does not give information about the environmental processes producing it. Environmental concentration is the result of complex distribution and transformation patterns that can produce significant modifications, in time and space, of the real exposure, either in terms of the situation prior to measurement or to future evolution. Moreover, environmental monitoring is a retrospective approach, showing a picture of the present situation and indicating that contamination is underway. This produces a large uncertainty in the environmental management of potentially dangerous substances, without any preliminary control of unwanted effects on a large scale. In other words, monitoring *a posteriori* could allow, if possible, to plan treatments but not any type of prevention.

Therefore, the monitoring of a potentially polluted environment should be the final point and not the starting point of an environmental study. Thus, a proper evaluation of the presence of potential pollutants in an ecosystem, or in one of the various environmental compartments (air, water, soil, biota), should be based on preliminary knowledge including the following:

- study of the territory and the human activities therein, in order to detect the possible presence of contaminants, and to evaluate the amounts produced and introduced into the environment, and to describe uses and discharge patterns;
- characteristics of the chemical substances and their main physicochemical properties of environmental relevance;
- main features of environmental partition, distribution, and transport of potential pollutants;
- transformation and degradation processes.

All of this information should be used as input data for predictive approaches — to produce indications of the compartments at risk, of the predictable environmental concentrations, and of their possible changes in time in function of uses and environmental processes. This could be the right approach in planning environmental monitoring, in order to verify the reliability of predicted concentration trends and distribution patterns.

On these bases, data produced through precisely aimed environmental monitoring could be a means of better understanding environmental distribution patterns, reconstructing the processes that determine specific environmental concentrations, and extrapolating them in space and time. Moreover, monitoring data may play an important role when basic physicochemical data are not precisely defined or in the presence of specific and/or not well known environmental situations. In other words, in this context, monitoring data are essential for the validation and improvement of the predictive capability of environmental models. On the other hand, a predictive approach is essential for preventive interventions, whereas environmental monitoring is obviously an *a posteriori* study.

The need for prevention and, as a consequence, for predictive *a priori* approaches, was recognized not only by the scientific community but also by politicians and legislators. In the late '70s two important regulations were promulgated: the Toxic Substances Control Act[1] in the U.S. and the Directive on Dangerous Substances[2] in the European Economic Community (EEC). Both require a preventive hazard assessment for the premarketing of chemical substances.

Thus, the concept of environmental protection through preventive management of chemical substances rather than through *a posteriori* control of emissions became more and more important, and this has contributed to the rapid development of predictive ecotoxicology in the last 15 years. With regards to pesticide risk evaluation, it should be noted that the cost of developing the data needed for registration of a new pesticide continues to escalate and is now estimated to be $20 to 40 million (U.S.)[3] At the same time, the cost of computation has decreased exponentially. Thus, predictive approaches appear to be extremely useful and promising tools to replace expensive field research or, at least, to enhance the understanding and extend the interpretation of experimental data.

THE ROLE OF EXPOSURE IN HAZARD ASSESSMENT AND RISK EVALUATION

One of the fundamental concepts of ecotoxicology is that of hazard assessment. The concept of hazard assessment was defined for the first time by Cairns et al.[4] as follows: "a hazard assessment must ultimately be based upon sound scientific judgement applied to knowledge of the expected environmental concentration of the material and of the material's toxicological properties." In other words, a hazard assessment derives from a comparison between the effects that a chemical substance can produce on living organisms or on ecosystems and the level of exposure. Effects could be quantitatively represented by an environmental quality criterion (QC) or by a safe concentration (NOEL: No Observed Effect Level), and exposure by a predicted environmental concentration (PEC) (Figure 1).

Environmental concentration is the result of partition and physical transport patterns, determining the spatial distribution of the chemical, and of mechanisms of reactivity to physical, chemical, and biological factors, determining its persistence in time. During a hazard evaluation procedure, increasingly accurate estimates can be produced of biological effects and of the environmental concentration. The sequential evaluation procedure should be carried out to the point where the accuracy of the information is sufficient to allow a considered judgmental decision about the hazard (Figure 2).

It should be noted that, at least for preliminary hazard assessment, exposure could not be related to a specific environmental condition. A hazard assessment can be produced at the screening level, for comparative purposes. In this case, a theoretical exposure level, without reference to a real environment, can be evaluated by means of simple predictive models applied to standard systems assumed as representative of a hypothetical general environment.

Figure 1   Scheme of the contribution to hazard assessment of the evaluation of biological effects (NOEL: No Observed Effect Level) and of environmental exposure (PEC: Predicted Environmental Concentration).

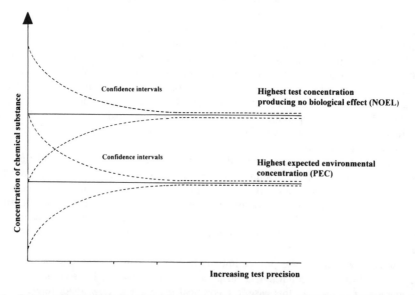

Figure 2   Relationship between Predicted Environmental Concentration (PEC) and No Observed Effect Level (NOEL) in function of increasing test precision (Adapted from Cairns, J., Jr., K. L. Dickson, and A.W. Maki, *Estimating the Hazard of Chemical Substances to Aquatic Life,* Philadelphia, ASTM, 1978, pp. 191–197, with permission.

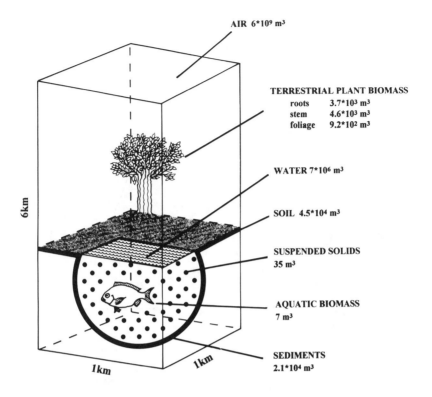

AIR $6*10^9$ m³

TERRESTRIAL PLANT BIOMASS
roots $3.7*10^3$ m³
stem $4.6*10^3$ m³
foliage $9.2*10^2$ m³

WATER $7*10^6$ m³

SOIL $4.5*10^4$ m³

SUSPENDED SOLIDS
35 m³

AQUATIC BIOMASS
7 m³

SEDIMENTS
$2.1*10^4$ m³

6km

1km   1km

Figure 3   The "standard unit of world" of the fugacity approach. Figure represents the unit of world of the Mackay model[5] modified by Calamari et al.[6] for the inclusion of terrestrial plant biomass. The various compartments are not in scale.

An example is the standard "unit of world" proposed in the fugacity approach[5] representing a surface (1km²) corresponding to more or less 1/500,000,000 of the real world with quite realistic volumes of the different environmental compartments (Figure 3).

A completely different approach is needed for a risk evaluation. Risk is associated with the presence of a chemical substance in a specific environment and in relation to a specific target. It represents, in probabilistic terms, the possibility for the target to be exposed to dangerous levels of a chemical substance. In this case, the evaluation of environmental concentration must be more precise and site specific. In predictive terms, suitable models should be applied, adapted to the specific environment, and experimentally validated.

THE BEHAVIOR OF CHEMICAL SUBSTANCES IN THE NATURAL ENVIRONMENT

DISTRIBUTION AND FATE IN THE ENVIRONMENTAL COMPARTMENTS

A chemical substance, either natural or xenobiotic, discharged in whatever environmental compartment (air, water, soil) follows a specific biogeochemical cycle that determines its transport and distribution. This distribution depends not only on diffusion and transport patterns within the single compartment, but also on partition processes among the various compartments. Moreover, within each compartment, the chemical is subject to degradation or, more generally, transformation processes. A scheme of the main mechanisms regulating distribution and fate of a chemical in the environment is shown in Figure 4.

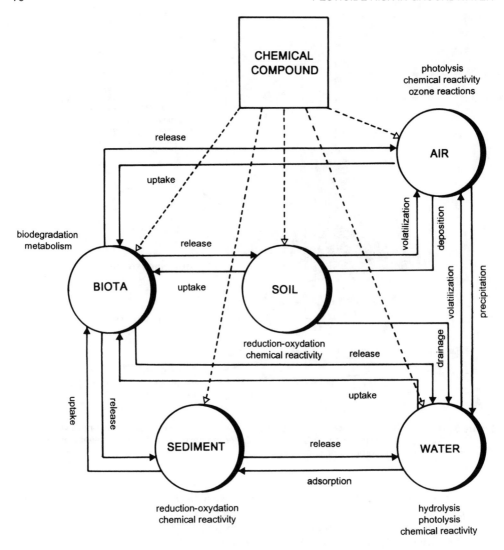

Figure 4   Transport and transformation patterns regulating the distribution and fate of organic chemicals in the environment.

It is obvious that the biogeochemical cycle of a chemical and its environmental fate depend on the characteristics of various compartments and on other environmental parameters. On the other hand, when the environmental conditions are fixed, the environmental distribution and fate of a chemical substance depend on its properties.

In particular, at least in ideal conditions of equilibrium, steady state, and well mixing, distribution among the various compartments is regulated by some physicochemical properties of the chemical substance, representing essentially partition coefficients among the different phases. It follows that a few molecular parameters regulate environmental partitioning of organic substances and, as a fully preliminary step, a rough evaluation can be based on single molecular properties. For example, it is evident that a very soluble substance will partition mainly in water, whereas a very volatile one will go principally into the air. A more precise description of the role of molecular properties will be given in the following pages.

Moreover it must be remembered that many environmental factors will highly affect molecular parameters. It is obvious that, for instance, temperature influences vapor pressure and water solubility more than other parameters such as Kow. A more important role can be played by pH in the case of dissociable compounds.

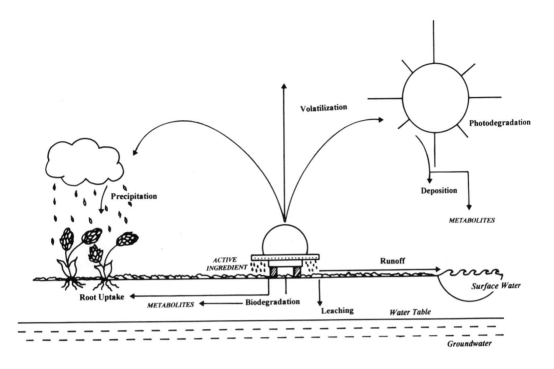

Figure 5 Main distribution and fate patterns of a pesticide.

To summarize, the environmental fate of chemical substances is the result of a sum of partition, mass transport, and degradation processes depending on environmental and molecular factors strictly related to one another. As for persistence, it depends on several transformation processes typical of different environmental compartments, as shown in Figure 4. In this case too, the fate of a chemical substance is a function of environmental factors (temperature, irradiation, microbial activity, etc.) and of the intrinsic properties of molecular stability with respect to various transformation processes.

THE BEHAVIOR OF A CHEMICAL SUBSTANCE IN THE SOIL ENVIRONMENT

Soil is a complex matrix consisting of air, water, mineral matter, and organic matter, with biomass in it. Soils vary enormously in their composition and texture and consist of various layers of different properties. The upper layer, which is more directly involved in the input of chemical substances and where distribution and fate processes are more complex, is the vadose or unsaturated zone. This means that the pore spaces in the soil or rock materials are not fully filled with water.

Below the vadose zone, there is the saturated zone, where pore spaces are fully filled with water. The top of the saturated zone is the water table, corresponding to the level to which water will rise at atmospheric pressure in a hole dug in the earth.

The term "groundwater" generally refers to the water below the water table. Therefore, the transport of chemical substances into groundwater depends essentially on mechanisms occurring in the vadose zone. A typical soil in the vadose zone may consist of 50% solid matter, 20% air, and 30% water, by volume. The organic content of the solid matter may be typically 1 to 5%. Soil is generally covered with vegetation, which stabilizes the soil and prevents erosion by wind or water action. The fate and transport of chemicals in the soil environment are controlled by complex processes and dynamic interactions (Figure 5). In accordance with Donigian and Rao,[7] soil processes can be grouped in the following five categories:

- Transport: chemicals applied on soil can move essentially by volatilization to the atmosphere, by runoff and erosion to surface waters, and by leaching to groundwater; leaching of chemicals to groundwater is primarily a liquid-phase process involving water movement and associated dissolved solute.
- Sorption and partition: partitioning of a chemical between the sorbed and dissolved phases is a critical factor in determining how rapidly the chemical will leach; partition and sorption depend on physicochemical properties of the substance, regulating the affinity for the organic or mineral component of the soil.
- Transformation/degradation: the key processes determining the persistence of a chemical in the soil environment are, in rough order of priority, biodegradation, chemical hydrolysis, oxidation-reduction reactions, and photolysis. Since light does not penetrate into the soil, photochemical transformations play a significant role only at the surface of the soil. Biological transformations are the main degradation pathway in the soil layer where a relevant bacterial community is active, indicating up to some tens of cm deep.
- Volatilization: can be defined as the loss of a chemical in vapor form from soil or plant surfaces. It depends on physicochemical properties of the substance and on environmental parameters (temperature, wind, etc.). Moreover, for pesticides, a significant role is played by use and application patterns (sprayed, applied on the soil surface, etc.).
- Plant processes: the plant roots allow uptake of the chemical from the soil into the plant, which may be considered as a complex system where chemical substances enter, are transported, are sorbed into organic matter and bioaccumulated, and are metabolized and degraded. Thus, in agricultural and silvicultural areas, plant processes may play a significant role in the fate of pesticides.

The ultimate distribution of chemicals in the soil is determined by dynamic, temporal, and spatial interactions among these processes.

DESCRIPTIVE AND PREDICTIVE APPROACHES
AT DIFFERENT DEFINITION LEVELS

In recent years, simulation and predictive approaches have been increasingly produced and applied to evaluate exposure to environmental pollutants. In particular, a wide variety of approaches has been developed to assess the distribution and fate of pesticides in the environment. These approaches vary tremendously in their objectives, their capabilities, their utility, and their level of complexity. Most of them have been developed for research or educational purposes and relatively few were originally developed with the aim of practical application, to predict true concentrations in a real environment. According to Jones,[8] environmental models could be applied to study pesticide behavior in the following four ways.

Educational — The behavior of an agricultural chemical in the environment is a complex process affected by many variables. However, in modeling simulations, the effects of single variables can be readily studied. Modeling simulations can be used to help determine which variables are the key factors in specific situations.

Generalize Field Research — Field research studies provide information on the behavior of an agricultural chemical under the conditions of the study. However, even at the same site, the movement and degradation in soil can change from year to year due to changes in climatic conditions, application procedures, or agricultural practices. A modeling approach may led to an understanding of the role and influence of various factors and to an extrapolation of the results of field studies to other environmental conditions.

Estimate Leaching Potential — One of the uses of unsaturated zone models is to estimate the potential of a pesticide to move through the soil profile and into groundwater. Screening simulations can be used to identify areas where more detailed modeling and/or field work needs to be performed.

Develop Management Practices — Modeling approaches can be extremely useful in developing management practices for minimizing the environmental impact of pesticides and, in particular, their movement to groundwater, to protect drinking water supply. The use of simulation models may indicate the optimal use of pesticides, the establishment of well-setback distances, and the identification of vulnerable areas to be protected.

The wide variety of simulation and prediction approaches, developed at different levels of complexity, and the multiplicity of possible uses determine the need for a careful selection of the suitable model for each particular purpose. A classification of models proposed by De Coursey[9] is as follows:

Screening Models — These are models that do not have extensive input data requirements and are relatively easy to use. The quality of output from such models varies considerably depending upon the theory underlying their development. The output of screening models may not be numerically quantitative values, but qualitative descriptions (e.g., pollution potential high or low). They can be used for classification of substances or sites, and their primary function is preliminary assessment.

Research Models — Research models simulate the physical processes as accurately as possible using physically based numerical structures. Such models require large amounts of input data and can be time consuming and costly to perform. They are generally complex and can require extensive knowledge of the physical system and long computer run times. Research models are not designed as prediction models because it is generally not possible to obtain all of the input data required to drive the model accurately enough, other than in a research environment. Validated research models are useful tools in the study of field data and for extrapolating information from a limited number of experiments to different crop and environmental conditions.

Planning, Monitoring and Assessment Models — These models fall between the two extremes of screening and research models and are used primarily for prediction. They have a very wide range of complexity from simple regression to complex process-based structures. Users of these models must make a statement of purpose for which the model is to be used and apply this statement in assessing the alternative models available. Features such as input data requirements, model structure, difficulty in identifying parameter value, and the quality of data used in model development should all be considered in the selection process. One should never use a more complex model than necessary.

The classification proposed by De Coursey[9] leaves a wide margin of uncertainty and vagueness in the last type of models which are the most important for practical and applicative purposes. In practice, it could be very important to define in more detail the difference between screening approaches and site-specific models. A screening approach can be defined as a method of classification, based either on very simple property evaluations and algorithms or on more or less complex models, able to produce a hazard ranking of chemical substances. In function of the level of complexity of the method used, the output can be a rough indicative and qualitative classification (e.g., high, medium, or low affinity for the water compartment) or a quantitative result useful for a more precise numerical ranking of compounds. A screening approach must be based on sound theoretical fundamentals that represent the guarantee of their validity. An experimental validation can be performed only through more or less environmentally unrealistic laboratory experiments.

In any case, a screening approach can never be used to predict environmental concentrations. The results produced, even if quantitative, are frequently independent of an environmental scenario or refer to a common hypothetical standard environment taken as reference for comparative purposes (e.g., the standard unit of world of the fugacity model, see page 87). In other cases, screening approaches can be applied to specific environmental scenarios, but, even here, results cannot be used as realistic predictions of environmental concentrations but only for hazard ranking.

Screening methods are very useful for practical purposes (e.g., for premarketing hazard assessment of chemical substances), but, at least in the case of more comprehensive approaches (e.g., multicompartmental models), they are important tools that contribute to the understanding of environmental distribution and fate processes. In this sense, they can be considered as a research instrument.

On the other hand, site-specific approaches can be defined as methods able to produce realistic environmental concentrations, with a level of approximation and accuracy depending upon the characteristics of the model. Few models have been originally designed for site-specific realistic prediction of concentrations of chemical substances in the various environmental compartments, but the recent development of studies on the application of simple evaluative models has demonstrated the high versatility of these approaches. These models were designed mainly as a "tool for thinking", i.e., as a means for interpreting and understanding the general trends regulating environmental distribution. Nevertheless, in spite of their lack of ecological realism, they have been frequently adopted and applied to real environmental conditions, and the experimental validations have demonstrated an acceptable predictive capability. On a relatively small scale (some hundreds of hectars) predicted and observed values were generally in agreement within an order of magnitude.[10] The difference between screening and site-specific models is not related to the complexity of the approach. Many simple models can be used as screening methods if applied to a standard "unit of world" and can produce enough reliable results if adapted to site-specific conditions. This is the case, for example, of the models derived from the fugacity approach (see page 87). On the other hand, a complex model such as the Pesticide Root Zone Model (PRZM)[11] is originally defined as a screening model, even if referred to a specific field condition.

An overview of the main approaches utilized for the description and prediction of pesticide distribution and fate, applicable either for screening or for site specific purposes, will be given in the following chapters. Various levels of complexity will be described, from the simple qualitative evaluation of molecular properties to complex predictive models.

THE ROLE OF MOLECULAR PROPERTIES

As previously stated, a few molecular properties regulate the partitioning of a chemical substance into environmental compartments. Thus, in comparative terms, assuming environmental parameters as constant, environmental distribution can be evaluated, as a first approximation, on the basis of these properties. It is a rough qualitative evaluation but, for preliminary screening purposes, can easily produce useful information. Therefore, it is necessary to describe the main molecular properties, in order to define their environmental meaning and their role in the classification schemes of chemical substances.

Water Solubility

Water solubility (S) quantifies the affinity of a substance for the water compartment. Solubility of the most common organic substances ranges from complete miscibility (ethanol, acetone) to levels as low as 10^{-6} g/L (DDT, hexachlorobenzene, dioxins) or even less for very hydrophobic compounds. Very roughly, the affinity for the water compartment can be described as follows:

- if water solubility is higher than 1 g/L, this property is the main driving force, and the affinity for water is always very high;
- between 1 and 10^{-2} g/L, affinity for water is high if other properties do not play a major role as driving forces (e.g., very high volatility);
- between 10^{-2} and 10^{-3} g/L, affinity for water is generally intermediate and strongly affected by other parameters;
- between 10^{-3} and 10^{-5} g/L, water affinity is generally low, except for particular cases;
- below 10^{-5} g/L, the substance is highly hydrophobic.

Vapor Pressure

Vapor pressure (VP) indicates volatility and, therefore, the affinity for the air compartment, even if this is better quantified by the Henry's law constant. For substances of ecotoxicological interest and in particular for pesticides, which generally show relatively low volatility, the most commonly used unit to measure vapor pressure is the Pascal (1 Pa = 0,0075 mm Hg), and the range is generally between 10^3 and 10^{-9} Pa. Vapor pressure higher than 1 Pa generally indicates high volatility, whereas, below 10^{-6} Pa, the air affinity is very low. Intermediate values are highly influenced by other physicochemical properties.

Henry's Law Constant

Henry's law constant (H) could be generally expressed as the ratio between vapor pressure and water solubility (H = VP/S = Pa m^3 mol^{-1}). In practice, H represents, but for a constant, a partition coefficient between air and water. In fact the concentration of a substance in air at saturation (Ca) is expressed by the following equation:

$$Ca = VP/RT \qquad (1)$$

where R is the gas constant (R = 8.314 Pa m^3 mol^{-1} $^\circ$K^{-1}), and T is the absolute temperature (T $^\circ$K = 273 .15 + t $^\circ$C).

Taking into account that the concentration in water in a saturated solution corresponds to the solubility:

$$Cw = S \qquad (2)$$

the air/water partition coefficient is represented by the following equation:

$$Kaw = Ca/Cw = VP/SRT = H/RT \qquad (3)$$

and

$$H = Kaw \, RT \qquad (4)$$

Therefore, H can be assumed as an index of the affinity for the air compartment. Henry's constant usually ranges between 10^5 and 10^{-9} Pa m^3 mol^{-1}, but pesticides show values above 10 only in a few cases. Values higher than 10 are always indices of very high air affinity, whereas below 10^{-4} the air affinity is low or negligible. Intermediate values are affected by other parameters.

Octanol/Water Partition Coefficient

The octanol/water partition coefficient (Kow) quantifies the lipophilicity of a substance and is therefore assumed as an index of the ability to pass through biological membranes and to bioaccumulate in living organisms. Thus, it is utilized as a measure of the affinity for the biota. For most pesticides Kow ranges from 10^6 and 10^{-2}.

A widely utilized classification scheme, even if not officially accepted, is the following:

bioaccumulating substances	log Kow > 3.5
low bioaccumulation potential	3 < log Kow < 3.5
nonbioaccumulating substances	log Kow < 3

Even in this case, the role of other physicochemical parameters should be evaluated. Moreover, it must be taken into account that Kow indicates only a "potential" for bioaccumulation, independent from other biological factors affecting the uptake and release of chemicals in living organisms (metabolism, excretion, etc.).

Octanol/Air Partition Coefficient

The role of the octanol/air partition coefficient (Koa) as an important environmental parameter for the evaluation of bioaccumulation in plants from air was recently shown.[12]

Octanol/air partition coefficient can be obtained as follows:

$$Koa = Co/Ca = (Co/Cw) \times (Cw/Ca) = Kow/Kaw \qquad (5)$$

From equation (3) it follows that:

$$Koa = (Kow/H) \, RT \qquad (6)$$

Therefore, at room temperature (25°C = 298.15°K), Koa can be calculated according to the equation:

$$Koa = (Kow/H) \times 2479 \qquad (7)$$

$$\log Koa = \log Kow - \log H + 3.39 \qquad (8)$$

Theoretically, on the basis of variability ranges of Kow and H, log Koa could range from 20 to –4.

In practice, taking into account interactions among parameters (for example high Kow values correspond to low S and, as a consequence, to relatively high H values), usually log Koa ranges from 2 and 10. In general, values above 8 indicate high bioaccumulation potential, below 4 the affinity for plants is very low.

Organic Carbon Sorption Coefficient

The organic carbon sorption coefficient (Koc), usually assumed as an index of soil affinity, represents, in practice, the sorption coefficient for the organic carbon of the soil, and it is strictly related to Kow.

The most commonly utilized equations for the calculation of Koc from Kow are those proposed by Karickhoff:[13,14]

$$\log Koc = \log Kow - 0.21 \qquad (9)$$

or

$$\log Koc = \log Kow - 0.39 \qquad (10)$$

This definition of Koc indicates a coefficient depending only on partition properties of a substance between two phases (organic carbon and water) and independent of other possible processes of sorption or other interactions with the inorganic matrix of the soil. The true partition coefficient between soil (or other similar compartments such as sediment or suspended solids) and water (Kp = Cs/Cw) should generally take these interactions into account. Nevertheless, for nonpolar and nonionized substances, it could be reasonably assumed that affinity for the soil compartment depends mainly on Koc and on the amount of organic carbon in the soil. Thus Kp can be calculated as follows:[15]

$$Kp = Koc \times Foc \qquad (11)$$

where Foc is the fraction of organic carbon in the studied environmental compartment (soil, sediment, suspended solids). This approach is appropriate for hydrophobic pesticides (i.e., nonionic substances with a water solubility less than $10^{-3} M$), but may be suitable, practically speaking, even for compounds that are slightly polar and too water soluble to be considered hydrophobic.[16]

Nevertheless, an alternative approach may be desirable for some pesticide–soil combinations. For example, the Koc approach may be inappropriate for very soluble substances or for soils or sediments having a sufficiently high (e.g., > 40) clay/organic carbon ratio. In this case, the clay contribution to pesticide sorption could play a significant role. An approach to predicting Kp values on the basis of the relationships of Kp to either the organic carbon content or the specific surface (SS) of the soil, was developed by Pionke and De Angelis.[17]

The specific surface is calculated as follows:

$$SS = 100 \, (\%OC) + 2 \, (\% \, Clay) + 0.4 \, (\% \, Silt) + 0.005 \, (\% \, Sand) \qquad (12)$$

Table 1 Affinity of an Organic Chemical for the Different Environmental Compartments According to the Main Molecular Parameters

Affinity for the Compartments	Water WS g/L	Air H Pa m³/mol	Soil log Koc	Animal Biomass log Kow	Vegetal Biomass log Koa
Very high	> 1	> 10	> 5	> 5	> 8
High[a]	$1–10^{-2}$	$10–10^{-1}$	5–4	5–3.5	8–7
Average[a]	$10^{-2}–10^{-3}$	$10^{-1}–10^{-2}$	4–2	3.5–3	7–5
Low[a]	$10^{-3}–10^{-5}$	$10^{-2}–10^{-4}$	2–1	3–1	5–4
Very low	$< 10^{-5}$	$< 10^{-4}$	< 1	< 1	< 4

[a] Influenced by other parameter values.

On the basis of the Pionke-De Angelis method, Kp can be calculated as follows:

$$Kp = m \, SS \tag{13}$$

where m is a fitted coefficient of linear regression based on experimental Kp for 35 pesticides.

Returning to the Koc approach, since the amount of organic carbon in most soils in agricultural area ranges from 1 to 5%, it can be roughly assumed that log Koc values higher than 5 indicate great affinity for soil, whereas values below 1 indicate low affinity, independently of the role of other molecular parameters.

In conclusion, a first rough and qualitative approximation of environmental distribution, based on the examination of the properties of a single molecule, is shown in Table 1. It should be remembered that the boundaries of various classes of affinity, in particular at intermediate levels, are empirically determined and must be assumed as highly subjective.

LEACHING INDEXES AND RANKING SYSTEMS

A further step, quite more advanced than the simple evaluation of single molecular parameters, is represented by comparative indexes and ranking systems. This approach requires few input data, either molecular properties or environmental parameters, and is based on simple algorithms that cannot be assumed as true models. These indexes produce nonquantitative values that allow the comparison of several compounds and the hazard ranking for one or more environmental compartments. In particular, most of these systems were produced specifically for groundwaters and therefore allow a classification of the leaching capability of chemical substances. A selection of leaching indexes, among the most common and currently utilized, is shown in Table 2. The main characteristics and the algorithms utilized for the calculation are reported.

Leaching indexes can be classified as follows:

- Indexes based only on molecular properties. In general they are based on Koc, half life in soil ($t_{1/2}$) and, in some cases, other physicochemical properties, such as water solubility or vapor pressure. They allow a generic comparison of chemical substances, apart from use patterns and environmental properties. Their main advantage is their extreme simplicity of application and interpretation. Examples of this kind of indexes are GUS[18] and LI.[19] It is to be noted that the reliability of the extremely simple algorithms of the GUS index was confirmed by a procedure of application of a complex model such as GLEAMS.[20]
- Indexes that, besides molecular properties, utilize environmental parameters related in particular to soil and groundwater. These indexes refer to some specific characteristics of a site area, thus they could be assumed as containing, in some way, a rough "site specific" meaning, and, therefore, they could be more suitable than the former if applied to a particular territory. They could be utilized as generic screening instruments for various compounds if applied to a hypothetical "territory unit" with fixed characteristics. On the other hand, they could be suitable for a comparison not only of the leaching capability of different chemicals but also for the assimilative capability of different sites. Examples of these kinds of indexes are AF,[21] DRASTIC,[22,23] and TC.[24]

Table 2   Leaching Indexes and Ranking Systems

Index and Source	Formula	Type
Hazard Ranking System (HRS) Caldwell et al.[25] Canter et al.[23]	$SM = \dfrac{1}{1.73}\sqrt{Sgw^2 + Ssw^2 + Sa^2}$ Sgw, Ssw, Sa, are the scores indicating the risk, calculated for groundwater, surface water and air	Classification system for chemical substances discharged in the environment (surface water, groundwater, air)
Attenuation Factor (AF) Rao et al.[21]	$AF = \dfrac{M_2}{M_0} = \exp(-B)$ $B = \dfrac{0.693\, tr}{t_{1/2}}$ M_2 = amount of pesticide reaching groundwater M_0 = amount of pesticide applied on soil tr = time needed for the transport through soil $t_{1/2}$ = half life	Calculation of pesticide transport from surface soil to groundwater; used for preliminary hazard evaluations of pesticides
DRASTIC Aller et al.[22] Canter et al.[23]	Contamination potential calculated through DRASTIC is a sum of relative scores and weights obtained by adding seven different factors, groundwater Depth, groundwater Recharge rate, characteristics of the Aquifer, characteristics of Soil, Topography (slope), Impact of the vadose zone, Conductivity of the aquifer	Score for the evaluation of groundwater contamination potential using hydrogeological information
Leaching Index (LI) Laskowsky et al.[19]	$LI = \dfrac{S \cdot t_{1/2}}{VP \cdot Koc}$ S = water solubility $t_{1/2}$ = half life VP = vapor pressure Koc = organic carbon sorption coefficient	Index for the evaluation of groundwater contamination potential
Tc (Travel time) Jury et al.[24]	$Tc = \dfrac{TH \cdot RF \cdot L}{q}$ TH = water content of soil, RF = retardation factor, L = groundwater depth, q = groundwater recharge rate	Allows the calculation of time needed to reach the water table through water leaching transport
GUS (Groundwater Ubiquity Score) Gustafson[18]	$GUS = \log t_{1/2}(4 - \log Koc)$ $t_{1/2}$ = half life Koc = organic carbon sorption coefficient	Allows the classification of chemicals in function of groundwater contamination potential

- Indexes requiring application rates as additional input data. This is obviously important information, taking into account that different active ingredients can be applied at the level of kg/ha (e.g., triazine herbicides) or of g/ha (e.g., pyrethroids insecticides, sulfonylureas herbicides), in relation to their agronomic effectiveness. Among the indexes listed in Table 2, only HRS[23,25] accounts for application rates. In reality, HRS is not a simple leaching index but a comprehensive index of environmental risk, taking into account several factors, such as toxicologic evaluations, targets, etc. On the other hand, all indexes listed at the points "a" and "b" could easily be modified by including a multiplicative load factor based on agronomical application rate.

Notwithstanding their simplicity, even these indexes can present some problems related to the availability of reliable input data, particularly in relation to soil half-life, needed for almost all the algorithms described.

PARTITION ANALYSIS AND RELATED MODELS

To predict the environmental distribution and fate of chemical substances, Baughman and Lassiter[26] introduced the concept of the evaluative model, with the aim of developing a quantitative approach for the evaluation of exposure. According to these authors, evaluative models "incorporate the dynamics of no specific environment but are based on the properties of stylized environment of hypothetical pollutants for which we specify (rather than measure) inputs." In the ensuing years, many publications appeared on the same subjects.[27–32]

In general, models related to the original approach proposed by Baughman and Lassiter are extremely simple and easy to handle; they require few input data and are based on partitive properties of the chemical substance. This means that the main input data needed are partition coefficients among the various environmental compartments. Several models based on partition analysis were developed at different levels of complexity, from simple standard screening models to more complex site-specific approaches, with more or less high predictive capability.

The Fugacity Approach

Among the various methods based on the partition analysis, one of the most promising is the fugacity approach proposed by Mackay and coworkers.[5,15,33] Fugacity is an old concept of physical chemistry[34] introduced at the beginning of this century as a criterion for equilibrium between phases. Accordingly to Clark et al.,[35] "fugacity is analogous to chemical potential as it pertains to the tendency of a chemical to escape from a phase (e.g., from water). It is expressed in units of pressure (pascals) and is essentially a partial pressure exerted by the chemical in each medium. If a chemical attains concentrations in various media that are in equilibrium, its fugacity is equal in these media."

At the low levels corresponding to environmental concentrations, fugacity is linearly related to concentrations in environmental compartments (moles m^{-3}) through a proportionality constant, the fugacity capacity Z (moles m^{-3} Pa^{-1}), that can be calculated for each compartment and can roughly be defined as the potential capability of an unit volume of a certain compartment to "retain" (by absorption, bioaccumulation, etc.) the chemical substance.

By knowing the environmental capacity for each compartment, calculated as functions of partition coefficients, and, assuming a constant fugacity in the different phases at equilibrium, the mass distribution and relative concentrations in various environmental compartments can be calculated. In its standard form, the fugacity model refers to a "unit of world" of 1 km^2 corresponding to about 1/500x10^6 of the real world (Figure 3). Originally, the unit of world was divided into six environmental compartments (air, water, soil, sediments, suspended solids, and aquatic biota). A further compartment, the terrestrial plant biomass, was then added to the unit of world.[6,36] The fugacity model can be applied at different levels of complexity, which can be schematically described as follows:

Level I — description of the distribution among compartments at equilibrium and steady state in a closed system, assuming no degradation and a single immission of the chemical substance. The model requires only basic physicochemical properties of the molecule (molecular weight, water solubility, vapor pressure, octanol–water partition coefficient) as input data.

Level II — as level I with the evaluation of persistence or residence time of a chemical. Transformation rate constants for various degradation processes in different environmental compartments (photolysis, hydrolysis, biodegradation, etc.) and advection rates are needed as additional input data.

Level III — simulates the levels of input and output of the chemical substance and circulation among compartments in a nonequilibrium but steady-state condition. Additional data needed are emission rates of the chemical substance in the different environmental compartments, background concentrations of the chemical substance in the various media, transport coefficients between adjacent compartments, and from each compartment and the surrounding environment (D values).

Level IV — as level III for unsteady-state conditions. Level IV does not require additional input data but only an improvement of calculation due to the need of differential equations instead of steady-state mass balance equations.

All levels of the original fugacity model can be easily applied, with limited computer requirements, in its standard form to a theoretical unit of world and adopted as a useful screening method. For those purposes, the fugacity model was suggested by the OECD[37,38] and in Europe by some national official organizations[39] as a procedure for the hazard evaluation of chemical substances. Moreover the extreme versatility of the fugacity model, allows its application in a site-specific form, substituting the standard unit of world with the characteristics of a real environment. Obviously, in this case, additional input data are the properties of the experimental system such as:

- volumes of environmental compartments
- meteorologic and climatic characteristics
- soil and sediment characteristics
- hydraulic balance and other advection parameters.

Conceptually, the site-specific application of the fugacity approach can be performed at different spatial scales, from the field scale to the regional level.[40] At present, the predictive capability of the model in real environmental conditions was experimentally validated on a relatively small scales, on the order of some hundreds of hectars.[41] On a larger scale, the model can be assumed to be an extremely useful method to evaluate distribution trends and to study and understand environmental fate patterns. Its predictive capability and its reliability to produce realistic environmental concentrations on a regional scale still need experimental validation.

A regional scale attempt of validation of a level III fugacity model was carried out by Sinkkonen and Di Guardo[42] with the simulation of the behavior of some chlorinated hydrocarbons in the Bothnian Bay, comparing the modeled results with measured concentrations in several environmental media (air, water, fish, pine needles). Besides the various levels of the fugacity model, the basic fugacity concept was then applied to several relatively more complex multimedia models, such as the Quantitative Water Air Sediment Interactions (QWASI) developed to predict distribution in lakes and streams[43,44] and the Spatial Multimedia Compartmental Model (SMCM) proposed by Cohen et al.[45] for a rapid screening-level prediction of the multimedia partitioning of organic chemicals in the environment on a regional scale.

More details about the various aspects of the fugacity models and other models related to the general fugacity concept can be found in the original papers or in a comprehensive overview of the fugacity approach.[46] Here the particular application of the fugacity approach to the soil environment will be described more precisely.

Applications of the Fugacity Approach to the Soil–Water System

Various models, directly related to the fugacity approach, have been produced for the prediction of the presence of pesticides in surface and groundwater. As for surface water contamination

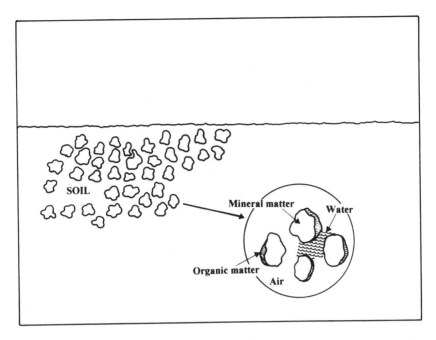

Figure 6 Scheme of the surface soil structure (Adapted from Mackay, D., Multimedia Environmental Models, The Fugacity Approach, Boca Raton, FL, Lewis Publishers, 1991, p. 257.)

through runoff from agricultural areas, two models (AgriFug and Soil Fug) were recently developed and validated.[10,47] To evaluate the risk to groundwater due to pesticide leaching, a surface soil model was developed[48] as a simplified version of a previous model of Jury et al.[24] In the model, the soil matrix is treated as four phases: air, water, organic matter, and mineral matter (Figure 6). Partition is determined by calculating, for each phase, a capacity (Z) value in function of partition coefficients, as for the original fugacity model. At equilibrium, a common fugacity (f) applies to all soil phases, and partition can be calculated as described by a Level I fugacity model. In particular conditions (heavily contaminated soils subject to spills or immediately after a direct heavy application of pesticides), fugacity can exceed the vapor pressure. This means that the capacity of all phases to "dissolve" the chemical is exceeded. Essentially the "solubility" of the chemical in the soil is exceeded. In this case, phase separation of pure chemical will occur, and its behavior cannot be described by the partition model.

As for transport and transformation, three processes are considered: degrading reactions, volatilization, and leaching. In each case, the rate is characterized by a transport (D) value. Specific D values can be calculated if a rate constant, k_i, is known for a specific phase. For example a water leaching rate (mm/day) can be derived from rainfall or irrigation and converted into a total water flow rate Ge (m³/h), which is combined with the water Z value to give the leaching D value as follows:[48]

$$De = Ge\ Zw \tag{14}$$

This assumes that the concentration of chemical in water leaving the soil is equal to that in the water in the soil, i.e., a local equilibrium situation. The model allows the calculation of the loss of a pesticide, in particular by leaching, from a surface soil layer, assumed as relatively homogeneous in structure and properties, whose thickness is a function of several factors, such as agricultural practices, tillage, etc.

As for the general fugacity model, the surface soil model can be applied to a fixed scenario, taking as preliminary assumptions some environmental characteristics (volume fractions and densities of various phases, organic carbon content, temperature, etc.) or to more realistic scenarios corresponding to real agronomic conditions.

Overview of Different Groundwater Models

Besides the simple fugacity approach, a number of more complex models have been developed in the last decade. Some examples of the most important models developed to predict pesticide transport in the unsaturated zone will be described in this chapter. The aim is not to give an exhaustive review of groundwater models, but only to present some typical cases of increasing complexity. The selected examples are the following:

- Pesticide Analytical Solution — PESTAN;[49]
- Seasonal Soil compartment model — SESOIL;[50]
- Groundwater Loading Effect of Agricultural Management Systems — GLEAMS;[51]
- Pesticide Root Zone Model — PRZM;[11,52]
- Leaching Estimation and Chemistry Model — LEACHM.[53,54]

The main characteristics of each model are shown in Table 3.

PESTAN was developed as a relatively simple screening model requiring few input data and without detailed descriptions of the environmental scenario. For example, soil–water flux may be estimated from simple water balance computations using long-term weather records. The model assumes a one-dimensional steady water flow, in a homogeneous soil profile, with constant hydraulic properties. With these assumptions, PESTAN could hardly account for the variability in the upper unsaturated zone. Thus, the model could be more suitable for the more homogeneous lower layer of the unsaturated zone, below the root level. Moreover, the model does not account for vapor-phase transport. Therefore, it cannot be applied to volatile compounds. In spite of its limitations, the model has been used extensively by the EPA Office of Pesticide Programs and has proved to be a useful tool for screening assessment to evaluate the potential for pesticide contamination of groundwater.[6]

The SESOIL model is defined as a long-term simulation of leaching in the soil environment. Its main characteristic is the "seasonal" approach. This means that it calculates the distribution in the soil compartment at the end of a relatively long period (e.g., month, year). Therefore, most input information is based on statistical description on a long-term basis. For example, the hydrologic cycle is based on a statistical representation of the water balance and does not require the description of rain events. The soil environment is divided into various layers (at least four) having uniform properties, but differing from each other. Therefore, there is a more detailed and realistic description of soil, in comparison with the former model. Nevertheless, for the hydrogeologic cycle, a homogeneous soil column is assumed. Thus, the model can be applied with some restrictions to sites exhibiting large vertical variations in soil properties. As for PESTAN, SESOIL has limited applicability for volatile compounds. According to Watson and Brown,[56] the model can be successfully applied to generic environmental conditions. Application on a site-specific basis requires careful calibration. For its "seasonal" character and long-term prediction capability, SESOIL can be considered as a useful tool for management purposes.

GLEAMS was developed as a modification of the original CREAMS (Chemicals, Runoff, and Erosion from Agricultural Management Systems) model[57] to study pesticide movement to groundwater. The model is a field-scale approach requiring a relatively detailed and accurate description of the environmental system. For example, hydrology should be described on a daily basis. Various soil layers with different properties are also included. CREAMS and GLEAMS do not have particular limitations for volatile compounds and can also be used for foliar-applied chemicals and not only for compounds directly applied on soil. These models proved especially useful in making relative comparisons of pollutant loads from alternate management practices.[58]

The PRZM approach requires a large amount of detailed input data. Precise meteorologic and hydrologic data are needed on a daily basis. An accurate soil description is needed for a number of soil layers, in function of the nonhomogeneous distribution of soil properties. The measure of the complexity of the PRZM model is given in Table 4, showing a list of needed input data. A number of field applications demonstrated the reliability of the PRZM model for management and research purposes.[7,59]

Table 3 Main Characteristics of Some Groundwater Models

Model	Characteristics	Fluid Flow Processes	Solute Processes	Computational Requirements	References
PESTAN	Screening model, homogeneous soil profile	Average from annual data, steady water flow	Advection, sorption, degradation; not suitable for volatile compounds	low	49
SESOIL	Management model, layered soil (homogeneous layers)	Statistical and seasonal data	Advection, ad/desorption, degradation; not suitable for volatile compounds	moderate	50
GLEAMS	Management model, layered soil	Transient hydraulic model (tipping bucket), surface runoff, percolation	Advection, sorption, degradation	moderate	51
PRZM	Management model, layered soil	Transient hydraulic model (tipping bucket), leaching, surface runoff	Advection, ad/desorption, degradation, plant uptake, foliar loss	moderate	11,52
LEACHM	Research model, layered soil	Transient hydraulic model (Richards equations); considers upwards flow	Advection, ad/desorption, degradation, metabolites	high	53,54

Adapted from Matthies, M., *Chemical Exposure Prediction*, Boca Raton, FL, Lewis Publishers, 1993, pp. 103–114.

Table 4   Input Data Needed for PRZM Model

Meteorologic Data

- daily rainfall or irrigation amounts
- daily evaporation measured by means of evaporimeter
- daily mean temperature
- hours of sunlight
- snowmelt

Hydrologic Data

- conversion factor by means of evaporimeter
- conversion factor for snowmelt
- factors of universal erosion equation
- basin surface area
- mean runoff duration after intense rain
- runoff curve for previous conditions

Pesticide Data

- organic carbon sorption coefficient (Koc)
- water solubility
- number of treatments
- application date
- application rate
- type of application and depth of incorporation
- for foliar application: foliar extraction coefficient and extraction parameters
- pesticide/soil partition coefficient for each soil layer
- half-life for each soil layer
- foliar decay rate

Soil Data

- total number and description of different layers
- apparent diffusion coefficient
- saturated water content
- field capacity water content
- wilting point water content
- bottom boundary flexibility
- bulk density
- organic carbon content
- thickness, apparent density, hydrodynamic dispersion and initial water content for each soil layer
- soil texture for each soil layer
- depth of sampling
- initial level of pesticide in each compartment

Crop Data

- crop surface covering
- maximum leaf dry weight
- root density distribution
- maximum rooting depth
- maximum water uptake capacity
- sorption efficiency
- pesticide uptake relationship
- crop production system
- soil tillage effects

Still more complex is the LEACHM model, developed essentially for research purposes. If input data are adequate, the output of the model includes the following series of data:[59]

1. Hydraulic conductivity and water content for each layer of the soil at the matric potential values of 0, −3, −10, −30, −100, and −500 kPa.
2. Cumulative totals and mass balances of water and pesticide. This includes the amount of pesticide initially in the soil profile, currently in the profile, the simulated change, additions, losses, and a composite mass error.
3. A summary by depth of soil–water content, matric potential, soil–water flux between layers, evapotranspiration, pesticide mass, pesticide concentration for both individual species and total pesticide, and the chloride concentration.
4. A summary by depth of root density, water uptake, and pesticide uptake.

A field application, to evaluate the fate of aldicarb, described by Wagenet and Hutson,[53] indicates an excellent agreement between predicted and observed values. Nevertheless, as for PRZM, the complexity of the approach and the difficulty of obtaining reliable input data represent a severe limitation to the practical applicability of this kind of model.

VALUE AND LIMITATIONS OF VARIOUS APPROACHES

All described approaches, from simple molecular properties to complex models such as PRZM or LEACHM, can be extremely useful, at different levels, in studying the distribution and fate of pesticides and in evaluating exposure for hazard assessment and risk evaluation. Obviously one must be fully aware of the kind of results that each approach could produce, as a function of the objectives to be obtained:

- rough indication of main trends of environmental distribution and of compartments at risk;
- preliminary screening and hazard ranking of chemicals for priority list purposes;
- understanding environmental processes and quantifying the main driving forces that regulate distribution and fate patterns;
- producing realistic predictions of site-specific environmental concentrations.

In particular, the last two objectives need a more careful selection of the suitable approach.

According to Barnthouse,[60] the real issue in determining the utility of ecological models in hazard assessment and risk evaluation is not validity, i.e., the capability of the model to make precise predictions, but credibility. The credibility of a model depends on a sound theoretical basis and determines its usefulness in understanding environmental processes, while punctual validity is strongly affected by occasional conditions, such as the reliability of input data and the accuracy of environmental description.

A sound conceptual basis of a model, independent of its quantitative validation in specific field conditions, would permit an understanding of environmental patterns and an evaluation of the main driving forces, a result probably as important as a precise quantitative prediction. For all models described above, the credibility and conceptual soundness is not in doubt, and, in most cases, also the punctual validity has been experimentally demonstrated. The ecological realism and the predictive capability increase, at least in theory, with the increase in the complexity and accuracy of environmental description. In reality, the major limitation of complex models is the need for detailed and precise input data, in particular for the description of the environmental scenario. This limitation is due not only to practical and economic reasons. Obviously a detailed and reliable data set requires much effort, with a substantial investment of human and economic resources. A more conceptual limitation depends on the variability of environmental conditions on a large scale, making a precise and detailed description impossible. Therefore, the more complex a model, the more reduced is the spatial scale of its practical applicability. Complex models can give very good predictions at the field scale, but they fail on a larger scale.

Carsel et al.[61] describe an excellent agreement between predicted and observed concentrations of the pesticide metalaxyl in validation experiments of PRZM in two experimental areas of 3.5 and 0.6 ha, respectively. Another validation experiment of PRZM is described by Hedden[62] for aldicarb and metolachlor on a 3.9 ha experimental area.

There are no theoretical obstacles to applying such models on a wider area, but the variability of environmental parameters makes it practically impossible to obtain a description as detailed and accurate as the models require. Simple evaluative models, which require few input data and an approximate description of the environmental scenario, are able to produce less precise results, but their versatility allows their application on relatively nonhomogeneous areas and, therefore, on a larger scale.

Examples of field validation of simple runoff models derived from the fugacity approach indicate an acceptable predictive capability (within an order of magnitude) in experimental areas up to more than 1000 ha.[10] An attempt of validation on a larger scale with this kind of model indicates promising results on the relatively heterogeneous drainage area of a small creek, with a surface of about 100 km^2.[63]

At this point the question is how precise a predictive value should be for use in predicting the exposure for hazard assessment and risk evaluation? If we consider the range of variability of monitoring data due to space and time changes of environmental concentrations and due to the various sources of error of experimental measurements (sampling, storage, analysis), it is reasonable to state that a predictive capability within one order of magnitude can be an acceptable result, at least for preliminary hazard assessment. Obviously, this is provided that the *credibility* of the model, i.e., its conceptual soundness, has been proved.

In conclusion, simple models capable of producing, with a moderate effort, indications as to the general trends of environmental distribution and fate, detection of the compartments at risk, and approximate evaluation of the range of concentrations in various matrices, appear at present to be the best instruments, realistically applicable, for the management of pesticides. On the other hand, it goes without saying that the predictive capability of models is strongly dependent on the availability of reliable input data.

THE PREDICTION OF PERSISTENCE

Persistence is one of the most important parameters needed as input data for all approaches developed to predict the environmental distribution and fate of chemicals. In particular, to assess the hazard from pesticides to groundwater, the soil half life, due to the main reaction and transformation processes that can occur in the soil compartment, is an essential requirement.

The availability of reliable data on persistence is, at present, one of the weakest points in the prediction of the environmental fate of chemicals. This kind of data is very seldom available in the literature and, when found, its usefulness and reliability must be carefully checked. As an example, a lot of data available on residence time for pesticides in soil refer to a comprehensive disappearance of the molecule due to various processes, including advection or other transport patterns, and not only on reaction and transformation processes. Thus, they are completely useless for persistence evaluations in evaluative models. Much work has been performed on biodegradation of chemicals, but, in general, the interest of microbiologists is more in defining metabolic patterns than in producing rough numbers to quantify degradation time, suitable for utilization in evaluative models.

Persistence of pesticides in soil depends on the intrinsic stability of the molecule and on a quantity of environmental variables. Climatic conditions, physicochemical and pedologic properties of soil and biological characteristics (temperature, moisture, pH, soil texture and composition, presence of adapted bacterial populations, etc.) highly affect the half-life in soil. In some cases, relatively precise data, referring to different degradation processes that may occur in the different soil layers, are needed to predict the risk for groundwater. For example, the half-life of atrazine due mainly to biodegradation in the surface soil layer is about some tens of days, whereas in deep soils and in groundwater, hydrolysis and other abiotic degradation processes determine a half-life of some years.[64] Therefore, for reliable site specific predictions, the knowledge of degradation patterns related to a particular environmental scenario, obtained through the results of *ad hoc* experiments, could be necessary. Nevertheless, at least for screening purposes, approximate information on persistence depending on the intrinsic stability of the molecule, independent of environmental variables, can be enough. In this case too, the lack of useful experimental data is coupled with predictive instruments that are still unsatisfactorily inefficient, even if it is generally recognized that a Quantitative Structure-Activity Relationships (QSAR) approach can produce approximate but useful results.

Studies on the relationships between the structure of a chemical substance and its persistence are relatively old. Some general information, based on qualitative relationships between chemical structure and biodegradability, was produced very early on. Well-known examples are the studies related to the problem of detergents in the 1950s and 1960s, indicating the highly different behavior of linear and branched alkyl chains of surfactants.[65] The problem of qualitative structure-biodegradation relationships has been extensively tackled by Alexander and coworkers.[66-68] Studying mainly aromatic and aliphatic hydrocarbons, they demonstrated that the presence of particular functional groups could increase or decrease the biodegradability of a molecule. According to

Table 5 Different Approaches to the Study
of Structure-Biodegradability
Relationships

Qualitative	Quantitative
Boolean classifications	Modeling
Ranking studies	QSAR
Inductive relationships	

Keunemann et al.[69] the various approaches to predict biodegradability can be divided into some fundamental types as shown in Table 5.

Qualitative approaches are, in some cases, simple boolean classifications or rough screening methods capable of classifying chemicals as biodegradable or persistent, without a ranking in function of their degradability potential. A further qualitative approach is represented by ranking methods developed for congeneric classes of chemicals in general. This approach is based on the assumption that, on the same parent skeleton, the presence of some substituent or of some molecular feature (e.g., halogens on an aromatic ring, double bonds, etc.) can change the persistence of the molecule. Finally, a more advanced qualitative approach to describe biodegradation is represented by inductive relationships that take into account several structural features of chemical substances and elaborate them by means of statistical techniques, such as discriminant analysis. This approach was used by Enslein et al.[70] and by Niemi et al.[71] to classify with enough reliability a large number of chemical substances on the basis of their biochemical oxygen demand (BOD) (Table 6). Quantitative predictive methods based on a QSAR approach could be more useful in producing input data for evaluative models. Recently, many attempts have been made to obtain quantitative relationships in order to predict degradability from physicochemical properties or other molecular descriptors.

A review of the state of the art of the application of the QSAR approach to the prediction of persistence is given in Parsons and Govers[72] and by Vasseur et al.[73] for biodegradation and by Macalady and Schwarzenbach[74] for chemical transformation. Advances in the application of QSAR to environmental fate are proposed in a book edited by Hermens and Opperhuizen.[75] The interest in this field of research is rapidly increasing and promising results have been obtained; nevertheless, the practical applicability of the QSAR approach to the prediction of persistence is not yet comparable to those now attained for other aspects of chemical behavior (e.g., for aquatic toxicology) and further research is needed.

An interesting attempt to find a relatively quick and homogeneous method that can be used for comparing a large number of molecules and therefore is useful at least for a relative ranking of their "instability" has recently been proposed by Tremolada et al.[76] These authors use mass-spectrometry-derived data as a possible method for a preliminary approximate prediction of environmental persistence of organic substances. Different methods of interpreting the fragmentation patterns of

Table 6 Structural Features Associated with Chemicals Readily Degradable in a 5d BOD Test

Description of Association	Range of Half-Life (days)	Number of Chemicals	
		Correctly Predicted	Incorrectly Predicted
One halogen substitution on an unbranched chemical	<12	4	0
One cyano substitution on an unbranched chemical	<10	3	0
Aldehydes	2–11	15	0
Hydrocarbons	3–17	6	0
Alcohols, esters, amines	2–16	64	13
Acids	3–12	29	2
Amino acids	2–5	13	0
Sulfonates	2–17	15	3
Substitutes benzene rings with Kow<2.18	2–16	21	0
Biphenyl and two or fewer hydroxy-substitutes polyaromatics	<15	4	0
Cyclic chemicals consisting only of C, O, N, and H	2–15	21	2
Two aromatic rings (e.g., naphtalene and amino-naphtalene	<15	2	0

Adapted from Niemi, G. J., G. D. Veith, R. R. Regal, and D. D. Vaishnav, *Environ. Toxicol. Chem.*, 6, 515–527, 1987.

organic substances in mass spectrometry were evaluated, and the results obtained for several classes of pesticides were discussed in relation to known persistence data. Despite many limitations, it was concluded that this could be a promising approach and that, ideally, one should be able to compare the mass-spectrometry data with experimental results on degradation obtained in homogeneous environmental conditions.

NEEDS FOR RESEARCH IN EXPOSURE PREDICTION

As discussed, the rapid development of predictive approaches in the last 15 years allow us to make reliable enough evaluations of the exposure of chemicals, and in particular of pesticides, for management purposes. Nevertheless, many questions are still open and many important topics need further research. The various aspects of chemical exposure prediction have been reviewed in a recent book[77] and four main areas requiring research have been identified.

1. *Physicochemical data base.* In recent years a number of methods for experimental measurement and for theoretical evaluation or calculation of molecular characteristics of environmental relevance have been developed. The data base of reliable physicochemical properties is rapidly growing. Nevertheless, the information is not yet complete, and there is still the need for data on these properties, which are fundamental as input in all predictive approaches.
2. *Prediction of degradation.* This item has already been described as particularly relevant in the previous pages.
3. *Partition coefficients and kinetics of the distribution phenomena.* Most predictive approaches, and in particular simple evaluative models, are based on partition properties between the various environmental compartments. Value and limitations of the partition approach for the description of transport between compartments (soil/air, soil-sediments/water, soil/plant, etc.) should be carefully evaluated, taking into account the behavior of chemicals with different characteristics (e.g., polar and apolar). The role of time in partition/distribution processes should also be investigated.
4. *Models: validation and mass balances.* Guidelines are needed for the applicability of models in different situations and to establish a framework to determine whether a model is appropriate or not. It is necessary to develop and validate models capable of describing biogeochemical cycles of chemical substances, which include mass balance at different spatial scales. A coordinated effort is needed to develop guidelines for models and field validation to assess the influence of environmental variability on the predictive capability of models.

To this end, the Program on Environmental Toxicology (ENVITOX) sponsored by the European Science Foundation has formed a Working Group on Chemical Exposure Prediction based on the collaboration of several European research groups specializing either in the development and application of models or in producing environmental data. The objective of the Working Group is to collect and compare available information for the evaluation and validation of predictive models.

REFERENCES

1. U.S. EPA. Toxic Substances Control Act, Fed. Reg., 43, 4108, Washington DC (1978).
2. EEC Council Directive 79/831, *Off. Jour. Eur. Comm.*, L 10/259 (1979).
3. Wauchope, R.D. and M. J. Duffy. "Introduction to the symposium on role of modeling in regulatory affairs," *Weed Technol.* 6: 670-672 (1992).
4. Cairns, J., Jr., K.L. Dickson, and A.W. Maki. "Summary and conclusions," in *Estimating the Hazard of Chemical Substances to Aquatic Life*, J. Cairns, Jr., K.L. Dickson, and A.W. Maki, Eds. (Philadelphia: ASTM STP 657, 1978), pp. 191-197.
5. Mackay, D. "Finding fugacity feasible," *Environ. Sci. Technol.* 13: 1218-1223 (1979)

6. Calamari, D., M. Vighi, and E. Bacci. "The use of terrestrial plant biomass as a parameter in the fugacity model," *Chemosphere* 16: 2359-2364 (1987).

7. Donigian, A.S. and P.S.C. Rao. "Overview of terrestrial processes and modeling," in *Vadose Zone Modeling of Organic Pollutants,* S.C. Hern and S.M. Melancon, Eds. (Boca Raton, FL: Lewis Publishers, 1987), pp.3-36.

8. Jones, R.L. "Use of modeling in developing label restrictions for agricultural chemicals," *Weed Technol.* 6: 683-688 (1992).

9. De Coursey, D.G. "Developing models with more detail: do more algorithms give more truth?" *Weed Technol.* 6: 709-715 (1992).

10. Di Guardo, A., D. Calamari, G. Zanin, A. Consalter, and D. Mackay. "A fugacity model of pesticide runoff to surface water: development and validation," *Chemosphere.* 28:511-531 (1994).

11. Carsel, R.F., L. A. Mulkey, M. N. Lorber, and L. B. Baskin. "The pesticide root zone model (PRZM): a procedure for evaluating pesticide leaching threats to ground water," *Ecological Modeling* 30: 46-69 (1985).

12. Paterson, S., D. Mackay, E. Bacci, and D. Calamari. "Correlation of the equilibrium and kinetics of leaf–air exchange of hydrophobic organic chemicals," *Environ. Sci. Technol.* 25: 866-871 (1991).

13. Karickhoff, S.W. "Semiempirical estimation of sorption of hydrophobic pollutants on natural sediments and soils," *Chemosphere* 10:833-849 (1981).

14. Karickhoff, S.W., D.S. Brown, and T.A. Scott. "Sorption of hydrophobic pollutants on natural sediments," *Water Res.* 13:241-248 (1979).

15. Mackay, D. and S. Paterson. "Calculating fugacity," *Environ. Sci. Technol.* 15:1006-1014 (1981).

16. Green, R.E. and S.W. Karickhoff. "Sorption estimates for modeling," in *Pesticides in the Soil Environment: Process, Impacts and Modeling,* H.H. Cheng, Ed. (Madison, WI: Soil Science Society of America. Inc., 1980), pp. 79-101.

17. Pionke, H.B. and R.J. De Angelis. "Method for distributing pesticide loss in field runoff between the solution and adsorbed phase," in *CREAMS, a Field Scale Model for Chemicals, Runoff, and Erosion from Agricultural Management Systems,* USDA Conservation Res. Rep. 26 (Wasington, DC: USDA, SEA) pp. 607-643.

18. Gustafson, D.I. "Ground water ubiquity score: a simple method for assessing pesticide leachability," *Environ. Toxicol. Chem.* 8:339-357 (1989).

19. Laskowsky, D.A., C.A.I. Goring, P.J. MacCall, and R.L. Swann. "Terrestrial environment," in *Environmental Risk Analysis for Chemicals,* R.A. Conway, Ed. (New York: Van Nostrand Reinhold Co., 1982) pp. 198-240.

20. Goss, D.W. "Screening procedure for soils and pesticides relative to potential water quality impacts," *Weed Technology* 6:701-708 (1992).

21. Rao, P.S.C., A.G. Hornsby, and R.E. Jessup. "Indices for ranking the potential for pesticide contamination of groundwater," *Proc. Soil Crop Sci. Soc. Fla.* 44:1-8 (1985).

22. Aller, L., T. Bennett, J. Lehr, and R. Petty. "*DRASTIC: A Standard System for Evaluating Ground Water Pollution Potential Using Hydrogeologic Settings,* U.S. EPA, R.S. Kerr Environmental Research Laboratory, Ada, OK, U.S. EPA Report 600/2/2-85/018 (1985).

23. Canter, L.W., R.C. Knox, and D.M. Fairchild. "Pollutant source prioritization," in *Ground Water Quality Protection,* L.W. Canter, R.C. Knox, and D.M. Fairchild, Eds. (Boca Raton, FL: Lewis Publishers, 1987) pp. 562.

24. Jury W.A., W.S. Spencer, and W.I. Farmer. "Behavior assessment models for trace organics in soils: I Model description," *J. Environ. Qual.* 12:558-566 (1983).

25. Caldwell, S., K.W. Barrett, and S.S. Chang. "Ranking System for Releases of Hazardous Substances," in *Proceedings of the National Conference on Management of Uncontrolled Hazardous Waste Sites.* (Silver Spring, MD: Hazardous Material Control Research Institute, 1981), pp.14-20.

26. Baughman, G.L. and R.R. Lassiter. "Prediction of Environmental Pollutant Concentration," in *Estimating the Hazard of Chemical Substances to Aquatic Life*, J. Cairns, Jr., K.L. Dickson, and A.W. Maki, Eds. (Philadelphia: ASTM STP 657, 1978), pp. 35-54

27. Haque, R., Ed. *Dynamics, Exposure and Hazard Assessment of Toxic Chemicals* (Collingwood, MI: Ann Arbor Science Publisher Inc., 1980) p. 496.

28. Hutzinger, O., Ed. *The Handbook of Environmental Chemistry. Reaction and Processes, Vol. 2, Part B* (Berlin: Springer-Verlag, 1980) p. 205.

29. Neely, W.B. *Chemicals in Environment. Distribution: Transport, Fate, Analysis* (New York: Marcel Dekker Inc., 1980) p. 242.

30. Neely, W.B. and G.E. Blau, Eds. *Environmental Exposure from Chemicals. Vol. 1* (Boca Raton, FL: CRC Press Inc., 1985) p. 245.

31. Sheehan, P., F. Korte, W. Klein, and P. Bourdeau, Eds. *Appraisal of Test to Predict the Environmental Behaviour of Chemicals*, SCOPE 25 (New York: John Wiley & Sons, 1985) p. 400.

32. GSF. *Environmental Modelling for Priority Setting among Existing Chemicals*, (Munchen: Gesellschaft fur Strahlen und Umweltforschung mbH, 1985).

33. Mackay, D. and S. Paterson. "Fugacity revisited," *Environ. Sci. Technol.* 16:654-660 (1982).

34. Lewis, G.N. "The law of physico-chemical change," *Proc. Amer. Acad. Sci.* 37:49-55 (1901).

35. Clark, T., K. Clark, S. Paterson, D. Mackay, and R.J. Norstrom. "Wildlife Monitoring, Modelling and Fugacity," *Environ. Sci. Technol.* 22:120-127 (1988).

36. Bacci, E., D. Calamari, C. Gaggi, and M. Vighi. "Bioconcentration of organic chemical vapors in plant leaves: experimental measurements and correlation," *Environ. Sci. Technol.* 24:885-889 (1990).

37. Mackay, D., S. Paterson, A. Di Guardo, and F. Wania. "Generic and regional fugacity models levels I, II, III. OECD workshop on the application of simple models for environmental exposure assessment," Berlin (1991).

38. OECD. "Workshop on the application of simple models for environmental exposure assessment," Berlin (1991).

39. Lange, A.W. "The experience of a national competent authority in hazard assessment from the data provided by industry as it relates to ecotoxicology," in *Registering New Chemicals in Europe*, P.L. Chambers and C.M. Chambers, Eds. (Ashford, Ireland: JAPAGA, 1991) pp. 89-104.

40. Mackay, D., S. Paterson, and W.Y. Shiu. "Generic models for evaluating the regional fate of chemicals," *Chemosphere* 24:695-717 (1992).

41. Bacci, E., A. Renzoni, C. Gaggi, D. Calamari, A. Franchi, and M. Vighi. "Models, field studies, laboratory experiments: an integrated approach to evaluate the environmental fate of atrazine (s-triazine herbicide)," *Agriculture, Ecosystem and Environment* 27:513-522 (1989).

42. Sinkkonen, S. and A. Di Guardo. "Simulation of chlorinated hydrocarbons distribution in Bothnian Bay using a level iii fugacity model: bioconcentration in salmon and long range transport," *Proceedings of the Workshop on Evaluation of Fate and Exposure Models*, European Science Foundation, November 9–12, 1992, Dübendorf- Zurich.

43. Mackay, D., M. Joy, and S. Paterson. "A quantitative water, air, sediment interaction (QWASI) fugacity model for describing the fate of chemicals in lakes," *Chemosphere* 12: 981-997 (1983).

44. Mackay, D., S. Paterson, and M. Joy. "A quantitative water, air, sediment interaction (QWASI) fugacity model for describing the fate of chemicals in rivers," *Chemosphere* 12:981-997 (1983).

45. Cohen, Y., W. Tsai, S.L. Chetty, and G.J. Mayer. "Dynamic partitioning of organic chemicals in regional environments: a multimedia screening-level modeling approach," *Environ. Sci. Technol.* 24:1549-1558 (1990).

46. Mackay, D. *Multimedia Environmental Models, The Fugacity Approach.* (Boca Raton, FL: Lewis Publishers, 1991) p. 257.

47. Di Guardo, A., D. Calamari, G. Zanin, M.J. Cerejeira, and A. Consalter. "AGRIFUG: Previsione della contaminazione di acqua superficiali con un modello sequenziale di fugacità," *Ingegneria Ambientale* 22:386-395 (1993).

48. Mackay, D. and W. Stiver "Predictability and environmental chemistry," in *Environmental Chemistry of Herbicides Vol II*, R. Grover and A.J. Lessna, Eds. (Boca Raton, FL: CRC Press, 1991) pp. 281-297

49. Enfield, C.G., R.F. Carsel, S.Z. Cohen, T. Phan, and D.M. Walters. "Approximating pollutant transport to ground water," *Ground Water* 20:711-722 (1982).

50. Bonazountas, M. and J. Wagner. *SESOIL: a Seasonal Soil Compartment Model* (Cambridge, MA: Arthur D. Little, Inc. 1984).

51. Leonard, R.A., W.G. Knisel, and D.A. Still "GLEAMS: groundwater loading effects of agricultural management system," *Trans.* ASAE, 30:1403-1418 (1987).

52. Carsel, R.F., C.N. Smith, L.A. Mulkey, J.D. Dean, and P. Jovise. *User's Manual for the Pesticide Root Zone Model (PRZM)*, U.S. EPA, Athens, GA, U.S. EPA Report 600/3-84-109 (1984).

53. Wagenet, R.J. and J.L. Hutson. "Predicting the fate of nonvolatile pesticides in the unsaturated zone," *J. Environ. Qual.* 15: 315-322 (1986).

54. Wagenet, R.J. and J.L. Hutson. *LEACHM: A Finite Difference Model for Simulating Water, Salt and Pesticide Movement in the Plant Root Zone. Continuum. Vol. 2. Version 2.0.* (Ithaca, NY: New York State Water Resources Inst., Cornell Univ., 1989).

55. Matthies, M. "Transport and behavior in soil" in *Chemical Exposure Prediction* D. Calamari, Ed. (Boca Raton, FL: Lewis Publishers, Inc., 1993) pp. 103-114.

56. Watson, D.B. and S.M. Brown. *Testing and Evaluation of the SESOIL Model* (Palo Alto, CA: Anderson-Nichols and Co., Inc., 1984) p. 56.

57. Knisel W., Ed. *CREAMS: A Field-Scale Model for Chemicals, Runoff, and Erosion from Agricultural Management Systems,* U.S. Department of Agriculture, Conservation Research Report No. 26, (1980).

58. Leonard, R.A. "Movement of pesticides into surface waters," in *Pesticides in the Soil Environment: Process, Impacts and Modeling,* H.H. Cheng, Ed. (Madison, WI: Soil Science Society of America. Inc., 1990) pp. 303-350.

59. Wagenet, R.J. and P.S.C. Rao. "Modeling pesticide fate in soil," in *Pesticides in the Soil Environment: Process, Impacts and Modeling,* H.H. Cheng Ed. (Madison, WI: Soil Science Society of America. Inc., 1990) pp. 351-400.

60. Barnthouse, L.W. "The role of models in ecological risk assessment: a 1990s perspective," *Environ. Toxicol. Chem.* 11:1751-1760 (1992).

61. Carsel, R.F., W.B. Nixon, and L.B. Ballantine. "Comparison of pesticide root zone model predictions with observed concentrations for the tobacco pesticide metalaxyl in unsatured zone soils," *Environ. Toxicol. Chem.* 5:345-353 (1986).

62. Hedden, K.F. "Example field testing of soil fate and transport model, PRZM, Dougherty Plain, Georgia," in *Vadose Zone Modeling of Organic Pollutants.* S.C. Hern and S.M. Melancon. Eds. (Boca Raton, FL: Lewis Publishers, 1987) pp. 81-102.

63. Barra Rios, R., M. Vighi, and A. DiGuardo. "Prediction of runoff of chloridazon and chlorpyriphos in an agricultural watershed in Chile," *Chemosphere,* 30, 485-500, (1995).

64. Wethje, G., L.N. Mielke, J.R. Leavitt, and J.S. Schepers. "Leaching of atrazine in the root zone of an alluvial soil in Nebraska," *J. Environ. Qual.* 13: 507-513.

65. Swisher, R.D. *Surfactant Biodegradation* (New York: Marcel Dekker Inc., 1970).

66. Alexander, M. and B.K. Lustigman. "Effect of chemical structure on microbial degradation of substituted benzenes," *J. Agr. Food Chem.* 14:410-416 (1966).

67. Dias, F.F. and M. Alexander. "Effect of chemical structure on the biodegradability of aliphatic acids and alcohols," *Appl. Microbiol.* 22:1114-1121 (1971).

68. Hammond, M.W. and M. Alexander. "Effect of chemical structure on microbial degradation of methyl substituted aliphatic acids," *Environ. Sci. Technol.* 6:732-739 (1972).

69. Keunemann, P., P. Vasseur, and J. Devillers. "Structure biodegradability relationships," in *Practical Applications of Quantitative Structure Activity Relationships (QSAR) in Environmental Chemistry and Toxicology,* W. Karcher and J. Devillers, Eds., (Dordrecht, The Netherlands: Kluver Acad. Publ., 1990) pp. 343-370.

70. Enslein, K., M.E. Tomb, and T.R. Lander. "Structure activity models of biological oxygen demand," in *QSAR in Environmental Toxicology,* K.L.E. Kaiser, Ed. (Dordrecht, The Netherlands: D. Reidel Publishing Co., 1984) pp. 89-107.

71. Niemi, G.J., G.D. Veith, R.R. Regal, and D.D. Vaishnav. "Structural features associated with degradable and persistent chemicals," *Environ. Toxicol. Chem.* 6:515-527 (1987).

72. Parsons, J.R. and H.A.J. Govers. "Quantitative structure-activity relationships for biodegradation," *Ecotoxicol. Environ. Saf.* 19:212-227 (1990).

73. Vasseur, P., P. Kuenemann, and J. Devillers. "Quantitative structure biodegradability relationships for predicting purpose," in *Chemical Exposure Prediction*, D. Calamari, Ed. (Boca Raton, FL: Lewis Publishers, Inc., 1993) pp. 47-62.

74. Macalady, D.L. and R. Schwarzenbach. "Predictions of chemical transformation rates of organic pollutants in aquatic systems," in *Chemical Exposure Prediction*, D. Calamari, Ed. (Boca Raton, FL: Lewis Publishers, Inc., 1993) pp. 27-46.

75. Hermens, J.L.M. and A. Opperhuizen. *QSAR in Environmental Toxicology. IV*, (Amsterdam: Elsevier, 1991) p. 705.

76. Tremolada, P., A. Di Guardo, D. Calamari, E. Davoli, and R. Fanelli. "Mass-spectrometry-derived data as possible predictive method for environmental persistence of organic molecules," *Chemosphere* 24:1473-1492 (1991).

77. Calamari, D., Ed. *Chemical Exposure Predictions*. (Boca Raton, FL: Lewis Publishers, Inc., 1993) p. 233.

CHAPTER 4

Groundwater Vulnerability to Pesticides: An Overview of Approaches and Methods of Evaluation

Giuseppe Giuliano

CONTENTS

INTRODUCTION

Groundwater is the most important source of drinking water supply in many countries especially in plainlands, often amounting to over 90% of total tapped water. The introduction of stringent limits for pesticide concentrations in water destined for human consumption has focused attention, mainly in Europe and the U.S., on groundwater contamination by such compounds.[1,2]

A frequent coincidence between cultivated areas, where pesticides are applied, and their occurrence in the underground aquifers extensively used for drinking water has been observed. As a matter of fact, in such areas, groundwater often reveals traces or significant concentrations of pesticides that limit its use.

The long-accepted hypothesis that soil and the deeper unsaturated layers could always constitute an effective defense against penetration by pesticides is thus questionable. Extensive research has been developed on the conditions contributing to groundwater contamination by organic agrochemicals, focusing on the roles played by the different elements of the underground environment and on the behavior of the compounds. An effective assessment of the susceptibility of groundwater bodies to pollution by pesticides is the necessary basis to evaluate the contamination hazard associated with using these substances on land surface in order to identify measures attenuating such risk, as well as to delineate reliable monitoring programs.

0-87371-439-3/95/$0.00+$.50
© 1995 by CRC Press, Inc.

THE VULNERABILITY CONCEPT

Groundwater vulnerability is a general concept that identifies the natural susceptibility of underground water bodies to be affected, to some extent, by contaminating substances migrating from a pollutant load imposed on land surface. In this framework, pollutants are expected to move only along natural pathways; no shortening of route or bypass due to artificial means (e.g., injection wells in aquifer) is considered in the definition of the water body condition. A high degree of vulnerability may increase the chance of groundwater quality deterioration with negative impact on water use.

This susceptibility strongly depends upon the features of the underground environment affected by pollutants migration, from the land surface down to the water table of the aquifer and, ultimately, to the capture well of groundwater. Along an ideal route or pathway, contaminants pass through different underground zones, each one characterized by the occurrence of sets of dominant and subsidiary processes.

At least two zones may be schematically recognized, the unsaturated (vadose) zone above the water table and the saturated aquifer. However, in the unsaturated (vertical) profile, it is useful to single out the uppermost part near the land surface, i.e., the soil–root zone, for its specific features and role in groundwater protection. Pollutants migrate along the natural pathways by effects and interactions of physical, chemical and biological processes, which are highly influenced by the intrinsic properties of these substances. The degree of vulnerability of a groundwater body is also dependent upon the type of pollutant. Structurally, vulnerability may be intended as an integrated behavior of the underground zones (soil, unsaturated thickness, aquifer) with respect to a migrating pollutant. Such behavior is controlled by a large number of mechanisms, parameters, and properties, among which we can group those related to the nature of the underground (lithologic composition of layers, hydrogeologic structure, and flow regime) to the reactivity potential of the matrix and to the chemodynamics of the compounds. Vulnerability appears also to be dependent on the morphoclimatic conditions of the area concerned and the modes of pollutant application on land, which, respectively, determine the direct hydraulic and pollutant loadings on the water body.

From a functional standpoint, groundwater vulnerability may be schematically thought of as a combination of different aspects:

- the possibility of pollutants to penetrate into the saturated aquifer, which is connected to the hydrogeologic features of the unsaturated zone;
- the load attenuation capacity of the layers above the water table, resulting from interactions between pollutants and the matrix;
- the dilution capacity of the aquifer once pollutants have penetrated it, mainly depending on (saturated) flow conditions and on water resource renewal.

These aspects may be evaluated in terms of either qualitative attributes of intensity or numerical variables expressing transit time and residual concentration at any step of migration. However, the magnitude of pollutant load available to move downward and the net recharge rate determine the actual possibility of a pollutant to extend to the capture point. In the assessment of vulnerability conditions, the type of pollution source also has to be considered because of its effects on the spatial domain concerned and on the impact features.

For point pollution conditions (such as spills, landfills, etc.) vulnerability evaluations may be referred to some limited volume of the ground beneath the polluting source and downward where the typical "plume" of migrating leachate may develop.[3] The site parameters may be determined on field-scale, and the pollutant impact on the receiving water body is defined on the basis of pollutant pathway monitoring.

In the case of contamination from agrochemical use on land, the pollutant source is typically of an areal type, while the volume of groundwater body affected is quite large and correlated to the extent of the surface where the load is imposed. Natural heterogeneity in the intrinsic properties of soils and underground geologic materials and spatial variability of physical-chemical processes are the leading features of the domain. Detection of actual conditions by field investigation is not an easy task; homogenizing and averaging the local conditions is essential to attain a workable representation of the real world. Thus, vulnerability may be intended as an index of the overall behavior of the groundwater body as a sort of statistical mediation of many local conditions, which are largely undetectable.

Given the above framework, it appears that the general groundwater vulnerability concept shows serious limitations if not in the context of a contamination scenario, then defined in terms of environmental features and of types of pollutants. Difficulty of obtaining sound evaluations extends to effective deterioration of groundwater quality, or pollution level, if reference is not made to water quality criteria or standards related to a given substance.

A large variety of operative definitions and methods have been developed for assessing groundwater vulnerability. They differ in approach, but are similar in structure and scope; consequently, data needs and computational requirement may be substantially different.

Some procedures are empirical and include a few parameters considered crucial for explanation, at a screening level, of the contamination hazard for groundwater of different compounds. Other procedures are physically based, resulting in formulations that schematize underlying processes and simulate pollutant movement and behavior in the underground. Others, moreover, are hydrogeologically based, focusing on potential macrobehavior of groundwater bodies with regard to pollutant penetration. Because of the complexity of the overall process and the numerous disciplinary expertises involved, approaches to vulnerability problems are often sectorial, limited either to some segment of the pollutant route or to specific features of load attenuation or to compound behavior. However, this complexity and variety of aspects makes the search for an integrated approach important. Within such a framework the most relevant elements may be identified and used for appropriate evaluations, taking into account the type of pollutants, the environmental situation concerned, and the evaluation goals.

A REFERENCE FRAMEWORK FOR PROCESSES AND PARAMETERS

The principal features of groundwater vulnerability may be understood following the likely routes of migration of pollutants, from land surface to a groundwater body, and evaluating the most relevant controlling factors. When a pollutant load moves downward through the underground environment, it is affected by physical, chemical, and biological processes to different extents in each underground zone. It can result in a decrease of the total load, a redistribution of concentrations in time, and a retardation in pollutant transfer with respect to water flow. These effects are different expressions of the comprehensive phenomena of load attenuation.

The relationship between input load to each zone and the output from it describes the lumped behavior of that zone with respect to the retardation-elimination process and, consequently, its efficiency in protecting groundwater under given climatic conditions and pollutant application mode on land.

Pollutant migration in the unsaturated profile will be reviewed first, discussing the aspects related to both water flow and compound mobility. In the unsaturated zone, the open spaces represented by the interconnecting pores of the matrix contain both water and air. Water is the main carrier of dissolved pollutants. It moves vertically within small soil pores at a low rate, depending on the temporary balance among precipitation, infiltration, water binding, runoff, and water uptake by plants, under the combined effects of the matrix potential and gravity gradients.[3] The unsaturated flow is controlled by hydraulic conductivity, k, which is, in turn, dependent on the matrix pressure head (Ψ) and the moisture content (θ), also a function of the head. This means that hydraulic conductivity decreases with a decrease in water content or with an increase in pressure head (negative). The relationships between θ and Ψ, k and Ψ, k and θ are hysteretic and experimentally determined.

After a rain storm or an irrigation stage, a downward movement and moistening of deeper levels of soil occur. Upward movement may take place in the uppermost levels by means of evaporation and evapotranspiration. At a given time, the flow condition in the unsaturated zone may be macroscopically represented by the moisture profile, i.e., the vertical distribution of volumetric moisture content $\theta(z)$. The wetting front moves downward at a rate that is controlled by hydraulic properties of the soil (porosity, permeability, bulk density) and by the antecedent soil moisture content. The unsaturated zone is affected by time-variant recharge inputs of different magnitudes. If the recharge is long and large enough, the movement will proceed to the water table. The infiltration water may take a rather long time to reach it, i.e., a week, a month, or much more.

The unsaturated one-dimensional (vertical) flow may be described by a number of similar equations that differ in assumptions and simplifications of the real world. Under constant conditions, a generalized Darcyan scheme relates volumetric flow, q, to soil water pressure head, Ψ, along the soil depth, z, times unsaturated conductivity, $k(\Psi)$. In macro terms the travel rate, V, is determined by Darcy's water velocity, v, and the volumetric water content:

$$V = \frac{v(\Psi)}{\theta}$$

The saturated condition, with $\Psi = h$ (hydraulic head), $\theta = n$ (effective porosity) and $k = k(\Psi)$ = constant may be intended as a particular case of unsaturation. It is frequently referred to as the worst condition for the unsaturated vertical flow of pollutants.

Because of the slow movement of water in the unsaturated medium, which is furthermore limited to smaller pores, and of favorable geochemical conditions, relevant chances for interaction between solid and liquid phase occur in this underground zone. The different processes ultimately result in trapping, decomposing, and transforming the dissolved contaminant, thus attenuating the load.

Transport of pollutant solute throughout the unsaturated profile is commonly described by the classical Fickian type of convection-dispersion mechanisms. The relationship between solute concentration and depth (vertical transport) is controlled by different transport components: convective water flux; dispersive flux; interaction between the liquid and solid phases of soil, mainly represented by sorption and exchange processes; and pollutant decay.

Under the assumption of homogeneous media for constant conditions of flow, the transport may be expressed by:

$$R\frac{\partial c}{\partial t} = D\frac{\partial^2 c}{\partial z^2} - v\frac{\partial c}{\partial z} - \lambda$$

where c is the soil solution concentration, v is in the porewater velocity, D is the dispersion coefficient (assumed to be limited to the longitudinal one)[4] equal to velocity times a constant, dispersivity, characteristic for each type of medium (hydrodynamic dispersion is assumed to be the only mechanism contributing to dispersive flux), R is the so-called retardation factor, a parameter accounting for the effect of the sorption and exchange processes. Equilibrium type interaction (instantaneously achieved) and linearity between solution and sorbed concentrations are assumed for convenience. Lastly, λ is a term accounting for the loss of a substance from biochemical or chemical degradation processes and is related to concentration.

The expression of the factor R relates the actual velocity of water flow to the transport velocity of an absorbed compound through a combination of different parameters of the porous media and the substance:

$$R = \frac{V_w}{V_s} = 1 + \frac{K_D(1-n)\rho_s}{\theta}$$

where V_w and V_p are the velocities for water and absorbed compound respectively, n is the matrix porosity, ρ_s is the grain density of the matrix, θ is the moisture content, K_D is the partition coefficient of the pollutant (which may be approximately estimated from the corresponding coefficient for organic carbon K_{oc}). R, in the assumption of simple linearity between solute and sorbed concentrations, becomes independent of concentration.

Many other formulations for solute transport have been developed to consider specific aspects or to overcome simplifying assumptions on diffusive flux, equilibrium and linear sorption, multicomponent flow, etc. Their discussion lies outside the scope of the present paper. Notwithstanding the apparent superiority of more refined approaches to solute transport in unsaturated zones, their practical use is still in the preliminary phase.

The relative importance of the convective and dispersive terms in the transport equation depends on the prevailing conditions of flow in the different sections of the unsaturated profile. Percolation down to the water table is generally dominated by gravitational convective flow, while diffusive contribution is restricted mainly to interaction situations. Infiltration flow in the upper sections may largely be affected by diffusive phenomena availing itself of the uneven distribution of water and solute concentrations in the matrix.

The flow conditions in the unsaturated zone are influenced by climate and weather conditions, primarily in terms of annual infiltration. However, much attention needs to be paid to hydraulic conditions generated by the important recharge events.

Field conditions are normally related to variations in moisture content due to normal hydraulic load and approaching retention capacity after prolonged drainage. Under heavy hydraulic loads, saturated flow conditions are attained, and flow becomes dependent on effective porosity and saturated hydraulic conductivity. That will result in a dramatic reduction in transit times, compared to unsaturated conditions, and a bulk transfer of pollutant downward to the water table. Due to the large differences in physicochemical properties of pollutants, contaminants, similar behavior for environmental mobility should be considered.

Water solubility is, in principle, a relevant feature for estimating mobility but, due to the commonly high rates of pesticides, it does not seem to be a limiting factor. The behavior of pesticides in soils is strongly affected by processes of (reversible) adsorption and desorption, the most relevant aspect of interaction between a compound and the soil matrix. Adsorption is brought about by organic constituents, by ion exchange with minerals, and by polar interaction. Matrix components that seem to play a control action are clay mineral content, organic matter, and hydroxides.

Attenuation of pollution load depends also on biological degradation of compounds and their transformation into metabolites, but disappearance can take place through chemical processes, such as hydrolysis, oxidation, and neutralization with reduction of original solubility and/or generation of less soluble residues. Finally, persistence of compounds plays a major role in mobility due to the normally long times in which they remain in the soil. The physicochemical interactions between compounds and matrix have been extensively investigated. Several expressions proceeding from a variable set of assumptions and constraints are available.

Despite the most convenient linearity assumption, adsorption-desorption processes are generally nonlinear and have chemically controlled kinetics (multicomponent transport). Instantaneous equilibrium mechanisms are also questioned in favor of diffusion-controlled nonequilibrium sorption. Moreover, interaction characteristics and parameters have been developed by laboratory experimentation. Their application and validity at field scale are often questioned. More difficult to evaluate are biotransformation phenomena.

The uppermost, or soil–root, zone of the unsaturated profile is the site where a very consistent part of the elimination-attenuation-transformation processes of the penetrating pesticide load occurs. The amount of pesticide leached at the base of the soil zone should greatly depend on the intensity of the interaction between the chemodynamic properties of the substance and the physical properties of the soil. For persistent, low-reactive pollutants, a simple time lag in transfer downward may be the leading feature of the migration process.

The important structural defense of groundwater against penetrating pesticides represented by the soil zone always occurs in situations of diffuse agricultural pollution. Sometimes the horizon may be bypassed when specific land treatments are applied, such as deep ploughing and furrow digging. The efficiency of soil zone control over pesticide leaching may be greatly diminished by the development of so-called preferential flow in the soil.

The flow and chemical transport of pollutants commonly assume that the soil zone is porous and homogeneous, but, in reality, this is not the case. Soil at the field scale is heterogeneous both in its vertical and horizontal dimensions. This heterogeneity may be related to biological activity (presence of roots, insect passages, bioperturbations, etc.), to textural features (drying cracks in clay-like soils) or to the intrinsic soil structure (macropores, aggregates, fissures, etc.).[5,6] The occurrence of preferential paths may also be linked to specific hydraulic situations, such as fingering in a wetting front, presence of water repellent zones, etc. Such preferential paths in soil allow for a rapid downward movement of pesticide solutions with negative effects on the elimination and transformation processes, because full interaction with the microporous matrix is impeded or, in any case, less enhanced.

The importance of preferential transport in leaching pesticide through soils has stimulated the search for alternative evaluation and the use of models that associate two domains; one is characterized by a soil matrix in which an unsaturated darcyan type flow occurs, the other consisting of a single macropore or a network of macropores through which water primarily flows under the influence of gravity.

Another specific mechanism that decreases the action of the soil filter and increases leaching is the transport of pesticides by means of some "carriers" made up of fine clay colloids and particulate organic matter. Such a mechanism may occur in coarse-textured or fissured soils.

Pesticide leached at the base of the soil–root zone enters an unsaturated environment extending downward to the water table. Here, processes causing elimination and/or attenuation generally explicate at a lesser rate. This is related to a much smaller content of dispersed clay, mineral, and organic matter and a reduced bacterial activity. Thus, the mobility of pesticides is generally higher in the lower part of the profile than in the upper soil zone. The degree of attenuation depends greatly on the flow regime and residence time in the unsaturated thickness.

The downward advancement of water and pollutants may be contrasted by less permeable, or impermeable, lenses and layers frequently present in the unsaturated profile. Above these, a temporary perched water accumulation may form. However, these aspects already pertain to a hydrogeologic view of the underground structure.

Generally speaking, the occurrence of deep (impermeable or low permeable) clay-like layers is the best barrier to pollutant migration from the surface. These layers (aquiclude or aquitard) may confine, in the hydraulic sense, the groundwater bodies with effects on the flow conditions,[3] while the active recharge area is to be found displaced upward, with respect to imposed load surfaces.

In this hydrogeologic framework, it is important to note that the occurrence of a consolidated geologic material, such as limestone or sandstone, in the underground is frequently associated with the presence of a network of fissures or fractures (sometimes, in the case of karstified formations, turning to channels) that increases the downward accessibility of pollutants. In such a situation, they may penetrate to depths much deeper than predicted for porous media.

Though background theory and models on transport in fractured media (or with dual type porosity, like sandstones) have been developed; due to the complexity of natural conditions, they are based on very simplifying assumptions, and their use in practice is still under examination.

The final segment of the contaminant route is the penetration into the saturated zone or aquifer and its migration toward a discharge or capture area. When the pollutant load, remoduled along the unsaturated profile to both space and time distribution, enters the aquifer, it is further attenuated by any one of a number of factors, primarily hydraulic. Dilution appears to be the most relevant as it is the final feature of flow conditions occurring in the aquifer.

The geochemical attenuation of pollutant loads here depends both on the chemical composition of the groundwater and the presence of reactive components in the aquifer matrix. Groundwater is generally less oxygenized at a certain saturated depth, while environmental bounding conditions, expressed by factors such as acidity, redox potential, and trace elements, may strongly affect compound mobility. Reactive materials such as clay, organic matter, and lime, generally do not attain significant levels. Groundwater is also poor for bacterial nutrition. From all these aspects, it may be argued that the attenuation and elimination processes are generally slight and difficult to quantify in this underground zone.

At low concentrations of contaminant, as those normally expected in aquifers due to diffuse pesticide inputs, dilution is a function of flow rate and transport processes. The overall effect of hydrodynamic dispersion is to spread or dilute the advective contaminant moving through the porous medium. Occasionally its contribution may become relatively important. The expressions that characterize saturated flow and transport processes are well established under different hydraulic conditions (unconfined or confined aquifers, porous or fractured media, etc.).(see, e.g., Bear and Verrujt).[7] The flow field is determined by the features of the aquifer, namely the type of hydrostructure and flow network, the hydraulic properties of the matrix, the boundary conditions expressed by geometric limits, and recharge-discharge relationships. Several parameters, such as dispersion coefficients, are necessay to describe transport conditions besides flow velocity. Their experimental determination is not always easy and effective for reproducing real contaminant migration, especially in the case of diffuse contamination such as that of pesticides. Other aspects affecting dilution are related to the overall behavior of the aquifer, which may be expressed by the renewal capacity of the water resource hosted and/or the transit time from the recharge area to the discharge or capture zone.

Most of the procedures and models concerning evaluation of pollutant (pesticide) movement along natural pathways refer to homogeneous media and incorporate constant value parameters for a given matrix and/or compound. In the real world, heterogeneous media, time variability, and dependency of chemical transport parameters on environmental conditions are the primary features, and they strongly affect the capability of the above tools to explicate field scale problems. Analysis, either of the variability and dependency of transport parameters (such as the partition coefficient) on physical and chemical characteristics of the environment or of consistency of equilibrium-type assumptions in their evaluation are beyond the scope of this paper. For these aspects, the reader should consult the extensive literature on the subject.

One aspect, however, that needs to be discussed in some detail, because of its prominent role in the evaluation of the migration process, concerns the spatial and time variability of the parameters. The effects of this variability on water flow and transport of pesticides appear to be significant, rendering the results obtained for definite conditions questionable and suggesting that deterministic solutions of equations may not accurately describe processes in natural fields. The problem of uncertainty and spatial variability is particularly relevant in making an assessment of the effects of pesticide leaching on groundwater, taking into account the regional features of contamination phenomena and the interest for area averaged results.

Alternative approaches to the deterministic formulation of processes were developed in a stochastic model resorting to Monte Carlo techniques (generation of PDF) applied to transport parameters or incorporating uncertainties into the flow models.[8-10]

Groundwater contamination has been examined, so far, considering the most prominent physical and chemical processes that occur along the natural pathway in a disaggregate mode. The pertinent parameters controlling mechanisms and processes, and possibly influencing vulnerability conditions have been discussed.

A completely different approach to the problem of vulnerability evaluation integrates the unsaturated and saturated zones into a unitary or systemic view. According to a classical definition, a system is a functional entity, acting as a correlated set of elements, processes, and aspects, whose behavior may be evaluated by input–output relationships. The system's behavior is analyzed as a whole without entering into the details of processes and mechanisms, but the response may be correlated to the most important features that characterize the system in terms of structure setting and functional operation. Such a correlation is expressed either empirically or conceptually, but it is not necessarily a causative scheme.

Interest is now focused on the evaluation of the functional response of a hydrogeologic system[11] to a contamination event. The response is defined by referring to a conceptual scheme that relates the hydrogeologic attributes or features of the system and expected pollution potential. Such behavior may be influenced by travel time, flux, and attenuation potential associated with a pollutant event or situation, expressed empirically or qualitatively. Here, we are in the field of the hydrogeologically based vulnerability.

Three-dimensional distribution of properties and features over entire hydrogeologic units or geologic formations becomes important when determining the system's macrobehavior. Specific processes concerning pollutant behavior and attenuation by physicochemical reactions are not explicitly considered. As a matter of fact, hydrogeologic vulnerability assessment may be largely independent of pollutants. No direct estimate of pollutant impact (leaching concentrations, load reaching the aquifer, etc.) is explicitly obtained equally.

The prominent features affecting the functional response of the hydrogeologic system and, thus, vulnerability, concern the following:

- characteristics of soil and underground, mainly lithology, which regulate accessibility to the aquifer;
- depth to water table, which determines, in conjunction with the above feature, the transit time from the land surface to the aquifer;
- groundwater saturated flow, which expresses hydrodynamic behavior;
- recharge/discharge conditions, which control renewal of average groundwater reserve.

Soil may be evaluated with respect to infiltration capacity and to filtering capability. These features may be simply expressed by means of texture and organic matter content. Potential pollution may increase with the degree of primary permeability, related to the percentage of sandy and silty components in the soil texture and to the occurrence of shrinking clay creating a secondary vertical permeability by means of cracks when dried. The organic matter content may be taken as a lumped indicator of pollutant attenuation.

Lithologic and formational features of the intermediate media control accessibility to groundwater. The degree of primary permeability and/or fracturing and the occurrence of layered sequences may be used to describe behavior in relation to potential pollution. A high degree of primary permeability in unconsolidated formations (mixtures of sand, silt, gravel, clay) and of fracturing in consolidated formations (sandstones, igneous rocks) result in a high pollution potential. In limestone sometimes a network of solution cavities and fractures occurs, greatly increasing the potential.

Sequencing of layers of different lithologic composition frequently occur; pollution hazard is largely influenced by the presence of beds of clay material, which play a relevant role in the geologic setting of the underground. Thick, continuous layers of clay may isolate or confine the groundwater body from the upper porous formations protecting it from surface infiltration of pollutants.

Depth to groundwater has to be considered as a primary factor because it determines the space/ time dimension through which a contaminant must travel before reaching the exposed downward (unconfined) aquifer. By correlation, it expresses the maximum opportunity for a pollutant to interact with the unsaturated media. A water table aquifer located closer to the surface is necessarily much more exposed to pollution than a deeper one where the transit time of infiltration water is consistently higher. However, the transit time is affected by matrix permeability, frequence of bedding planes, and tortuosity of paths. Flow conditions in the aquifer are expressed by hydraulic conductivity or transmissivity parameters connected to porosity and fracturing of saturated media and by flow field features in terms of velocity and flowline patterns.

The hydrodynamic function expresses the ability to move pollutant downwater from sites of incidence on the aquifer and to dilute loads, thus reducing pollution potential. Given a certain gradient, higher rates of groundwater flow derive from high values of conductivity and transmissivity controlled by primary porosity and fracturing, and by the thickness of the aquifer. A divergent flow field pattern appears to favor dilution more than a convergent or parallel one. The pattern is strongly controlled by formational and geostructural features of the aquifer.[11]

Recharging conditions are controlled by climate and the characteristics of soil and unsaturated media. The amount of water that infiltrates (net recharge) downward is the vehicle for pollutants. For unconfined aquifers, recharge is generally direct, and the pollution potential is greater than that for confined aquifers. These latter are protected by top impermeable layers, and they are recharged laterally or by very low filtration from other aquifers (leakage). When polluted recharge occurs, the higher the rate, the higher the pollution potential for the aquifer. However, the ratio of vertical recharge to the aquifer flow has to be considered, because it determines the dilution rate that can be achieved. Another consideration regards the renewal of the average groundwater reserve as a consequence of the ratio of recharge to the reserve volume. High rates of renewal allow for a rapid flushing of the contaminants stored in the body, thus diminishing pollution potential.

The features or parameters discussed, or others, are used for evaluating hydrogeologic vulnerability according to different empirical schemes or procedures. They often lead to classifications of hydrogeologic systems referring to specific aspects of functional behavior with respect to potential contamination. The most frequent parameters considered include lithology and fissuring of the unsaturated layers, depth to water table, hydraulic conductivity and lithology of the saturated aquifer, recharge rate, flow field patterns, renewal rate of groundwater resources, and others.[12] Table 1 lists the parameters involved in vulnerability evaluation. Lumped approaches similar to those mentioned above hydrogeologic systems have been developed for soils. Behaviour is characterized in terms of some parameters related to pollutant migration throughout soils, sometimes with specific reference to pesticides. The main parameters considered concern the type of unsaturated flow, distinguishing between unconsolidated and consolidated materials, occurrence of by pass flow, occurrence of an impermeable or low permeable layer at a small depth, organic matter content, and agronomic factors. Empirical procedures or classification schemes may be used to combine the effects of the parameters for the evaluation of vulnerability of soils to penetrating pollutants.

EVALUATION OF VULNERABILITY

The different factors previously examined can be recognized, to a greater or lesser extent, in many formulations and procedures used for assessing groundwater vulnerability and related aspects. These originate from different scientific backgrounds; sometimes they are specifically task oriented and can be considered limited views of the more general problem of groundwater vulnerability.

Very simple and somewhat implicit estimations of vulnerability may be found in some screening formulations proposed to classify tendencies of pesticides to leach. They are often applied in a context of estimating of ecotoxicologic risk. In some of them, only very few physicochemical properties of the substances are considered for estimating their mobility, independently of underground environment features,[13,14] used as a proxy indicator of leaching potential to groundwater. Others refer to a simplified structure of the underground, distinguishing in the upper crop root zone from the intermediate vadose zone above the water table. Variables that specifically account for the nature of soil and the occurrence of an aquifer are included. An example of this type, is the Rao model,[15] which contains total porosity, volumetric water content (at field capacity), bulk density, organic carbon content as soil parameters, and some parameters related to crops, while groundwater is shortly characterized through depth to water table and recharge rate. In Rao's procedure, two indices, the attenuation factor and the delay factor, allow for a quantitative evaluation of the pesticide behavior through soil and intermediate unsaturated zone; the amount of pesticide reaching the water table and the relative time of travel quantify synthetically the contamination potential (relative) of the different substances to leach past soil zone and intrude into groundwater.

Similarly, the Jury index[16] is based on the simplified solution of a system of equations concerning the movement and degradation of pesticides in different environmental conditions. Pesticides are ranked on the basis of travel time by convective flow in soil water. No estimate of residual concentration is made.

In these procedures, attention is mainly concentrated on the soil, where the most important interactions with compounds take place. The behavior is highly simplified, and the data need is consequently very limited. The above models generally use uniform values for soil properties (water content, soil bulk density, and other properties), while flow is assumed to be constant with depth and time (the effects of effective rainfall and irrigation water on pesticide movement are not considered). Evaluations are given for standard scenarios (soils or site). These models are not designed to be predictive tools, and the ranking is not based on pesticide concentration in the groundwater. Information supplied with respect to groundwater vulnerability seems to be of a very preliminary character and largely insufficient.

More realistic procedures to evaluate the behavior of pesticides in the soil–root zone consist of a number of mathematical simulation models recently developed in the U.S. and Europe.[17-21] The models are based on physical concepts, that is, the advection/dispersion process for solute transport, reversible linear adsorption, first-order transformation kinetics, and passive plant uptake. Different options are available in some of them, such as temperature dependence on transformation, nonlinearity of sorption, application time of pesticide, and influence of water content.

Table 1 Vulnerability Degree of Hydrogeologic Complexes

E_E	E	A	M	B	B_B	Hydrogeologic Complexes and Settings Features
	■					Unconfined (water-table) aquifer in coarse- to medium-grained alluvial deposits, without any surficial protecting layer
■ (a | b)						Unconfined (water-table) aquifer in alluvial deposits: (a) streams free recharging groundwater body; (b) well or multiple-well system drawdowning water table under stream level (induced recharge)
	■					Confined, semiconfined (leaky), and unconfined aquifer with impervious (aquiclude) or semipervious (aquitard surficial protecting layer)
■						Aquifer in carbonate (and sulphate) rocks affected by completely developed karst phenomena [holokarst with high karst index (k.i.)]
	■ (a | b)					Aquifer in highly fractured [high fracturing index [(f.i.)] limestones with k.i. low or null: a) depth to water <50 m b) depth to water >50 m
		■ (a | b)				Aquifer in highly fractured (but not cataclastic) dolomite with k.i. low or null: a) depth to water <50 m b) depth to water >50 m
		■	■			Aquifer in medium- to fine-grained sand
			■			Aquifer in fissured sandstone or/and noncarbonatic cemented conglomerate
				■		Aquifer in fissured plutonic igneous rocks
			■	■		Strip aquifers in bedded sedimentary sequences (shale–limestone–sandstone flysch) with hgihly variable diffusion rate layer by layer
		■				Aquifer in highly clivated volcanic rocks and nonweathered plutonic igneous rocks with high f.i.

Table 1 (continued)   Vulnerability Degree of Hydrogeologic Complexes

E$_E$	E	A	M	B	B$_B$	Hydrogeologic Complexes and Settings Features

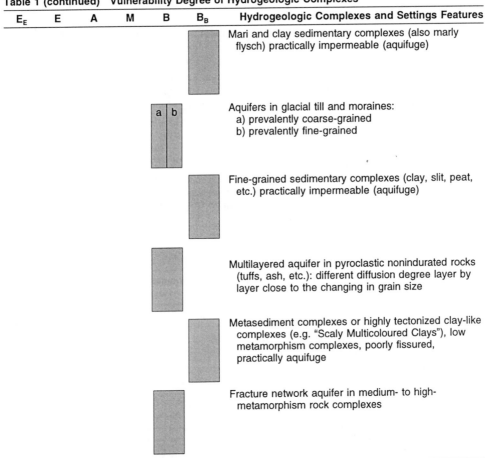

Mari and clay sedimentary complexes (also marly flysch) practically impermeable (aquifuge)

Aquifers in glacial till and moraines:
a) prevalently coarse-grained
b) prevalently fine-grained

Fine-grained sedimentary complexes (clay, slit, peat, etc.) practically impermeable (aquifuge)

Multilayered aquifer in pyroclastic nonindurated rocks (tuffs, ash, etc.): different diffusion degree layer by layer close to the changing in grain size

Metasediment complexes or highly tectonized clay-like complexes (e.g. "Scaly Multicoloured Clays"), low metamorphism complexes, poorly fissured, practically aquifuge

Fracture network aquifer in medium- to high-metamorphism rock complexes

E$_E$ = Extremely High; E = Very High; A = High; M = Medium; B = Low; B$_B$ = Very Low or Nil

Mass balance is computed mathematically obtaining the amounts of substance respectively degraded, absorbed in soil, leached at the base of the root zone, or occurring at a given depth. These models, typically monodimensional (vertical), simulate pollutant behavior on a fine time scale (e.g., daily) and use a large number of parameters that are empirically or experimentally estimated. Concerning the vulnerability aspects, most of these models are useful for estimating the potential of soils to leach pollutant at limited depths (generally, less than 1 m and not greater than 5 m); for greater thicknesses of the unsaturated zone, results seem to become less reliable. However, they tend to restrict the evaluation to the first functional aspect described in the chapter dealing with the vulnerability concept.

A crucial point for the application of these simulation models is the validation phase in terms of the influence of variability of parameters and of field tests. The data available for ranges of soils, pesticides, and climatic conditions are very limited. Results, however, have to be evaluated over sufficiently long times of simulation related to recharge rate and mass emission beyond the soil zone.

Validation studies sometimes indicate that results greatly depend on estimated values of input (chemodynamic) parameters, showing significant differences from field tests. It is argued that estimates of parameters referring to standard soil types are often not representative of the commonly tested sandy soils; some assumptions about processes in the model are easily invalidated in natural field conditions; other transport mechanisms not contemplated in the schematization (preferential flow, movement of adsorbed colloids, etc.) are likely to occur. It is interesting to note that these models have been used to classify the leaching potential of different compounds as well as to evaluate the relative susceptibility of different soils to leaching, with reference to specific compounds over large regions.[22]

In terms of vulnerability evaluation, the lower unsaturated zone, extending downward to the water table, might be considered as an extension of the soil zone. Transfer of pollutant in this zone can be still modeled using convective/diffusive mechanisms with components that account for sorption and decay processes. Notwithstanding the progresses in computing solutions, the practical utility of such tools is still questioned. Problems caused by spatial and temporal variability in hydraulic properties, flow conditions, and nonequilibrium processes affecting transport become relevant when evaluating the reliability of results. Models of different structure complexity are available. They require large sets of input data, and validation tests are difficult to perform because of the absence of suitable data collected along the profile or just at the water table, before the aquifer flow begins to dilute.

Recognizing the difficulty of obtaining valid data and questioning the practical usefulness of applying such advanced formal procedures to a relatively unexplored environment, sometimes the analysis of the lower unsaturated zone is carried out by means of simple expressions such as the retardation factor,[15] transit time,[16] or simplified models.[23]

An estimate of vertical flow may be obtained from a calculation of Darcy flux, under saturated or unsaturated conditions, depending on the matrix and recharge conditions, equalizing it in the latter case to final infiltration capacity.[24] The transit time through the unsaturated profile related to degradation time may also be understood as a measure of vulnerability.

Simulation procedures are extensively used to predict the movement of a pollutant once it entered a saturated flow field. A large variety of models, at different degrees of refinement, are available in the literature.[25] Estimates of parameters are more efficient in the saturated zone than in the unsaturated zone; validation can be carried out because of the possibilities of gathering data from sets of wells. The reliability of these models is generally good if the distribution of pollutant input is perfectly known. The problem of estimating this input, however, remains largely unsolved because of the above mentioned difficulties and uncertainties in assessing migration in the unsaturated zone.

Recently, complex procedures have been designed using jointly different submodels concerning soil, the unsaturated zone, and the aquifer, in order to evaluate the fate of agrochemicals and the associated contamination risk of groundwater.[26] Theoretically, such procedures could satisfy the entire functional set illustrated in the chapter dealing with the vulnerability concept but they are related mainly to functions a) and c). They appear to be the best tool for evaluating the actual vulnerability of groundwater because they quantify pollution potential in terms of physical magnitude, which can be compared with water quality criteria. However, the relevant amount of data needed makes them very complicated to use in practice.

The models previously described generally simulate the pollutant migration along the vertical dimension in the unsaturated zone and over a bidimensional space in the saturated aquifer, making reference to a finite differences or finite volumes discretization. For a regional assessment of pesticide loadings and movement in groundwater such tools would have to be coupled to some procedure that would take into account the spatial variability of soil and media properties or input variables (recharge, loadings, etc.). Some examples of regionalized evaluation, by means of conventional zoning or geographical information systems (GIS) are available.[10,27] Sometimes, such procedures are used for evaluating uncertainty too, by means of Monte Carlo techniques.

Different methods have been developed to evaluate hydrogeologic vulnerability. Integrated classifications of hydrogeologic systems may be based on formational and structural criteria of aquifers and on recharge-flow conditions. An extensive classification of this type used for preparing intrinsic vulnerability maps at different scales[28] is given in Table 2.

Accordingly, the behavior of hydrological complexes is globally estimated by taking into account different hydrogeologic aspects, i.e., setting features of lithologic complexes, the occurrence of protection layers in the unsaturated zone, the structural situation of the aquifer (confined, semiconfined, free surface), the flow conditions, and water table depth. A similar methodology proposed in France[29] takes into account the possibility of pollutants passing through unsaturated layers, spreading into the aquifer, and persisting in the saturated zone in relation to the recharge rate. Hydrogeologic complexes are characterized on the basis of a six-class ranking scheme. The

Table 2   Some Methodologies and Related Data Used for the Evaluation of Hydrogeological Vulnerability

Methodology and Reference	Surface			Soil				Unsaturated			Saturated			
	P	T	D	L	R	KS	CH	RR	NR	KU	S	F	H	K
Albinet, Margat (1970) (29)								•		•	•		•	•
Vrana (1968)									•			•		
Fenge (1976)			•					•	•	•		•	•	•
Josopait, Swerdtfeger (1976)								•	•	•			•	•
Fried (1987)									•	•				
Villumsen, Jacobsen, Sonderskov (1983)				•					•	•	•	•		•
Haertlè (1983)									•	•				
Vrana (1984)	•			•						•			•	
Subirana Asturias, Casasponsati (1984)								•		•	•		•	•
Engelen (1985)								•		•	•		•	
Zaporozec (1985)				•	•	•	•			•	•		•	
Breeuwsma et al. (1986)				•	•	•	•	•	•	•	•			•
Sotornikova, Vrba (1987)						•				•	•	•		•
Ostry et al. (1987)				•		•				•	•			
Goossens, Van Damme (1987)				•				•		•	•			•
Carter et al. (1987), Palmer (1988)				•	•	•					•			
Marcolongo, Pretto (1987)				•	•			•	•	•				
Foster, (1987, 1988)									•	•	•			
Schmidt, (1987)				•					•	•	•			
Civita (1988)									•	•	•		•	•
Aller et al. (1985, 1987)	•			•				•	•	•	•		•	•
Civita (1991)	•	•	•					•	•	•	•		•	•

Legend: P = precipitation; T = topography; D = drainage density; L = lithology; R = specific retention; KS = soil permeability; CH = physicochemical features; RR = linkage with surface network; NR = net recharge; KU = unsaturated permeability; S = depth to water table; F = piezometric fluctuation; H = hydrogeologic features of aquifer; K = hydraulic conductivity.

From Civita M. Proceedings of the 1st National Congress on Protection and Management of Groundwater, Marano sul Panaro, Italy, 1990.

controlling factors, such as flow rate, aquifer permeability and recharge rate are not explicitly evaluated by these methods.

Alternatively, prominent parameters may be independently mapped, and the aquifer may be divided into zones by intersection of these maps where simple vulnerability classifications are obtained in a semiquantitative mode on either value intervals of parameters or types of conditions. For example,[30] evaluation of vulnerability may be obtained by classifying, according to empirical scales, three main properties of aquifers: the hydraulic condition of groundwaters; the overall aquifer class, based on lithology and rock consolidation features; and the depth to water table. In this relatively simple procedure, emphasis is given to the hydraulic accessibility of groundwater and to the overall attenuation capacity of the saturated zone.

These methods are also suitable to produce very condensed classification schemes founded on very few leading properties, such as permeability of the unsaturated zone and depth to aquifer, in relation to some general environmental evaluation.[31] These empirical methods based on classified sets of hydrogeologic parameters, however, produce relative evaluations of vulnerability that may appear in some fashion too synthetic or condensed. In any case, they do not indicate the most important parameters in different situations and do not discriminate among them, unless the original methodologic background is altered, in relation to specific pollutant loadings. In these representation methods, correlation with the imposing load on the surface is limited to the overmapping of the spatial distribution of hazard sources. No inference on the effects on groundwater (pollutant concentrations) is drawn.

Extensively used, parametric procedures are based on a very limited set of independent parameters, classed or value distributed, depending on their specific role, which is crossrelated by means of different evaluation schemes, i.e., matrix, rating, point counting, etc. Weight or multipliers are sometimes used to increase the relevance of some aspects or parameters in relation to specific types or forms of pollution. Factor identification, assignment of weights, establishing scaling functions, development of numerical indices or classification schemes are all largely

Table 3 Assigned Weights for DRASTIC and Pesticide DRASTIC Features

Feature	Rating Range	General Weight	Pesticide Weight
Depth to water	1–10	5	5
Net recharge	1–9	4	4
Aquifer media	1–10	3	3
Soil media	1–10	2	5
Topography	1–10	1	3
Impact of vadose zone	1–10	5	4
Hydraulic conductivity of aquifers	1–10	3	2
Index range		23–226	26–256

subjective operations where sound professional judgment is needed to reach meaningful results. An exhaustive review of different methods for evaluating vulnerability is given by Civita.[32]

A very common empirical procedure is the numerical rating scheme called DRASTIC,[33] an acronym standing for the factors used in vulnerability evaluation: depth to water table (D), recharge rate (R), textural properties of the aquifer (A), soil properties (S), surface topography (T), impact on vadose zone (I), and hydraulic conductivity of the aquifer (C).

The procedure consists of the computation of a ranking index from quantitative factors which have been weighted and summed. Two different indices are proposed, a general index for point-source pollution and an index suitable for diffuse pesticides pollution. They differ in the weight selection and thus in the role given to the different factors in the different pollution situations. Each factor has been divided into ranges or significant types. The latter have been assigned a typical or variable rating for each factor.

The general additive model for determining potential vulnerability is

$$V.I. = D_R D_W + R_R R_W + A_R A_W + S_R S_W + T_R T_W + I_R I_W + C_R C_W$$

where subscripts R and W stand for rating and weight, respectively.

In Table 3, the weights assigned to DRASTIC factors are indicated. In the agricultural case, the most prevalent role played by soil media and topography and the lower importance assigned to hydraulic conductivity of aquifer should be noted. The DRASTIC procedure is a representative example of how relevant factors pertinent to the different aspects of potential migration processes may be associated to produce evaluations of vulnerability. In this hydrogeologically based method, great attention is paid to flow conditions, both in the unsaturated and saturated zones, and to other physical factors controlling potential movement of contaminants. The soil zone, although included, is not directly considered for its attenuation role.

Other procedures[34] focus more on attenuation potential, producing groundwater vulnerability evaluations as combinations of soil vulnerability, expressed by hydrologic factors (type of unsaturated flow, presence or lack of bypass flow) and organic matter content, and by depth to the aquifer or seasonally saturated layers. Then, the classification of soil vulnerability is compared with a classification of pesticides based on their relative mobility and persistence. For each combination of the two classifications, individual vulnerability assessments are made by means of a simple model expressing the likelihood of pesticide to travel to the aquifer. An interesting feature of the procedure is represented by the combination of two separate methodologies detailing with further vulnerability by means of a model (an attenuation factor) tailored to climatically representative ranges of soil water flux.

Empirical assessment methodologies have been developed for evaluating in a lumped form the groundwater hazard potential of different pollutant sources. The ranking index is based on factors concerning polluting characteristics of disposed materials besides those related to hydrogeologic and soil features of sites; sometimes parameters related to quality of groundwater are also taken into account. Ranges of numerical values are assigned to factors, according to weight, which are aggregated in the final ranking.[35-39] The results are relative classifications of hazard for different waste disposal sites.

In the field of pesticides, the methodologies may be applied by coupling the hydrogeologic and/ or soil parameters with factors expressing the intrinsic behavior of different compounds, such as mobility, to evaluate contamination hazard. Rao and Jury indices discussed previously and others similar[40] may also be defined as procedures of this type.

Susceptibility factors, expressed quantitatively, may further be used to compute the chance that some capture well has to be contaminated, once the source features have been defined. For example,[41] given the area and rate of application of the contaminant, chance may be estimated on the basis of soil type (with reference to organic matter content) in the recharge area, depth to water table, and groundwater flow (direction, velocity, saturated thickness) in the capture zone of the well, which in turn depends on the pumping rate.

FINAL CONSIDERATIONS

Groundwater vulnerability represents a comprehensive, integrated expression of interaction between migrating pollutants in the underground environment and hydrogeologic system behavior. A large number of factors, parameters, and processes are involved and play prominent roles in determining such interactions. They refer to the intrinsic properties of different underground zones through which pollutants migrate, to the interactions among the water-pollutant-matrix, to conditions of water flow from surface infiltration to the groundwater capture area.

The fate of pollutants and the behavior of hydrogeologic systems are natural phenomena that are very complex and not completely known or ascertainable. They can be analyzed, reproduced, and foreseen on the basis of more or less rigid and partial schematizations of control processes and environmental conditions. Such schematizations, on the one hand, should be sufficiently reliable and representative of the real world; on the other hand, they should be able to be treated and validated in an experimental context.

Natural processes and environmental behaviors are typically variable in space and not stationary in time. These properties have to be manipulated and/or simplified in order to be experimentally analyzed. Because of the complexity of the overall problem, analysis is relatively easier if referred to a point-source pollution where the pollutant source as well as the recipient underground body might be well defined, and the processes are characterized by strong gradients along main directions.

By contrast, the pesticide source is typically diffuse over space and intermittent in time while pollution is generally shown through small and random gradients resulting from intrinsic process variability. Thus, interest has to be shifted from a local or site view to a regional or areal framework. The volumes of the underground recipient water body, affected by contamination, are substantially large, and analysis must be developed in three-dimensional space. In such conditions, more attention should be paid to the behavior of the overall system in space and time under different input conditions than to the detailed response of a limited fragment of it with drastic analytical simplifications. The natural variability of the real system conditioning the reliability of response to contamination is a crucial problem that has strongly affected every kind of procedure.

The ultimate effects of the pollutant input on a groundwater body may be measured in terms of residual concentrations and loads at different steps of route and transfer time. A quantitative evaluation of vulnerability depends on the capacity in gaining consistent and representative estimates of these parameters. The presence of diffused sources of pollution, of strongly reactive compounds, and of large underground domains affected by pollution, such as those related to pesticides in agriculture, make quantitative estimates difficult to attain.

Despite the great availability of more or less advanced physically based procedures and models, the level of applicability and confidence in results appears to be, generally speaking, definitely questionable. A basic reason is the considerable amount of data needed to validate the analyses.

Much effort has been spent trying to overcome the limitations and uncertainties derived from the complexity of real systems and actual pollution events. Both the option of regionalizing the analysis and the explicit consideration of stochastic features of natural processes are being explored. However, these can be retained only as promising attempts that cannot attenuate or substitute the intrinsic question of reliability of schematization of natural systems and processes. Thus, in the present framework, it seems that what most of these methods can do is to express their results on behavior by ranking and rating the different conditions. At this point, a highly sophisticated evaluation procedure seems to be neither advisable nor necessary in order to attain a good result. They are more effective if used in relation to a specific contamination problem over a limited domain, e.g., to define the risk of attaining pollutant concentrations exceeding limits in a capture well. Here, the deterministic aspects of a real system may be assumed to be dominant and consequently the results will be consistent.

The system approach to vulnerability evaluation, by its very nature, does not permit an absolute measure by means of the above mentioned attributes. However, the procedures based on the global behavior of the system, taking into account hydrogeologic and hydropedologic features, seem to be more effective than the preceding procedures at a scale comparable to the domain of pollution phenomena. Good knowledge of the hydrogeologic framework and a sound identification of leading factors in the evaluation are unconditionally necessary.

The methodologies developed within this approach lead more or less directly to empiric classifications or ranking evaluations of the systems, sometimes divided into zones in relation to pollution conditions. This result might appear too simplistic or insufficient to characterize the real susceptibility of the systems to risk, especially as the procedures are mostly uncorrelated to the pollutant properties.

The conclusions of the different approaches, at first sight, might be considered as separate and independent. However, in light of more realistic analyses, a coordinated use of different procedures might be promising, because of the possibility of overcoming the drawbacks of each one.

Accordingly, the hydrogeologically based procedures may be defined as the qualitative framework for environmental conditions, expressed in terms of general system vulnerability, within which quantitatively oriented techniques can be applied to measure the relative degree of susceptibility, with reference to chemodynamically qualified pollutants. Examples of such integrated procedures have begun to appear in the scientific literature. Such an integration of different approaches and methodologies should also be carefully explored in connection to regulatory actions of pesticide use.

ACKNOWLEDGMENT

Publication #1092 of the National Groups for the Defense from Hydro-geologic Hazards (GNDCI) of the National Research Council of Italy. The work was carried out by the Research Unit 4.13 of the group set up by the Water Research Institute, National Research Council, Rome.

REFERENCES

1. Funari, E. and A. Sampaolo. "Erbicidi nelle acque potabili (Herbicides in drinking water)," 25: 353-362 (1989).
2. Hopman, R., C.G. van Beek, H.M.S. Ianssen, and L.M. Puijker. "Pesticides and drinking water supply in The Netherlands," KIWA Report 113 (1990), Nieuwegein, p. 176.
3. Freeze, R.A. and J.A. Cherry. *Groundwater,* (Englewood, NJ: Prentice Hall), p. 604.
4. Roberts, P.V., M. Reinard, G.D. Hopblins, and R.S. Summers. "Advection-dispersion-sorption models for simulating the transport of organics," in *Groundwater Quality,* Ward, C.H., Ed. (New York: J. Wiley & Sons, 1985), pp. 425-445.
5. Beven, K.J. and P. German. "Methodologies and water flow in soil," *Water Resources Res.* 18: 1311-1325 (1982).
6. Brusseau, M. L. and P.C.S. Rao. "Modeling solute transport in structured soils: a review," *Geoderma* 46: 169-192 (1990).
7. Bear, J. and A. Verrujt. *Modeling groundwater flow and pollution,* (Dordrecht: D. Reidel Publ. Co. 1987), p. 414.
8. Jury, W.A. and J. Gruber. "A stochastic analysis of the influence of soil and climatic variability on the estimate of pesticide groundwater pollution potential," *Water Resources Res.* 25: 2465-2474 (1989).
9. Boesten, J. J. T. L. and S.E.A.T. van der Zee. "Effects of soil heterogeneity on pesticide leaching to groundwater," *Water Resources Res.* 27 (12): 3051-3063 (1991).

10. Carsel, R.F., R.L. Jones, J.L. Hansen, R.L. Lamb, and M.P. Anderson. "A simulation procedure for groundwater quality assessment of pesticides," *J. Contaminant Hydrol.* 2: 125-138 (1988).

11. Castany, G. *Principes et Méthodologie de l' Hydrogéologie* (Paris: Unod Univ., 1982).

12. Civita, M. "La valutazione della vulnerabilità degli acquiferi all' inquinamento (The evaluation of aquifer vulnerability to pollution)," in *Proceedings of 1st National Congress on Protection and Management of Groundwater*, Marano sul Panaro, Italy (1990), vol. 3, pp. 39-86.

13. Laskowski, D. A., C.A. Goring, P.J. McCall, and R.L. Swann. "Terrestrial Environment," in Conway, R., Ed. *Environmental Risk Analysis for Chemicals* (New York: Van Nostrand Reinhold Co., 1982), pp. 198-240.

14. Gustafson, D. I. "Groundwater ubiquity score: a simple method for assessing pesticide leachability," *Environ. Tox. Chem.* 8: 339-357 (1989).

15. Rao, P.S.C., A.G. Hornsby, and R.E. Jessup. "Indices for ranking the potential for pesticide contamination of groundwater," *Soil Crop Sci. Soc. Fla.* 44: 1-14 (1985).

16. Jury, W. A., D.D. Focht, and W.J. Farmer. "Evaluation of pesticide groundwater pollution potential from standard indices of soil–chemical adsorption and biodegradation," *J. Environ. Qual.* 16: 422-428 (1987).

17. Addiscott, T.M. and R.J. Wagenet. "Concepts of soil leaching in soils: a review of modelling approaches," *J. Soil Sci.* 36:411-424 (1985).

18. Carsel, R. F., L.A. Mulkey, M.N. Lorber, and L.B. Baskin. "The pesticide root zone model (PRZM): a procedure for evaluating pesticide leaching threats to groundwater," *Ecol. Modelling* 30: 46-69 (1985).

19. Wagenet, R. J. and P.S.C. Rao. "Modeling pesticide fate in soil," in H. H. Cheng, Ed., *Pesticides in the Soil Environment: Processes, Inputs, and Modeling.* Book Series 2 (Madison, WI: Soil Science Society of America, 1990), pp. 351-399.

20. Boesten, J.J.T.L. and A.M.A. van der Lindern. "Modeling the influence of sorption and transformation on pesticide leaching and persistence," *J. Environ. Qual.* 20: 425-435 (1991).

21. Wagenet, R.J. and J.L. Huston. "Predicting the fate of nonvolatile pesticides in the unsaturated zone," *J. Env. Quality* 15: 315-322 (1986).

22. "Substainable use of groundwater." RIVM and RIZA Report no 600025 (1991).

23. Nofziger, D.L. and A. G. Hornsby. "A microcomputer-based management tool for chemical movement in soil," *Applied Agric. Res.* 1: 50-56 (1986).

24. Hillel, D. *Soil and Water: Physical Principles and Processes,* (New York: Academic Press, 1971) p. 288.

25. Anderson, M. P. "Movement of contaminants in groundwater: groundwater transport — advection and dispersion," in *Studies in Geophysics: Groundwater Contamination* (Washington, D.C.: National Academy Press, 1984) pp. 37-45.

26. Dean, J.D., P.S. Huyakorn, A.S. Donigian, A.K. Voos, R.W. Schanz, and R.F. Carsel. "Rustic: risk of unsaturated/saturated transport and transformation of chemical concentrations," EPA Report 600/3-89/048a (1989).

27. Loague, K. "The impact of land use on estimate pesticide leaching potential: assessment and uncertainties," *J. Contaminant Hydrology* 8: 157-175 (1991).

28. Civita, M. *Unified Legend For The Aquifer Pollution Vulnerability Maps* (Bologna, Italy: Pitagora Ed., 1988).

29. Albinet, M. and J. Margat. "Cartographic del la vulnerabilitè à la pollution des naffer d'eau soutarraine," *Bull. BRGM* 2,3,4: 13-22 (1970).

30. Foster, S.S.D. "Fundamental concepts in aquifer vulnerability risk and protection strategy," in *Proceedings of the International Conference on Vulnerability of Soil and Groundwater to Pollution.* Woordwijk, The Netherlands, 1990.

31. Haertlè, T. "Method of modeling and employment of EDP during the preparation of groundwater vulnerability maps," in *Proceedings of the Symposium on Groundwater in Water Resources Planning,* IAHS, 1983, pp. 1073-1085.

32. Civita, M. *Le carte di vulnerabilità degli acquiferi all'inquinamento: teoria e pratica.* (Bologna, Italy: Pitagora Ed., Quaderni di tecniche di protezione ambientale 31, 1994).

33. Aller, L., T. Bennet, J.H. Lehr, R.L. Petty, and G. Hackett. "DRASTIC: a standardized system for evaluating groundwater pollution potential using hydrogeologic settings," EPA Report 00/2-87/035, 1985.

34. Hollis, J.M. "Mapping the vulnerability of aquifers to pesticide contamination of the national/ regional scale," in Walken, A., Ed. *Pesticide in Soil Water. Current Perspectives. B.CPC Mono* 47: 165-175 (1991).

35. LeGrand, H.E. "System of reevaluation of contamination potential of some waste disposal sites," *J. Am. W.W.A.* 56: 959-974 (1964).

36. Hagerty, D.J., J.L. Pavoni, and J.E. Heer, Jr. *Solid Waste Management* (New York: Van Nostrand Reinhold, 1973), pp. 242-262.

37. Phillips, C.R., J.D. Nathwani, and H. Mooij. "Development of a soil waste interaction matrix for assessing land disposal of industrial wastes," *Water Res.* 11: 859-868, 1977.

38. Caldwell, S., K.W. Barrett, and S.S. Chang. "Ranking system for releases of hazardous substances," in *Proceedings of the National Conference on Management of Uncontrolled Hazardous Waste Sites*, Hazardous Material Control Research Institute (Silver Spring, MD: 1981), pp. 30-41.

39. Kufs, C. et al. "Rating the hazard potential of waste disposal facilities," in *Proceedings of the National Conference on Management of Uncontrolled Hazardous Waste Sites*, Hazardous Materials Control Research Institute (Silver Spring, MD: 1980), pp. 30-41.

40. Britt, J.K., S.E. Dwinell, and T.C. McDowell. "Matrix decision procedures to assess new pesticides based on relative groundwater leaching potential and chronic toxicity," *Env. Toxicol. Chem.* 11: 721-728 (1992).

41. Scheiber, T.D. and D.F. Lettenmaier. "Risk based selection of monitoring wells for assessing chemical contamination of groundwater," *Groundwater Monitoring Rev.* 4: 98-108 (1989).

Section III
Pesticides
and
Human Health

CHAPTER 5

Human Health Implications Associated with the Presence of Pesticides in Drinking Water

Enzo Funari

CONTENTS

INTRODUCTION

Surface and groundwaters can be contaminated by pesticides. In the case of surface waters, pesticide contamination is dependent on the season and generally does not last long. On the other hand, groundwater contamination, for the reasons previously discussed in this volume, usually is not season dependent and has a strong inertia. Moreover, a pesticide may became detectable in groundwater only after numerous agricultural applications over a period of years.

Therefore, human exposure to pesticides through drinking water is generally continuous in the case of groundwater and intermittent in the case of surface waters. The human health implications associated with these types of exposures can be evaluated by applying known procedure such as those of the World Health Organization (WHO) and the Environmental Protection Agency of the U.S. (USEPA).

The WHO recently defined the guideline values (GLs) for a number of agents, including many pesticides (Table 1).[1] These values refer to lifetime exposures. The USEPA has defined Health Advisories (HAs) for many pesticides for specific periods of exposure, ranging from 1 day to a lifetime (Table 2). In both cases, the most important sources of data are epidemiology and animal studies.[2]

Even though epidemiology is the only direct means for determining whether a chemical has produced a toxic effect in humans, only a few epidemiologic studies are available on substances to which human beings are exposed through drinking water, and these are not available for new substances. In addition, these studies are not without their limitations. The main one is that, at the generally low concentration of substances in drinking water, it is very difficult to detect any increase (if there is an increase) in human diseases and to associate this to a substance in drinking water instead of to exposures to this or other substances through air, food, etc. Accordingly, animal studies generally represent the main source of data utilized to assess the risk to human health

Table 1 Guidelines for Drinking Water Quality as Defined by WHO for Pesticides

Pesticide	Guideline (g/L)
alachlor	20[a]
aldicarb	10
aldrin/dieldrin	0.03
atrazine	2
bentazon	30
carbofuran	5
chlordane	0.2
chlorotoluron	30
2,4-D	30
2,4-DB	90
DDT	2
1,2-dibromo-3-chloropropane	1[a]
1,2-dichloropropane	20[b]
1,3-dichloropropene	20[a]
dichlorprop	100
fenoprop	9
heptachlor and heptachlor epoxide	0.03
hexachlorobenzene	1[a]
isoproturon	9
lindane	2
MCPA	2
mecoprop	10
metolachlor	10
metoxychlor	20
molinate	6
pendimethalin	20
pentachlorophenol	9[b]
permethrin	20
propanil	20
pyridate	100
simazine	2
2,4,5-T	9
trifluralin	20

[a] Substance considered to be carcinogenic; the guideline value is the concentration in drinking water associated with an excess lifetime cancer risk of 10^{-5} (one additional cancer case per 100,000 of the population ingesting drinking water containing the substance at the guideline value for 70 years).

[b] Provisional guideline value. This term is used for constituents for which there is some evidence of a potential hazard but where the health effect information available is limited; and/or where an uncertainty factor greater than 1000 is used in the derivation of the tolerable daily intake.

Reproduced, by permission, from: *Guidelines for drinking-water quality, Vol. 1, Recommendations.* Geneva, World Health Organization, 1993.

associated with exposures to chemicals through drinking water. Because of the many morphologic and physiologic analogies between laboratory animals and human beings, it is considered possible to use animal data to predict potential effects on human health.

PROCEDURES APPLIED BY WHO TO DERIVE GUIDELINE VALUES FOR PESTICIDES

The recent guidelines for drinking water quality were established by the WHO after a revision process that lasted some 5 years. They were derived from two procedures applied to substances that either exhibit a threshold for toxic effects or for which it is believed there is no threshold.

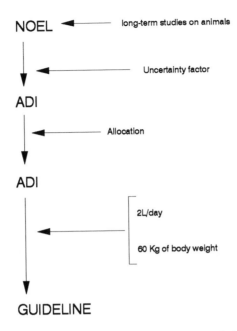

Figure 1   WHO method to derive a guideline value for treshold substances.

The threshold concept implies that a dose exists below which no adverse effects may be reasonably expected. This concept is based on the fact that the biological effects of a chemical depend on its concentration or on that of its metabolites in the target tissues or organs. This, in turn, is modulated by processes such as absorption, distribution, metabolism, and excretion. The threshold concept is perfectly reasonable in the case of many elements and compounds that are naturally present in human organisms at low concentrations and whose presence is essential for well-being. If these elements and compounds are present at high doses, they may cause adverse effects. The validity of this concept is also confirmed by experimental data.[3]

Toxicologic studies on laboratory animals are designed to obtain a dose-response relationship from which it is possible to derive the highest dose for which no significant effect is detected (No Observed Effect Level, NOEL). This experimental value is not necessarily the true No Effect Level (NEL), but it is within the threshold region.

For the reasons mentioned above as to the possibility of using animal data to predict potential risk to humans, an Acceptable or Tolerable Daily Intake (ADI or TDI) can be derived by applying an uncertainty factor from the NOEL. ADIs are established for food additives or pesticide residues that are present in food as a result of necessary technological or plant protection purposes. For contaminants, the term TDI is seen as a more appropriate term than ADI. Obviously, trace contaminants have no intended function, so the term "tolerable" is seen as a more appropriate term than "acceptable", signifying permissibility rather than acceptability.

The uncertainty factor takes into account intraspecies and interspecies variations, nature and severity of effects, quality of data, and the kind of population likely to be exposed. Once an ADI or a TDI has been calculated for a pesticide, it encompasses the whole intake from all the potential exposure sources. In order to derive the one that is due only to drinking water consumption it is necessary to define the proper proportion of the intake. For this reason, through monitoring data or the prediction of environmental fate, it is necessary to determine the potential exposure through foods, air, drinking water, etc. for each pesticide.

Finally, the guideline value is calculated by considering a consumption of 2 L of drinking water per day for an adult weighing 70 kg. In Figure 1, this procedure is schematized; in Figure 2 it is applied to molinate.

This procedure is not applied by the WHO in the case of carcinogenic compounds. For these compounds, it can be assumed that even a single biologically active molecule reaching the DNA may be the cause of a mutation in a somatic cell, and this is considered to be the initial event in the process of chemical carcinogenicity. Therefore, this process theoretically has no threshold,

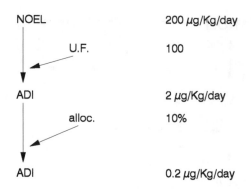

$$\text{guideline} = \frac{0.2 \, \mu g/Kg/day \times 70 \, Kg}{2 \, l} = 7 \, \mu g/l$$

Figure 2   Derivation of the guideline for molinate.

implying that a completely safe dose greater than zero does not exist. Mechanistic reasons can be proposed such as pharmacokinetics, DNA repair and immune surveillance to justify the exisistence of a threshold in most, if not all cases, but demonstrating this experimentally can be extremely difficult.[4]

On the other hand the linearity hypothesis seems to be confirmed by analysis of a number of relationships describing the quantitative trend of DNA adduct formation as a function of the dose of the active substance.[5] The linearity hypothesis is also consistent with the statistical analysis of many experimental carcinogenesis dose-response relationships.[6-8] For carcinogenic compounds as well, the evaluation of human health risk is essentially based on epidemiology and animal studies. However, the epidemiologic studies have the limitations mentioned above. At present, the only indisputable human evidence for carcinogenicity from a waterborne contaminant comes from data on arsenic.[9] A number of studies in several countries have given varying levels of correlation between the use of chlorinated waters and a range of cancers.[10-12] Therefore, most of the information for carcinogenic compounds also comes from studies on laboratory animals.

In these studies, high doses of the substances are used to compensate for the relatively small number of test animals and the subsequent low statistical weight. This may imply that tissue damage occurs, and, often, tumors are observed at the same time.[4] Moreover, a substance administered at high doses may be metabolized differently with respect to low doses.

Utilizing animal experimental data obtained at high doses to predict the human health risk associated with drinking water, in which compounds are generally present at low concentrations, makes some kind of extrapolation necessary. For this purpose, some mathematical models have been proposed that attempt to represent the biological mechanism of carcinogenesis. Nevertheless, their application yields results that vary over many orders of magnitude,[13] reflecting the gaps present in the knowledge of the carcinogenesis mechanisms. For this reason, the solution of using mathematical models that do not underestimate health risk seems reasonable and, at present, without valid alternatives.

The linearized multistage model is used by both the USEPA[4] and the WHO. In this model, a linear relationship between dose and effect is assumed, and it is one of the most conservative mathematical models.

The risk calculated for human beings is usually the upper 95% confidence interval, which means that there is only a 5% chance that the calculated risk will be exceeded at the dose given. Conversely, the lower 95% confidence interval is usually zero. The risk is given as one additional cancer in a population of 10^4, 10^5, 10^6 exposed for 70 years.

Essentially, the multistage model of the USEPA has generally been applied for substances classified into class 1 or 2A by the International Agency for Research on Cancer or ranked by the

WHO experts in these classes, and for which a genotoxic mechanism has been judged to be responsible for the carcinogenic activity. There is a problem for substances for which some evidence exists on their carcinogenic properties but without showing genotoxic activities. This has been dealt with by the WHO by considering them in the category of those having threshold effects. In these cases, an additional uncertainty factor has usually been applied.

PROCEDURES APPLIED BY THE USEPA TO DERIVE HAs
FOR PESTICIDES

The WHO guidelines for drinking water quality refer to long-term exposures. For situations relating to a spill and for specific exposures that are less than a lifetime, reference can be made to the USEPA, Office of Water, which developed a Drinking Water Health Advisory Program in 1978. The HAs are not federally enforceable but are reference documents with information on the main environmental and health aspects. HAs have been prepared for over 100 contaminants and several pesticides, including those examined in the National Pesticide Survey, which were considered likely to leach into groundwater throughout the U.S.[15]

HAs contain estimates of concentrations of a particular contaminant that would not cause adverse, noncancer health effects for specific durations of exposure. The risk assessments are developed for exposures up to 5 days, 14 days, a longer period of exposure that covers approximately 7 years or 10% of an individual's lifetime, and lifetime. Table 2 shows the HAs derived so far for pesticides as well as their carcinogenicity classification.

The procedure applied to derive a lifetime HA is similar to that used by WHO. More important differences rely on the definition of the other HAs. The first one concerns the type of the toxicologic studies required. The 1-day and 10-day HAs are derived from acute and subacute studies, in which animals are exposed for up to 7 days and from 7 to 30 days, respectively. The longer-term HA is calculated using subchronic animal data in which animals are exposed for 90 days, corresponding to approximately 10% of the animal's lifetime. Another relevant difference is that to derive both of the 1-day and the 10-day HA, the protected group is young children (weighing 10 kg). The longer HA is derived both for a 10-kg child and for an adult. Obviously, the lifetime HA refers to an adult person.

Again, different from the WHO procedure, in the cases of acute or subacute exposure to a substance, the USEPA does not consider other possible intakes, hence the relative source intake is not calculated.

SOME CONSIDERATIONS ON THE WHO GUIDELINES
AND THE USEPA HAs

The first comment on the WHO guidelines and the USEPA lifetime HAs is that, where a comparison is possible, these values are similar and, in the worst cases, they differ within one order of magnitude. The differences between GLs and HAs reflect some minor differences in the procedures and the possibilities of some differences in the data set available.

The procedure applied by both the WHO and the USEPA to derive the GLs and the HAs for threshold substances seems to be quite consolidated, even though the choice of uncertainty factors does not have a scientific basis and can be to some extent arbitrary. Its validity seems also proved by historical data on humans, even though some exceptions do exist. These refer to some minor diseases, for which the information gathered from animal studies does not always make it possible to predict human effects. Indeed, classical toxicologic studies using animals are poor at predicting gastrointestinal irritation, headache, nausea, and general malaise in humans as well as psychosomatic effects due to foul tasting water. These symptoms were recorded in the U.K. on the occasion of a spill of industrial chemicals, in spite of the fact that phenol, the major contaminant, and its by-products formed after the chlorination treatment, were present at levels well below those calculated as safe levels from the available animal data.[16]

Table 2   USEPA's Health Advisories for Drinking Water (1992).

	10-kg Child			70-kg Adult					
	1-day HA (mg/L)	10-day HA (mg/L)	Longer-term HA (mg/L)	Longer-term HA (mg/L)	RfD[a] (mg/kg/day)	DWEL[b] (mg/L)	Lifetime HA (mg/L)	mg/L at 10^{-4} Cancer Risk	Cancer Group[c]
Acifluorfen	2	2	0.1	0.4	0.013	0.4	—	0.1	B2
Alachlor	0.1	0.1	—	—	0.01	0.4	—	0.04	B2
Aldicarb	—	—	—	—	0.0002	0.007	0.001	—	D
Aldicarb sulfone	—	—	—	—	0.0002	0.007	0.001	—	D
Aldicarb sulfoxide	—	—	—	—	0.0002	0.007	0.001	—	D
Aldrin	0.0003	0.0003	0.0003	0.0003	0.00003	0.001	—	0.0002	B2
Ametryn	9	9	0.9	3	0.009	0.3	0.06	—	D
Ammonium sulfamate	20	20	20	80	0.28	8	2	—	D
Atrazine	0.1	0.1	0.05	0.2	0.005	0.2	0.003	—	C
Bentazon	0.3	0.3	0.3	0.9	0.0025	0.09	0.02	—	D
bis-2-Chloroisopropyl ether	4	4	4	13	0.04	1	0.3	—	D
Bromacil	5	5	3	9	0.13	5	0.09	—	C
Butylate	2	2	1	4	0.05	2	0.35	—	D
Carbaryl	1	1	1	1	0.1	4	0.7	—	D
Carbofuran	0.05	0.05	0.05	0.2	0.005	0.2	0.04	—	E
Carbon tetrachloride	4	0.2	0.07	0.3	0.0007	0.03	—	0.03	B2
Carboxin	1	1	1	4	0.1	4	0.7	—	D
Chloramben	3	3	0.2	0.5	0.015	0.5	0.1	—	D
Chlordane	0.06	0.06	—	—	0.00006	0.002	—	0.003	B2
Chlorothalonil	0.2	0.2	0.2	0.5	0.015	0.5	—	0.15	B2
Chloropyrifos	0.03	0.03	0.03	0.1	0.003	0.1	0.02	—	D
Cyanazine	0.1	0.1	0.02	0.07	0.002	0.07	0.001	—	C
2,4-D	1	0.3	0.1	0.4	0.01	0.4	0.07	—	D
DCPA	80	80	5	20	0.5	20	4	—	D
Dalapon	3	3	0.3	0.9	0.026	0.9	0.2	—	D
Diazinon	0.02	0.02	0.005	0.02	0.00009	0.003	0.0006	—	E
Dicamba	0.3	0.3	0.3	1	0.03	1	0.2	—	D
Dichloropropene (1,3-Dichloropropene)	0.03	0.03	0.03	0.1	0.0003	0.01	—	0.02	B2
Dieldrin	0.0005	0.0005	0.0005	0.002	0.00005	0.002	—	0.0002	B2
Dinoseb	0.3	0.3	0.01	0.04	0.001	0.04	0.007	—	D
Diphenamid	0.3	0.3	0.3	1	0.03	1	0.2	—	D
Diphenylamine	1	1	0.3	1	0.03	1	0.2	—	D
Diquat	—	—	—	—	0.0022	0.08	0.02	—	D
Disulfoton	0.01	0.01	0.003	0.009	0.00004	0.001	0.0003	—	D
Diuron	1	1	0.3	0.9	0.002	0.07	0.01	—	D
Endothall	0.8	0.8	0.2	0.2	0.002	0.7	0.1	—	D

Compound								
Endrin	0.02	0.02	0.003	0.01	0.0003	0.01	0.002	—	D
Ethylene dibromide (EDB)	0.008	0.008	—	—	—	—	—	0.00004	B2
Fenamiphos	0.009	0.009	0.005	0.02	0.00025	0.009	0.002	—	D
Fonofos	0.02	0.02	0.02	0.07	0.002	0.07	0.01	—	D
Formaldehyde	10	5	5	20	0.15	5	1	—	B1
Glyphosate	20	20	1	1	0.1	4	0.7	—	D
Heptachlor	0.01	0.01	0.005	0.005	0.0005	0.02	—	0.0008	B2
Heptachlor epoxide	0.01	—	0.0001	0.0001	1.3E-5	0.0004	—	0.0004	B2
Hexachlorobenzene	0.05	0.05	0.05	0.2	0.0008	0.03	—	0.002	B2
Hexazinone	3	3	3	9	0.033	1	0.2	—	D
Lindane	1	1	0.03	0.1	0.0003	0.01	0.0002	—	C
Malathion	0.2	0.2	0.2	0.8	0.02	0.8	0.2	—	D
Maleic Hydrazine	10	10	5	20	0.5	20	4	—	D
MCPA	0.1	0.1	0.1	0.4	0.0015	0.05	0.01	—	E
Methomyl	0.3	0.3	0.3	0.3	0.025	0.9	0.2	—	D
Methoxychlor	0.05	0.05	0.05	0.2	0.005	0.2	0.0431	—	D
Methyl parathion	0.3	0.3	0.03	0.1	0.00025	0.009	0.002	—	D
Metolachlor	2	2	2	5	0.15	5	0.1	—	C
Metribuzin	5	5	0.3	0.9	0.025	0.9	0.2	—	D
Naphtalene	0.5	0.5	0.4	1	0.004	0.1	0.02	—	D
Oxamyl (Vydate)	0.2	0.2	0.2	0.9	0.025	0.9	0.2	—	E
Paraquat	0.1	0.1	0.05	0.2	0.0045	0.2	0.03	—	E
Pentachlorophenol	1	1	0.3	1	0.03	1	—	0.03	B2
Picloram	20	20	0.7	2	0.07	2	0.5	—	D
Prometron	0.2	0.2	0.2	0.5	0.015[d]	0.5[d]	0.1[d]	—	D
Propachlor	0.5	0.5	0.1	0.5	0.013	0.5	0.09	—	D
Propazine	1	1	0.5	2	0.02	0.7	0.01	—	C
Propham	5	5	5	20	0.02	0.6	0.1	—	D
Simazine	0.07	0.07	0.07	0.07	0.005	0.2	0.004	—	C
2,4,5-T	0.8	0.8	0.8	1	0.01	0.35	0.07	—	D
Tebuthiuron	3	3	0.7	2	0.07	2	0.5	—	D
Terbacil	0.3	0.3	0.3	0.9	0.013	0.4	0.09	—	E
Terbufos	0.005	0.005	0.001	0.005	0.00013	0.005	0.0009	—	D
Trichloroacetic acid	—	2	4	13	0.04	1.3	1	—	C
Trifluralin	0.08	0.08	0.08	0.3	0.0075	0.3	0.005	—	C

a Reference Dose. An estimate of a daily exposure to the human population that is likely to be without appreciable risk of deleterious effects over a lifetime.

b Drinking Water Equivalent Level. A lifetime exposure concentration protective of adverse, noncancer health effects, that assumes all of the exposure to a contaminant is from a drinking water source.

c B1 — Probable human carcinogen (limited evidence in humans)
B2 — Probable human carcinogen (sufficient evidence in animals and inadequate evidence in humans)
C — Possible human carcinogen (limited evidence in animals in the absence of human data)
D — Not classifiable as to human carcinogenicity
E — Evidence of non-carcinogenicity in humans

d Under review.

For many years, substantial debate has arisen on the mathematical models that best represent the process of chemical carcinogenesis. In particular, many observations have been made on the many conservation assumptions of the multistage linearized model. However, as mentioned before, at present, due to the relevant gaps in the knowledge of the chemical carcinogenesis process, the application of a conservative model, which does not underestimate the human health risk, seems the most appropriate choice (as long as it is not excessively conservative).

The practical importance of the WHO GLs and the USEPA HAs is undoubtedly relevant. The WHO GLs are used in many countries as the legal concentration limits for agents in drinking water or as a basis for establishing such limits. The USEPA HAs do not necessarily coincide with the Maximum Contaminant Levels permitted in the U.S., but they are particularly useful as reference especially in cases of spills or in any kind of accident related to drinking water.

From this point of view, one limitation of the WHO GLs and the USEPA HAs is that they have been defined for a small number of pesticides as compared with the large number of pesticides used that are potential contaminants of drinking water supplies.

Another limitation of the WHO GL and the USEPA HAs becomes evident when there is exposure to a mixture of two or more substances; in these cases it is generally not clear what risk assessment procedure should be applied. This problem is quite complex, and there is no general solution. In addition, people can be exposed to chemicals by other routes as well. The problem of simultaneous exposure to two or more substances should be assessed on a case-by-case basis. In these cases, the risk assessment procedure should consider the toxicologic properties of the single substances and their possible interactions (synergism, additive effects, competition). These interactions depend on the number and the relative amounts of the chemicals present. In addition the degradation processes in water are generally different for contaminants, so that the concentration pattern may change with time.[8]

In spite of this complexity, some attempts have been made to estimate the error in risk assessment by applying the dose-additivity hypothesis to many mixtures. Using 53 pairs of industrial chemicals, it has been observed that the ratio between the predicted LD50 and the experimental LD50 was within 2 in most of the cases, with a maximum of 5. For these cases, the additivity hypothesis could be used by applying an uncertainty factor.[8] The additivity hypothesis has been checked also by using epidemiologic studies in which the effects of the simultaneous carcinogenic agents were examined.[17] The additivity hypothesis has been shown to be appliable — with the use of a small uncertainty factor — also in the case of simultaneous exposure to asbestos and smoke, which are synergistic.

The relationship between a quality criterion, such as a WHO GL or a USEPA HA, and a standard value depends on many factors. Among these factors, the most relevant are the technical and the economic ones. Decreasing exposure to a contaminant that is already at lower concentrations implies an economic cost. If this is considered to be worthless in the sense that it does not improve protection of human life, often this expenditure can be diverted to other social purposes.

The general philosophy of the EC, which derives from Directive 778/80 on the quality of drinking water, is not to allow humans to be exposed to chemicals that are present in the environment from anthropogenic activities. The maximum acceptable concentration of $0.1 \, \mu g/L$ for a single pesticide substantially means its absence from drinking water. This philosophy has received a wide consensus from the Green Movement and public opinion in many European countries. It is considered to be one of the trigger points to launch a strategy of protection and recovery of the environment.

On the other hand, this philosophy has also received strong opposition, particularly from the lobbies representing producers and users of pesticides. Indeed, some European countries still have not adopted the European Directive, and, in some other countries, such as the U.S., Canada, etc., the legal limits are established on the basis of a risk assessment procedure.

However, the debate concerning the relationship between the toxicologic limits and the legal ones goes beyond the drinking water issue; it also encompasses exposure to agents through food, air, etc. The influence of some of the elements that play an important role in this relationship is analyzed in depth in the last chapter of this book.

COMPARISON OF GLs AND/OR HAs WITH GROUNDWATER MONITORING DATA

As mentioned in the introduction of Chapter 1, one of the main uses of monitoring data is to establish whether there is a risk to human health associated with exposure to pesticides through drinking water and, if this is the case, to which pesticides. This evaluation can be done by means of a comparison between the pesticide levels and their relative GLs and/or HAs. Yet, using the data of Chapter 1, two main shortcomings affect this evaluation. The first refers to the fact that not even we know whether the monitoring data regard groundwater used for drinking purposes. The second refers to the limited possibility of this comparison, since GL and lifetime HA values have been defined only for a limited number of pesticides with respect to those for which monitoring data are available.

Hence, when this activity is possible, the mean levels are almost always within the GLs and/or lifetime HAs. With rare exceptions, these are exceeded by simazine or atrazine and rarely by 1,2-dichloropropane. On the contrary, aldicarb mean levels are often higher than the GLs and HAs. The maximum levels for many pesticides are often above the GLs and/or HAs. Among these, the ones considered to be of most concern appear to be aldicarb, chlortal-dimethyl (DCPA), dicamba, and 1,3-dichloropropene whose maximum levels often exceed the respective GLs and/or HAs of at least one order of magnitude.

As mentioned above, the human health implications associated with exposure to levels exceeding the WHO GLs and/or lifetime HAs cannot be defined as a general figure, but should be discussed case by case. Many elements should be borne in mind for these evaluations. The first one is related to the time of the exposure. The GLs and the lifetime HAs can be exceeded for short exposures. In these cases, depending on the length of the exposure, the exposure levels should be compared, for example, with 1 day, 10 days, or longer-term HAs. Another relevant element is the role played by other sources of exposure (air, foods, etc.). Indeed, the GLs and the lifetime HAs are derived by allocating a relatively small portion of the pesticide exposure to the drinking water consumption. This, in general, does not reflect the real/current situation of intake through drinking water and often represents a way to add further uncertainty factors in the definition of the GL or HA. In order to establish whether the intake of the pesticide through drinking water represents a risk, what is important is to check whether the ADI is exceeded or not in the actual conditions of exposure through drinking water and the other sources.

As for the carcinogenic risk, it seems that the monitoring levels are, in general, such that the cancer risk is below or much lower than 10^{-5}. Nevertheless, in some cases, the cancer risk is higher than 10^{-5} for alachlor, 1,2-dichloropropene, and especially ethylene dibromide as estimated by WHO and/or EPA.

ACKNOWLEDGMENTS

The author of this chapter is very grateful to John Fawell of the Water Resarch Center (UK) for his valuable advice. Particular thanks are also due to Bruce Mintz (USEPA) for his kindness in providing material for the preparation of the chapter.

REFERENCES

1. World Health Organization. *Guidelines for Drinking Water Quality*, 2nd ed., Vol. 1 Recommendations, Geneva (1993).
2. "Drinking Water Regulations and Health Advisories by Office of Water," U.S. Environmental Protection Agency — Washington, DC. 202-260-7572 (1992).
3. Taylor, J.R. and Friedman, L. "Combined Chronic Feeding and Three Generation Reproduction Study of Sodium Saccharin in the Rat," *Toxicol. Appl. Pharmacol.* 29: 154, abstract 200 (1974).

4. Fawell, J.K. and Young, W.F. "Assesment of the Human Risk Associated with the Presence of Carcinogenic Compounds in Drinking Water," *Ann. Ist. Sup. Sanità* 29:313-316 (1993).
5. Appleton, B.S., Hoel, D.G., and Kaplan, N.L. "A General Scheme for the Incorporation of Pharmacokinetics in Low-Dose Risk Estimation for Chemical Carcinogens: Example — Vinyl Chloride," *Toxicol. Appl. Pharmacol.* 55: 154-161(1982).
6. Food Safety Council. Quantitative Risk Assesment. *Food Cosmet. Toxicol.*, 18(6):11-733 (1980).
7. Zapponi, G.A., Bucci, A.R., and Lupi, C. "Reproducibility of Low-dose Extrapolation Procedure: Comparison of Estimates Obtained Using Different Rodent Species and Strains," *Biomed. l and Environ. Sci.* 1: 160-170 (1988).
8. Reichard, E., Cranor, C., Raucher, R., and Zapponi, G. "Consequence of Exposure: Dose-response Assessment" in *Groundwater Contamination Risk Assessment. A Guide to Understanding and Managing Uncertainties* (IAHS Publication N. 196, 1990) pp. 47-86.
9. Fawell, J.K. "The Impact of Inorganic Chemicals on Water Quality and Health," *Ann. Ist. Sup. Sanità* 29:293-303 (1993).
10. Fawell, J.K. and Fielding, M. "Identification and Assessment of Hazardous Compounds in Drinking Water," *Sci. Total. Environ.* 47:317-341 (1985).
11. "Chlorinated Drinking-water; Chlorination by-products; Some other Halogenated Compounds; Cobalt and Cobalt Compounds," IARC Monograph on the Evaluation of Carcinogenic Risks to Humans, IARC, Lyon, no. 52 (1991)
12. Morris, R.D., Audet, A.M., Angelillo, I.F., Chalmers, T.C., and Mosteller, F. "Chlorination, Chlorination by-products, and Cancer: a Meta-analysis," *Ann. J. Publ. Hlth* 82:955-962 (1992).
13. Rodricks, J. and Taylor, M.R. "Application of Risk Assessment to Food Safety Decision Making." *Regul. Toxicol. Pharmacol.* 3: 275 (1983).
14. "Guidelines for Carcinogenic Risk Assessment," USEPA Fed. Reg. 51(185): 33992-34003 (1986).
15. Zavaleta, J.O., Cantilli, R., and Ohanion, E.V. "Drinking Water Health Advisory Program," *Ann. Ist. Sup. Sanità* 29(2):355-358.
16. Jones, F. and Fawell, J.K. "Lessons Learnt from the River DEE Pollution Incident, January 1984. Public Health," in *Proceedings of the World Conference on Chemical Accidents CEP Consultant* (Edinburg: 1987), pp. 223-226.
17. Hammond, E.C., Selikoff, I.J., and Seidman, H. "Asbestos Exposure, Cigarette Smoking and Death Rates," *Ann. N.Y. Acad. Sci.* 330: 473-490 (1979).

Section IV
Possible Solutions

CHAPTER **6**

Techniques for the Purification of Groundwaters Polluted by Herbicides

Costantino Nurizzo

CONTENTS

INTRODUCTION

The purification of waters derived from groundwaters polluted by herbicides (e.g., atrazine, simazine, bentazone, etc.) can be attained in different ways, usually based on the following units:

- **adsorption on activated carbon** [mainly granular activated carbon (GAC) for groundwater];
- **oxidation** (mainly ozonation, alone or combined with hydrogen peroxide, UV rays or others);
- **membrane treatments** (reverse osmosis or others).

In any case, the experience gained with the aforementioned treatment options is presently quite different: some unit operations, such as adsorption on activated carbon, are well consolidated, while others led, until now, only to some industrial scale applications or are only at an experimental stage, used to foresee their feasibility and possible development capabilities.

UNIT PROCESSES

Here, the unit operations involved in the purification of drinking waters deriving from aquifers polluted by pesticides will be described, briefly reviewing their main characteristics.

Table 1   Main Properties of Activated Carbons

	PAC	GAC
Physical Properties		
• Surface area (m²/g)	—	600–1100
• Bed density[a] (g/cm³)	—	0.25–0.48
• Particle density[b] (g/cm³)	1.4–1.5	0.9–1.5
• Nominal diameter (mm)	0.15–0.30	0.9–1.7
• Effective size	—	0.6–1.0
• Uniformity coefficient	—	1.4–1.9
Other Properties		
• Iodine number[c] (mg/g)	600–1200	900–1000
• Abrasion number	—	70–85
• Ash (%)	3–6	0.5–8.0
• Moisture as packed (max. %)	3–10	1–2

[a] Backwashed and drained.
[b] Wetted in water.
[c] Express the capability to adsorb low molecular weight substances.

ADSORPTION ON ACTIVATED CARBON

Adsorption can generally be described as a mass transfer process capable of displacing some substances (adsorbables) from the fluid where they are diluted to a solid (adsorbent), on the outer and mainly on the inner surfaces, to which they are bound by local chemical-physical actions.

In the field of water purification, this process usually operates on low-concentration solute substances — mainly organic molecules — that are generally classed as **micropollutants**; for instance: volatile organic compounds (VOC), synthetic organic chemicals (SOC), trihalomethane (THM) precursors, etc. Activated carbon is the most popular adsorbent in the water treatment field, also because, not being a polar substance, interference of water on the process is negligible.[1] Activated carbons are prepared from carbon-rich substances (such as peat, lignite, wood, nuts, bones, etc.), and their production can be summarized as a two-step operation:[1,2]

1. During the first one, the basic stuffs, after a dehydration period at T ≈ 170°C, are **pyrolized** at T < 600°C in an oxygen-poor atmosphere to produce "amorphous coal", which contains residual organic substances;

2. These organic substances are then oxidized in an air/steam atmosphere (sometimes with CO_2) at T≈800 to 900°C; this step (**activation**) causes an extension of channels inside any single solid particle and the cleaning of these channels, with a marked increase in final porosity. This raises therefore the overall specific surface and the adsorptive capabilities.

In some cases, the process takes place using chemical dehydration products, but the obtained carbon is usually unsuited for drinking water purposes because of the risks related to chemical leaching.

The final structure of activated carbons is influenced by the ones of the initial stuffs and by the production processes, which affect the number, size, and distribution of pores inside the particles and therefore the adsorption capabilities. Table 1 summarizes some of the characteristics of activated carbons used for water and wastewater treatment.[2,5]

The adsorption process can be described by the simplified mass-transfer scheme shown in Figure 1, based on diffusional processes. The kinetics of the first transfer step depend on the existing concentration gradient and on the thickness of the liquid film; in the second step the process is driven by internal diffusion.

The adsorption capabilities of activated carbons can be modeled in different ways; for diluted solutions, such as groundwaters surely are, the most common mathematical approach is the well known **Freundlich model**, which, for a single contaminant, is given by:

$$q = k \cdot C_e^{1/n} \tag{1}$$

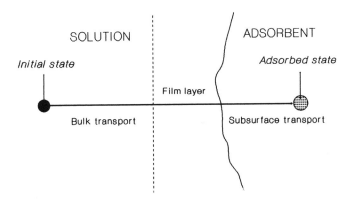

Figure 1 Mass transfer process in activated carbon.

which can be expressed also as:

$$\log q = \log k + n^{-1}\log C_e \tag{2}$$

where:

- $q = X/M$, is the carbon efficiency [g adsorbate/g carbon];
- C_e is the equilibrium concentration at a constant temperature of the single micropollutant [μg/L];
- k and n, are constants that depend on the adsorbed molecule and the tested carbon.

Figure 2 compares linearized log–log Freundlich isotherms of a GAC(F200), for some molecules (data deriving from single tests). As can be seen, there are no significant behavior differences between atrazine and simazine; for propazine, carbon efficiency (or loading) is slightly higher for high residual concentrations and quite lower for the low ones. This means that the tested carbon will not be very well suited for propazine, if very low final concentrations are requested, while it is, on the contrary, suitable in any condition for atrazine and simazine.

Adsorption on activated carbon is a stable process, but, in the case of waters polluted with different organics, competitive effects can arise, with the displacement of the previously adsorbed molecules (**desorption**) from GAC to effluent waters (this can also happen when a steep decrease of contaminant concentration in influent water[7] happens) with a strong and fast increase of effluent pollutant concentration and consequent possible health risks.

Practical Aspects of Activated Carbon Use

Activated carbons can be marketed as PAC (powdered activated carbon) or as GAC (granular activated carbon); generally speaking, the main difference between these two kinds of activated carbons is the fact that GAC can be reactivated when exhausted, while PAC is usually wasted after utilization. As a consequence, GAC is used in continuous flow filters (mainly for groundwater polishing treatments), while PAC use (generally on a seasonal basis) is restricted to the pretreatment of surface waters, taking advantage of the existing facilities provided for color and turbidity removal. In this case, the only extra machinery required is PAC stock silos and dosers; mixing, reaction, and settling tanks can be the same as used by the original plant and no special problem is to be expected in the rapid filters section. The only precaution to be taken is to check that mean residence times comply with the time required by PAC to reach equilibrium; in the meantime, adequate mixing levels (mean velocity gradient G = 50 to 70 s⁻¹) must also be secured. This last figure can be only partially respected in plants equipped with separate basins, being the standard mixing times in surface water treatment plants about 25 to 30 minutes;[8] on the contrary, layouts based on floc-blanket reactors are pretty suitable, being PAC's mean residence time in

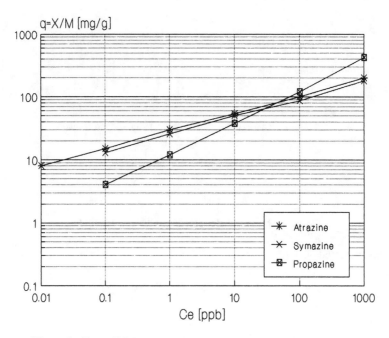

Figure 2   Freundlich isotherms for some herbicides on F200 carbon.

clariflocculation tanks long enough to secure that the equilibrium is reached.[5] Table 2 shows some data about PAC's capabilities to remove pesticides from surface waters;[5,10] it must be underlined that reported data refer to different experiences, and, therefore, PAC doses and removal levels are, not surprisingly, quite different; in fact adsorption capabilities in raw surface waters can be 50 to 93% lower than those obtained working on a lab scale with pure water.[5]

In GAC filters, a bed of activated carbon, usually downflow operated, is saturated by pollutants layer after layer (Figure 3); when the entire bed is exhausted, GAC needs to be reactivated. Possible layouts are generally based on multiple (parallel or series) filters; upflow or moving bed technology can also be utilized. In-series solutions are here encouraged to reduce the risks of unexpected desorption (see Figure 4).

The reactivation process, usually a thermal one, can be carried out at a regeneration station, generally outside the treatment plant; this practice causes an ideal GAC loss of 5 to 6% by weight,[4] but the real loss can sometimes reach 10% or more,[6,8] depending on the regeneration techniques used and on the current gas composition. Table 3 summarizes some of the main composition

Table 2   PAC Capabilities of Removing High Concentrations of Pesticide

Molecules	C_0 [μg/L]	C_e [μg/L]	PAC dose [mg/L]	Reference
Alachlor[a]	10.0	1.0	0.10	5
Atrazine[a]	10.0	1.0	0.24	5
Atrazine[b]	4.43	2.61	11.0	10
Atrazine[b]	8.11	1.86	18.0	10
Dieldrin[a]	10.0	1.0	3.0	5
Endrin[a]	10.0	1.0	1.8	5
Parathion[a]	10.0	1.0	2.5	5
Simazine[a]	10.0	1.0	0.3	5

[a] Lab tests performed with long contact times to reach ultimate equilibrium concentration (days).
[b] Data referring to real plants, with normal contact times.

Data derived from Najm, I. N. et al. *J. F.A.W.W.A.* 83(1), 65–76, 1991 and Miltner et al. *J. A.W.W.A.* 81(1), 43–52, 1989.

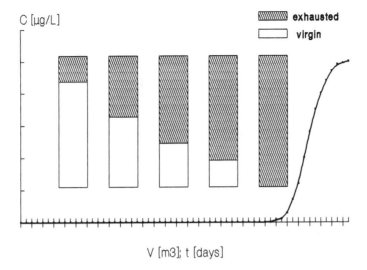

Figure 3 Exhaustion cycle for a GAC adsorber (breakthrough curve).

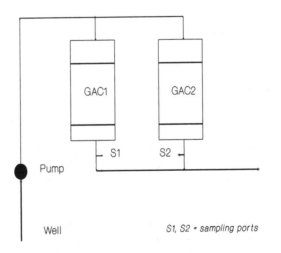

Figure 4 Schematic of a GAC simple plant (series layout preferred).

Table 3 Data on the Composition of Virgin, Spent, and Reactivated GAC

Type of Carbon	Composition (%)			
	Carbon	Hydrogen	Ash	Volatiles
Virgin Carbon	80.9	0.84	4.55	6.75
Spent Carbon	76.2	1.48	8.08	9.82
Reactivated Carbon	86.7	0.93	9.73	2.30

Data extracted from Smithson, G. R., "Regeneration of activated carbon; thermal, chemical, solvent, vacuum and miscellaneous regeneration techniques," in *Carbon Adsorption Handbook,* Ann Arbor Science, 1978.

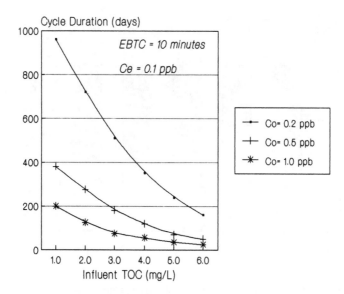

Figure 5 Raw water's TOC influence on cycle lenghts of a GAC filter for different influent atrazine concentrations. (Data extracted from Chramosta, H., De Laat, J., and Dore, M., "Effects of alkalinity and dissolved organic matter on the degradation of atrazine by O_3 and H_2O_2," in *Proceedings of the 10th Ozone Congress,* Monte Carlo, Monaco, 1991.)

figures of a GAC, at different operation stages.[6] Duration of single adsorption cycles depends on the bed depth, on the GAC type, on the TOC concentration in raw waters, on the filtration velocity, and mainly on the type of pollutants that are to be adsorbed, that can make considerable differences; for instance, in the case of atrazine, on the basis of laboratory simulations later confirmed by the experience gained during the last five years also in Italy, the standard cycle duration to take into account is 18 to 20 months, if the initial concentration does not exceed $C_0 = 0.5$ µg/L; for lower C_0, longer cycles durations can be expected (see Figure 5). Such long cycles require a countercurrent backwash, usually every 2 weeks, to reduce undesirable packing effects in the GAC bed.

On the contrary, operating on other organics (depending on their molecular structure and weight), the same carbon can display a different behavior (for instance chlorinated hydrocarbons such as TCE can exhaust a standard GAC adsorber in a few weeks). As it is quite common to detect different kinds of organics in groundwaters, the continuous control of SOC concentrations in influent waters is indispensable to avoid the risk of heavy leakages of some pollutants. Table 4 summarizes information about adsorption capabilities of a typical activated carbon for some common organic pollutants (cycle duration times are not given).[8]

Referring to the treatment of groundwater polluted only or mainly by herbicides, the usual layout[8] is based on rapid filters operating under pressure and filled with GAC fixed beds 2 to 3 m high (maximum vessel standard diameter 3.6 m); lower or higher layers can be provided within the range of 1.2 to 6 m. The geometric shape of the adsorber depends on the specific hydraulic load (SHL) (average filtration velocity), which is usually in the range of 5 to 20 $m^3/m^2{\cdot}h$; typical adsorption plants need an empty bed contact time (EBCT) of 10 minutes at least. Treating well

Table 4   Adsorption Capabilities of Activated Carbons for Some Organic Pollutants in Water

Class and Type of Organics	Adsorption Capability
Aromatic solvents (benzene, toluene, etc.)	Good
Pesticides (aldrin, endrin, chlordane, dieldrin, etc.)	Good
Herbicides (atrazine, simazine, propazine, etc.)	Good, very good
Aldehydes and ketones	Poor

Data derived from Nurizzo, C., "Purification of Groundwaters Polluted by Atrazine," ESF Workshop, Dublin, 1992.

Figure 6   Example of simple GAC plants for micropollutant removal at well's head in urban location around Milan (Italy).

adsorbable molecules such as herbicides, reference values can be as follows: EBCT = 10 to 20 min and SHL = 6 to 12 m/h.

At least two GAC filters are usually provided to balance leakage problems and to allow the GAC reactivation of a filter without stopping the operation (this is the case, for instance, of small plants installed at the well's head). An example of this solution is shown in Figure 6; the GAC station is operating just before the water distribution, this being the only treatment unit required for this kind of groundwater.

Some layouts are also based on fluidized beds with periodic extraction of the exhausted carbon — which is heavier than the virgin one — from the bottom of the vessels and its replacement with fresh carbon.

The importance of the previously mentioned parameters can be seen in Figure 7, where the specific throughputs of the same GAC (F100) are given for different herbicides as function of the bed depth;[14] the data refer to a filter operating at SHL = 10 m/h.

Another aspect of interest can be the influence that the raw water's TOC has on GAC cycle duration, even if this is much more a typical figure for surface water. However, some recent data

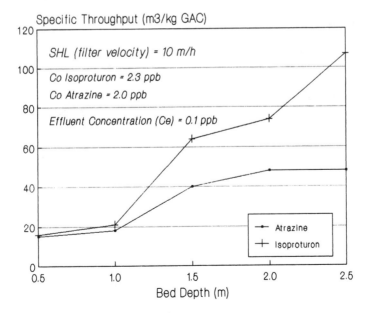

Figure 7   Specific throughputs of GAC filters as a function of bed's depth. (Data derived from Haist-Gilde, B., et al., "Removal of pesticides from raw waters," in Proceedings of the IWSA Special Conference AQUATECH92, Amsterdam, 1992.)

showed that, in some groundwaters, TOC can exceed 0.5 mg/L and sometimes be around 1 mg/L. The effect of TOC content on cycle length also depends on the initial atrazine concentration as it appears in Figure 5. The curves refer to three different atrazine initial concentrations (0.2, 0.5, 1.0 μg/L), for an equilibrium concentration C_e able to comply with pesticide EEC standard (0.1 μg/L) and for a filter operated with a short EBCT.

CHEMICAL OXIDATION

The capabilities of strong oxidants to break organic molecules in order to produce simpler and smaller ones can be of great interest for pesticide removal from waters used for drinking purposes. Great efforts were and are currently made to select appropriate oxidants and to check their applicability for a drinking water treatment; a great deal of work must be done in the field of oxidation by-products, which need to be recognized, characterized, tested for acute or chronic toxicity and finally checked for their influence on adsorption rates and GAC cycle durations.

Some new plants using strong and combined chemical oxidants to remove atrazine and other pesticides from raw or treated waters are now in operation utilizing registered industrial processes and many studies are currently in progress. A condensed review on the feasibility of this practice will be given in the next pages.

Some experimental works about chlorine capabilities to remove pesticides were done and data of analytical surveys on existing plants are also reported, but generally the results were poor.[10,18,19] Tests performed on atrazine, simazine, cyanazine, linuron, alachlor, and metolachlor rarely attained performances better than a 9 to 10% removal rate; in the case of carbofuran, a 24% removal rate is reported,[10] while very good results were obtained on metribuzin (over 90% removal),[10] operating with strong chlorine residuals (> 2.2 mg/L).

The use of ozone alone or possibly combined with hydrogen peroxide or UV rays appears to be, on the basis of some recent studies, a more promising way to break organic molecules, since many pesticides usually show a quite refractory behavior. Interest in this kind of treatment, followed by GAC adsorption, is presently very high in Europe, especially for groundwaters polluted with herbicides. For this reason, some considerations will be done, describing the proposed techniques and their possible utilizations, but having in mind the problems connected with oxidation by-products.

Ozone Oxidation

Ozone is a rather unstable and highly oxidant form of oxygen, thus capable of reacting with many substances; its practical solubility is 10 to 20 times higher than that of pure oxygen in a 0 to 30°C range and its half-life time is 165 minutes in distilled water, at 20°C.[20,21] Ozone decay in natural waters depends mainly on pH (at pH > 8, the ozone molecule dissociates into OH^-)[21] and on their natural concentration of oxidable substances, but it must also be noted that ozone oxidation efficiency can be strongly affected by HCO_3^- ions and by organic acids, which can consume the active free radicals.[12]

The O_3 decay reactions in distilled water can be described as follows:[22]

$$O_3 + H_2O \rightarrow HO_3^+ + OH^- \tag{3}$$

$$HO_3^+ + OH^- \rightarrow 2HO_2 \tag{4}$$

$$O_3 + HO_2^o \rightarrow HO^o + 2O_2 \tag{5}$$

$$HO^- + HO_2^o \rightarrow H_2O + O_2 \tag{6}$$

Ozone can be produced industrially using simple air, air added with oxygen and pure oxygen alone; the latter solution leads to simpler O_3 plant layout, as the air pretreatment steps are not necessary, but operation costs are usually higher. Generally speaking, the use of ozone in water treatment is better suited for large plants — to profit from scale effects and to take advantage of labor skills — since the complexity of ozone production plants is higher than the common one for other units of a water treatment plant.

The quantity of ozone produced is influenced by many parameters, but the following expression can well represent the ozone production:[23]

$$Q/A = k \cdot f \cdot \varepsilon \cdot V^2 \cdot \delta^{-1} \qquad (7)$$

where:

- Q/A is the production per unit surface [g/mm^2·h];
- f is the electric frequency [Hz];
- ε is the dielectric rigidity [V/mm];
- δ is the thickness of dielectric [mm];
- V is the electric potential [V];
- k is a constant.

Some of these parameters can be influenced by other quantities; for instance, ε of the gaseous flow can be influenced by particulates and moisture, and also by the operating pressure; another consequence of high moisture content is the production of nitrogen derived acids in the reactor, with heavy corrosion problems.[24] Another important quantity is the operating temperature inside the ozonator (ozone decay is higher if T > 35°C and refrigeration of both the gaseous flow and the ozonator itself is therefore very important).[21]

Reference data for industrial ozonators are as follows:[25]

- electric potential: V = 10 to 20 kV;
- ozone concentration in the outflowing gas: C = 12 to 18 g/m^3 (higher values can be reached with O$_2$ enriched air or with pure oxygen);
- electric frequency: the standard is f = 50 to 60 Hz, but, for large plants, a frequency converter can boost the value to f = 500 to 600 Hz;
- operating at standard frequency with natural air, the usual production of ozone is usually 50 to 100 g/m^2·h, with a total energy consumption of 20 to 30 Wh/g O$_3$ (60 to 70% of this amount of energy is consumed by the ozonator itself).

Ozone can be dissolved in water in different ways: gravity systems (usually multistage) are quite popular for disinfection, while oxidation of organic or inorganic substances can be performed either by gravity or under pressure facilities (in this case, ozone must be dissolved inside a mixing chamber — where the reaction begins — and the process will then continue in the following contact tower usually filled with high surface media to promote reaction).

Contact times are reported[26] to be around 15 minutes in order to allow a sufficient oxidation efficiency on many pesticides molecules, even if, in some cases, shorter contact times were shown.[8,26]

Great care must be devoted to the study of the ozonization by-products, which can have significant influence on the safe utilization of drinking water by humans. For this reason — and for economic ones, too — ozone doses must be contained to reduce the formation of aldheydes and ketones; another problem that can arise, if high O$_3$ doses are used, is the possible production of bromates, which takes place in bromide-rich waters.[27,28] It must also be underlined that proposed guidelines for bromates are fairly low (0.5 µg/L) and removal possibilities really poor.[27]

Some experimental works, which focused on ozone capabilities to remove or to transform the molecules of pesticides, were done and information will be given hereafter; in some cases the results were quite promising, but in other ones very poor efficiency was displayed. Operating on organic molecules other than pesticides, ozone was sometimes capable of improving the adsorptive capabilities of the following GAC units.[29]

More effective results on a wider range of organics in different operating situations were obtained by ozone activation, which can be performed using H$_2$O$_2$ or UV radiation. In this way, the generation of highly reactive — but poorly selective — hydroxyl radicals can be promoted,[29] helping to remove refractory organics. High concentrations of these radicals can reduce the trap-effect of alkaline (HCO$_3^-$ and CO$_3^-$ rich) waters on oxidation efficiency;[30] in the case of middle alkalinity waters (200 to 250 mg/L as $CaCO_3$) a considerable fraction of ozone dosage would be scavenged by bicarbonates.[12]

The common reference data for optimum operation in the case of triazines oxidation are as follows: ozone doses not exceeding 4 mg/L,[12] but prudently under 2 mg/L;[8] H_2O_2 to ozone dosage ratio about 0.4 to 0.5 mole per mole[12,30,31] (H_2O_2 must be injected before O_3, after strong dilution with water);[30] pH = 7.5.[26,30]

Some differences can, on the contrary, be found regarding the suggested contact times for effective removal rates; better results were obtained (Duguet et al.,[12] Ferguson et al.,28 Nurizzo,[8] etc.) if the contact times were around 15 minutes or more, while interesting data are reported by Brauch and Werner[32] for atrazine and terbutylazine, with a mere 5 minutes contact time, at pH = 7, even if higher removals were observed, increasing ozone contact times and ozone doses (up to 3.5 mg/L). In any case, negligible effects were reported for contact times exceeding 20 minutes.

In Figure 8, a comparison between the results obtained using ozone alone and ozone plus hydrogen peroxide is made (referring to atrazine and simazine, respectively);[30] as it can easily be noticed, the combined use of O_3 and H_2O_2 increases the removal of both compounds[12] (derived). The data were quite dispersed, but the interpolating lines report a significant trend; it can also be noted that, in the case of simazine, the better results displayed were probably due to a higher H_2O_2 to O_3 mass ratio.

MEMBRANE TREATMENTS

These treatments refer to two classes: pressure driven processes (reverse osmosis, nanofiltration, ultrafiltration) and electrically driven processes (electrodialysis). Of these, only pressure driven ones can be of some interest for pesticide removal, with the exclusion of the ultrafiltration process, whose membranes have large dimension pores being, therefore, not suited for molecules of this type. The process can be described as a pressure forced flow of water through a synthetic membrane, which rejects organic molecules and ions in the case of reverse osmosis (RO plants) and mainly organics in the case of nanofiltration (NF plants). A typical simplified scheme of these kinds of plants is shown in Figure 9.

In a membrane process, the purified water flux (permeate) mainly depends on the net driving pressure and on the membrane characteristics; a generally recognized expression for the specific flux is as follows:[34]

$$J_W = K_W \cdot P_{EFF} \tag{8}$$

where K_W is a constant, depending on the membrane type, and P_{EFF} is the effective driving pressure taking into account pump pressure, permeate pressure, and osmotic pressure. The brine flow is therefore obtained as the difference between the influent flow and the permeate flow, being:

$$Q_{INF} = Q_{PER} + Q_{BR} \tag{9}$$

where $Q_{PER} = J_W \cdot A_M$, and A_M is the membrane total effective area.

Reverse osmosis appears to be effective (see Table 5;[35] reported data refer to fairly high concentrations), but not really feasible due to high treatment costs and to combined, but usually undesired, salinity removal. It would probably be more attractive in the case of combined occurrence of atrazine and other pollutants, such as nitrates, nitrites, sulfates, chlorides, or other dissolved substances quite refractory to other treatments.

A new development of the process, now under evaluation, which can probably begin to modify the situation is nanofiltration, which is a particular type of reverse osmosis; in fact the method is the same, but the membrane is "more loose", allowing less expensive operation, with lower effective pressure and without strong changes in the water salinity.[36]

During a recent study,[34] the capability to remove pesticides and other micropollutants by nanofiltration was checked at a pilot plant. The plant was operated at 6 to 15 bars (Q_{INF}) using four different membranes; the removal results were very interesting, being, for instance, (depending on the membrane type) in the range of:

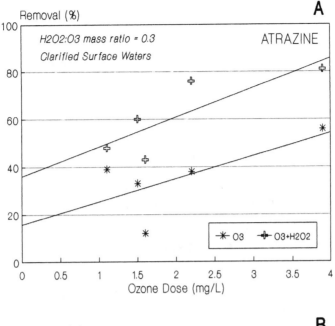

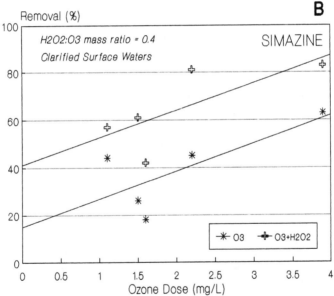

Figure 8   Removal of atrazine (A) and simazine (B), with ozone alone or ozone plus hydrogen peroxide (average contact time, 15 minutes). (Data derived from Duguet, J. P., Wable, O., Bernazeau, F., and Mallevialle, J., "Removal of Atrazine from the Seine River by the ozone–hydrogen peroxide combination in a full scale plant," in *Proceedings of the 10th Congress,* Monte Carlo, Monaco, 1991.)

- 80 to 98% for atrazine;
- 63 to 93% for simazine;
- 43 to 87% for diuron;
- 96 to 99% for bentazone.

These quite impressive results were obtained on pretreated water (two-stage chemical-physical section); in spite of this, severe membrane fouling was observed after a few days of operation (the previously cited driving pressure range is related to membrane progressive fouling). GAC adsorption followed the membrane treatment.

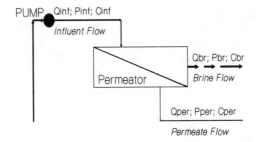

Figure 9   Schematic of a pressure-driven membrane unit (RO, NF).

Nanofiltration appears to be a quite promising technique that, however, needs further studies to improve the pretreatment operation and to check the real operation costs (chemicals, energy, membrane duration, etc.).

POINT OF USE/POINT OF ENTRY TREATMENTS

A great deal of problems come from rural districts where a not very large number of persons get their supplies for drinking water from a large number of sources, mainly groundwaters coming from poorly protected shallow aquifers. In this situation, the previously mentioned treatment options must be adapted case by case, being unfeasible to install small "standard" plants in this situation.

In many areas, POU (point-of-use) or POE (point-of-entry) plants could be a solution to this problem, also to avoid emergency practices such as packed water distribution. Generally speaking, these facilities can be based on:

- GAC filtration for micropollutants removal;
- RO treatment for: saline water treatment, micropollutants or specific ions removal, bacteria removal;
- AA (activated alumina) plants for fluoride removal;
- UV lamps for general disinfection.

Table 5   Removal of Selected Pesticides by Reverse Osmosis Using Different Membranes

Pesticide	Membrane type	C_0 [μg/L]	Removal [%]
Chlorinated Pesticides			
Aldrin	NS-100	142.3	100.00
	CA	142.3	100.00
Heptachlor	NS-100	505.4	100.00
	CA	505.4	100.00
Dieldrin	NS-100	321.3	100.00
	CA	321.3	99.88
Organophosphorous Pesticides			
Diazinon	NS-100	473.7	98.05
	CA	473.7	98.25
Malathion	NS-100	1057.8	99.65
	CA	1057.8	99.16
Parathion	NS-100	747.3	99.83
	CA	747.3	99.88
Miscellaneous Pesticides			
Atrazine	NS-100	1101.7	97.82
	CA	1101.7	84.02
Trifluralin	NS-100	1578.9	99.99
	CA	1578.9	99.74
Captan	NS-100	688.9	100.00
	CA	688.9	97.78

Data derived from Eisenberg, T. N. and Middlebrooks, E. J., "Reverse Osmosis Treatment of Drinking Water", Butterworths Publishers, Stoneham, MA (1986).

Units can be installed directly at the point of use (faucet) or in the basement of any single house to produce drinking-quality water. In Italy, a general purpose POU/POE plant is usually based on an RO unit, which is eventually connected to GAC and UV units; prefilters and postfilters are also common.

The main problems related to these kinds of plants, neglecting here the high water costs, are as follows:

- high risks coming from low-level or scarce technical survey of the plants operation;
- possible bacterial colonization of the GAC cartridges and the related need for effluent disinfection;[37]
- desorption risks from GAC filters in case of micropollutant changes (kind and number);
- possible high fouling risk of RO membranes and, therefore, pretreatment needs;
- possible worsening of effluent quality due to replacement delays of some parts of the treatment unit (GAC, membrane, etc.);
- expensive waste removal.[38]

The use of POE/POU (mainly POU) is presently promoted in Italy by the producing industries as a further safety factor, also in the case of public water supply, to cope with possible network failures. This situation is quite different from that reported existing in the U.S., where the USEPA does not view POE and POU as BAT (Best Available Technology) and accepts POU only as an interim solution.[38]

SOME EXPERIMENTAL RESULTS — TREATING SPECIFIC MOLECULES

As in many other laboratories, some research work is currently in progress at the Environmental Section of the D.I.I.A.R. of the Milan Politecnico, using ozone oxidation (followed by GAC filtration) with low ozone doses to reduce by-products formation. Among tested molecules the following herbicides were also used: atrazine, simazine, bentazone and molinate, mainly used for maize and rice weeding. Tests were carried out at an emergency water treatment plant capable of producing up to 20 m^3/h of drinking water; the schematic layout of this plant is shown in Figure 10, while Figure 11 shows a view of the same plant.

To check the effects of contact time (taking advantage of residual ozone to simulate the possible effect of a final storage tank), each sample was divided prior to the analysis; the first subsample was left undisturbed in a closed bottle for at least 90 minutes before the analysis, while the second one was immediately stabilized by reducing residual ozone with sodium sulfite; in this case, the real contact time did not exceed 4 minutes. Table 6 summarizes the findings obtained in a first set of field tests on different herbicides using two ozone doses with different initial concentrations (influent water was totally free from herbicides before the micropollutants injection); tests were carried out separately for each molecule. In the case of low contact times (< 4 minutes), atrazine and simazine displayed — as was expected — a quite similar behavior with poor herbicides removals using O_3 alone; better results were obtained with O_3 oxidation of molinate, while excellent ones were shown by ozone oxidation of bentazone, even for very short contact times (a few seconds). The situation was fairly different if long contact times (even without mixing) were allowed, also in the case of refractory molecules such as atrazine or simazine. In any test condition, the GAC filter was always capable of completely removing all kinds of micropollutants. Figure 12 summarizes the overall results obtained for atrazine at different initial concentrations (A sample was added with sodium sulfite, while sample B was left undisturbed).

These preliminary findings seem to point out that, even for low O_3 doses, some results can be obtained if sufficient contact times are possible. As a consequence, a simple layout modification that permits O_3 dosage before the system storage tank, followed by GAC adsorption on the storage tank effluent, can probably lead to interesting results for large plants allowing sufficient contact times for ozone oxidation; the storage tank must be closed to avoid ozone dispersion.

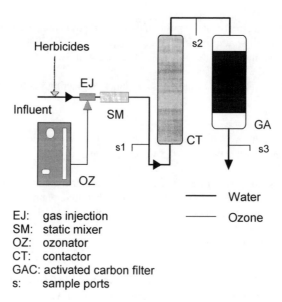

EJ: gas injection
SM: static mixer
OZ: ozonator
CT: contactor
GAC: activated carbon filter
s: sample ports

Figure 10   Scheme layout of the plant used for ozone oxidation tests on herbicides.

Figure 11   View of the demo-scale plant used at Novara (Italy); on the side the ozonator, in the background the GAC filter.

Table 6   Herbicide Removal Efficiency (avg. as %), Using
           Ozone in Low Doses at Different Contact Times,
           (D_1 = 1.3 mg O_3/L, D_2 = 1.7 mg O_3/L; T1 ≤ 4 min,
           T2 ≥ 90 min)[a]

Herbicide	Ozone Dose D_1		Ozone Dose D_2	
	T1	T2	T1	T2
Atrazine				
C_0 = 0.12 µg/L	18	81	25	83
C_0 = 0.52 µg/L	14	88	18	100
C_0 = 1.52 µg/L	22	53	24	60
Simazine				
C_0 = 0.12 µg/L	10	100	10	100
C_0 = 0.44 µg/L	45	87	59	93
C_0 = 1.42 µg/L	23	68	24	78
Molinate				
C_0 = 0.11 µg/L	68	100	72	100
C_0 = 0.40 µg/L	60	100	63	100
C_0 = 1.00 µg/L	48	98	61	100
Bentazone[b]				
C_0 = 0.10 µg/L	100	100	100	100
C_0 = 0.40 µg/L	100	100	100	100
C_0 = 1.80 µg/L	100	100	100	100

[a] A 100% removal indicates that molecule concentration is not detectable in the effluent.
[b] Results obtained just after premixing (3 to 4 sec); only the lowest O_3
dose tested (0.9 mg/L) showed a mere 67% removal in the pre-mixer
effluent, for the highest influent concentration (1.8 µg/L). After the
contact tower, the removal was complete in any case.

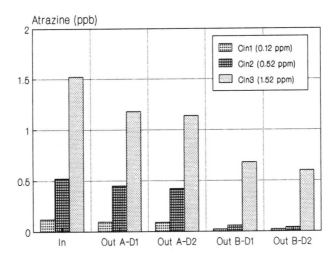

A: sulphite added; B: no sulphite added
D1= 1.3 mg O3/L; D2= 1.7 mg O3/L

Figure 12   Results of atrazine removal by ozone oxidation at different doses (D_1 = 1.3, D_2 = 1.7 mg
             O_3/L), in the 0.12 to 1.52 µg/L range of initial concentration (A: sulfite added sample to avoid
             residual ozone reaction inside the sample; B: undisturbed sample to simulate long contact
             time with residual ozone).

CONCLUSION

- When, as it happens in Italy, the water distribution is mainly based on local small authorities pumping from a few unconnected wells, the dispersion of plants leads to a sole possible solution: adsorption on granular activated carbon; in fact GAC gave good results in different operating situations, can be automatically operated, and needs moderate survey practices.
- Further knowledge must in any case be obtained about the possible risk of desorption when raw water shows a mix of different micropollutants and about the related cycle lengths.
- Another topic of great interest is the ability of GAC to become a good medium for bacterial growth; on one hand, this can also be helpful, but on the other bacterial count standards must be respected.
- PAC techniques can easily be applied to surface water treatment stations, where their feasibility is more marked in case of a seasonal pollution.
- Some new and promising processes are related to the use of O_3, alone or combined with hydrogen peroxide or UV rays, but always followed by GAC adsorption to remove harm coming from residual pollutant, reaction intermediates, or by-products. In fact, oxidation by-products are often unknown or not yet fully identified and characterized, but, in many cases, aldehydes and ketones were detected.[39,40] When ozone was used on triazines, desethyl-derivatives were also found.[32]
- Operating at contact times < 5 minutes, ozone alone (at a 1.7 mg/L dose) was only capable of reaching efficiencies <25% for atrazine and simazine, but >60% for molinate; bentazone was completely removed under the same conditions. For very long contact times, checked to simulate the behavior inside a storage tank, removal levels always > 60% were obtained for atrazine with even better results for simazine. In the case of molinate, results very close to 100% were shown in any condition; clearly bentazone was completely removed.[8]
- It must be underlined that ozone doses below 1 mg/L always performed quite poorly, except in some very favorable cases (bentazone, for instance).
- The use of technologies such as ozonization appears likely to be devoted to medium or large size stations; it is, in fact, quite unusual to use facilities with an ozone production lower than 5 kg O_3/h. This figure roughly means a daily flow of at least 30,000 to 40,000 m^3 (depending on ozone doses, type of pesticides, and ozone solubility) and, therefore, a served population of about 100,000 inhabitants.
- Other treatments that are capable of development are membrane methods such as nanofiltration, but some fouling-related problems and high costs are, at present, the main obstacles to a wide-range application.
- Ozone and membrane plants need skilled personnel for operation and maintenance, and membrane treatments do not fit very easily with high flow requirements.
- Construction and operation costs can be variable, depending on plant size and its general layout, on initial pesticide concentrations, and on possible copresence of other micropollutants; some information referring to the present situation on the Italian market will be given hereafter, as a rough indication:
  - in the common case of a small treatment station just at the well's site, the only solution considered here is GAC adsorption; the reference construction cost for a standard plant as described before would be, in this case, also on the basis of a work recently carried out by a Technical Committee of the Regione Lombardia,[41] in the range of 1,800,000 to 2,000,000 Italian Liras (ITL) for every m^3/h of treated nominal flow (for instance, a 50 m^3/h GAC plant would cost about 90,000,000 to 100,000,000 ITL). Present exchange rate is roughly 1,600 ITL/US $ (January 1995).
  - For large treatment stations, always based on GAC adsorption, reported data can clearly be different, due to common layout choices (usually gravity filters), control devices, etc. Construction costs per unit flow of a huge fully equipped GAC plant recently designed, are about 1,500,000 ITL.[42] Plants using ozone oxidation and GAC adsorption display a specific construction cost, some 30 to 40% higher.[12]

- Operation costs for plants at the well's site are quoted between 50 and 70 ITL/m^3, depending on plant size, initial pesticide (triazine) concentration, concurrent presence of other micropollutants, etc. In the case of larger plants (always referring to triazine removal from groundwaters), they can be in the 45 to 50 ITL/m^3 range. These figures take into account: energy, manpower, ordinary maintenance, periodic GAC reactivation every 24 months, and capital costs. In situations different from those common in Italy, operation costs can change and further modifications must be expected when O_3 plus H_2O_2 are used prior to GAC adsorption; data coming from a 3000 m^3/h French plant[12] give a total operation cost of 1.5 FF/m^3 (1 FF = 5.6 US $), but in this case one year GAC cycle and doses of 2 mg O_3/L and 0.6 mg H_2O_2/L were provided, respectively. Table 7 reports data from another French study comparing the treatment costs of different options with those expected for the adsorption units of the provided upgrade provided for one of the Firenze (Florence) drinking water treatment plants[30,42] (both of the Firenze plants treat the polluted waters of the Arno river).
- For POU plants (usually operating in a flow range of 50 to 100 L/d), costs of 1,500,000 to 2,000,000 ITL/unit (50 L/d flow) are common, and specific operation cost (mainly related to membrane change, roughly once each year) ranges around 15,000 to 18,000 ITL/ m^3.

In conclusion, the possible treatment options for groundwaters polluted by pesticides can be different, but, in any case, a process unit based on adsorption on activated carbon seems to be necessary to assure drinking water of proper quality.

In the very near future, a possible trend could be the following:

- GAC filtration as the sole specific unit for small plants and as the final unit of existing ones;
- Chemical oxidation plus GAC filtration for large new plants. Chemical oxidation (mainly ozone plus hydrogen peroxide) would operate as the first treatment step, while GAC filtration would be the final one to perform the removal of residual pollutants and oxidation by-products.

Table 7 Construction and Operation Costs for Atrazine Removal

- Mt. Valerien plant (avg. flow 2500 m^3/h, different treatment trains, C_{OMAX} = 1 μg/L of atrazine)[a]
- Firenze Anconella plant (upgrade works in progress, avg. flow 10,000 m^3/h, GAC adsorption, C_{OMAX} = 1 μg total pesticides/L)[b]

Treatment Train	Construction Costs	Operation Costs[c]
Mt. Valerien Plant (2,500 m^3/h)[d]		
GAC filtration	11,000,000 FF	0.52 FF/m^3
O$_3$ oxidation plus GAC filtration	39,000,000 FF	0.64 FF/m^3
O$_3$+H$_2$O$_2$ oxidation plus GAC filtration	40,000,000 FF	0.61 FF/m^3
Firenze Anconella Plant (10,000 m^3/h)[e]		
GAC filtration	15,400,000,000 ITL	46 ITL/m^3

Note: 1 US $ ≈ 1400 ITL; 1 US $ ≈ 5.6 FF (october 1992).

[a] Data derived from Duguet, J. P. et al. *Proceedings of the B.E.T.I.E.E. 92*, Toulouse, France, 1992, with permissions.
[b] Personal communication.
[c] Capital costs included.
[d] 12 months cycle for GAC.
[e] Provisional data with a 24-month cycle.

REFERENCES

1. Tchobanoglous, G. and Schroeder, E.D. *Water Quality: Characteristics, Modeling, Modification* (Reading, MA., Addison-Wesley, 1987).
2. Benefield, L.D., Judkins, J.F., and Weand, B.L. *Process Chemistry for Water and Wastewater Treatment* (Englewood Cliffs, N.J.: Prentice-Hall, 1982).
3. Chemviron Carbon. Technical Informations (1990).
4. Clark, R.M. and Lykins, B.W., Jr. *Granular Activated Carbon: Design, Operation and Cost* (Boca Raton, FL.: Lewis Publishers, 1989).
5. Najm, I.N., Snoeyink, V.L., Lykins, B.W., Jr., and Adams, J.Q. "Using Powdered Activated Carbon: a Critical Review," *J. A.W.W.A.* 83(1):65-76 (1991).
6. Smithson, G.R. "Regeneration of Activated Carbon: Thermal, Chemical, Solvent, Vacuum and Miscellaneous Regeneration Techniques," in *Carbon Adsorption Handbook,* Chemerinisoff, P.N. and Ellerbusch, F., Eds. (Ann Arbor, MI: Ann Arbor Science Publishers Inc., 1978).
7. Snoeyink, V.L. "Adsorption as a Treatment Process for Organic Contaminant Removal from Groundwater," in *Proceedings of the AWWA Seminar Organic Chemical Contaminants in Groundwater: Transport and Removal* (St. Louis, MO: AWWA 1981).
8. Nurizzo, C. "Purification of Groundwaters Polluted by Atrazine," ESF Workshop, Dublin (1992).
9. Weber, W. J., Jr. "Activated Carbon Treatment for Removal of Potential Hazardous Compounds from Water Supplies," in *Innovations in the Water and Wastewater Fields.* E.A. Glysson, D.E. Swan, and E.J. Way, Eds. (Stoneham, MA: Butterworth Publishers, 1985).
10. Miltner, R.J., Baker, D.B., Speth, T.F., and Fronk, C.A. "Treatment of Seasonal Pesticides in Surface Waters," *J. A.W.W.A.* 81(1): 43-52 (1989).
11. van Santfoort, F. "Test Accelerato su Colonna per la Progettazione di un Impianto a Carbone Attivo," *Inquinamento* 29(3): 96-100 (1991).
12. Duguet, J.P., Wable, O., Bernazeau, F., and Mallevialle, J. "Removal of Atrazine from the Seine River by the Ozone–Hydrogen Peroxide Combination in a Full Scale Plant," in *Proceedings of the 10th Ozone Congress,* Monte Carlo, Monaco (1991).
13. Faust, S.D. and Aly, O.M. *Chemistry of Water Treatment* (Woburn, MA.: Butterworth Publishers, 1983).
14. Haist-Gilde, B. et al. "Removal of Pesticides from Raw Waters," in *Proceedings of the IWSA Special Conference AQUATECH92* (Amsterdam, NL, 1992).
15. Davis, M.L. and Cornwell, D.A. *Introduction to Environmental Engineering* (New York: McGraw-Hill, 1991).
16. Speth, T.F. and Miltner, R.J. "Adsorption Capacity of GAC for Synthetic Organics," *J. A.W.W.A.* 83(2): 72-75 (1991).
17. Montgomery, J.M. *Water Treatment: Principles & Design* (New York, N.Y.: *Organic Chemical Contaminants in Groundwater,* John Wiley & Sons, 1985).
18. Cremisini, C. et al. "Valutazione degli effetti della disinfezione delle acque con ipoclorito sulle triazine ed altri erbicidi di interesse igienico-sanitario," *Acquae & Aria* 12: 581-588 (1990).
19. Smith, E.J., Jr., Renner, R.C., Hegg, B.A., Bender, J.H. *Upgrading Existing or Designing New Drinking Water Treatment Facilities,* PTR no. 198 (Park Ridge, N.J.: Noyes Data Corporation, 1991).
20. Venosa, A.D., Meckes, M.C., Opakten, E.J., and Evans, J.W. "Comparative Evaluation of Ozone Contactors," in *Proceedings of the 52nd W.P.C. Conference* (Houston, TX, 1979).
21. Rice, R.G. "Innovative Applications of Ozone in Water and Wastewater Treatment," *Innovations in the Water and Wastewater Fields* (Stoneham, MA.: Butterworth Publishers, 1985).
22. Maccarthy, J.J. and Smith, C.C."A review of Ozone and Its Applications to Domestic Wastewaters Treatment," *J. A.W.W.A.* 66(12): 718-725 (1974).
23. White, G.C. *Disinfection of Wastewater and Water for Reuse*, (New York: Van Nostrand Reinhold, 1978).
24. Masschelein, W.J. "Aspect Thermodynamiques de la Generation Industrielle de l'Ozone," in *Ozone et Ozonation des Eaux* (Paris: TECDOC-Lavoisier, 1991).
25. Degremont. "Memento Technique de l'Eau". (Paris-CEDEX, F, 1984).

26 Chramosta, H., De Laat, J., and Dore, M. "Effects of alkalinity and dissolved organic matter on the degradation of atrazine by O_3 and H_2O_2," in *Proceedings of the 10th Ozone Congress* (Monte Carlo, Monaco, 1991).

27. Kruithof, J.C. and Schippers, J.C. "The formation and removal of bromate," *Proceedings of the IWSA Specialized Conference, AQUATECH92* (Amsterdam, NL, 1992).

28. Ferguson, D.W., McGuire, M.J., Koch, B., Wolfe, R., and Aieta, M. "Comparing PEROXONE and Ozone for Controlling Taste and Odor Compounds, Disinfection By-products and Microorganisms," *J. A.W.W.A.* 82(4): 181-191 (1990).

29. Jun, Y. and Bao Zhen, W. "Efficacy and Mechanism of Removal of Organic Substances from Water by Ozone and Activated Carbon," *Water Sci. Technol.* 21: 1735-1737 (1989).

30. Duguet, J.P., Wable, O., and Mallevialle, J. "Le Couplage Ozone-Peroxide d'Hydrogène: Exemple d'une Innovation Technologique dans le Domaine des Eaux Potables," *Proceedings of the B.E.T.I.E.E. 92* (Toulouse, France, 1992).

31. Paillard, H., Brunet, R., and Dore, M. "Conditions Optimales d'Application du Système Oxidant Ozone-Peroxide d'Hydrogène," *Water Res.* 22(1): 91-103 (1988).

32. Brauch, H.J. and Werner, P. "Identification and Characterization of Ozonization By-products of Selected Pesticides, by Chemical Analysis and Bio-test," *Proceedings of the 10th Ozone Congress* (Monte Carlo, Monaco, 1991).

33. Bouillot, P. et al. "Efficiency of Different Possibilities of Making Use of Ozone, Alone or Coupled, as Regard the Refractory Organic Carbon Tranformation in Carbon Biodegradable by Bacteria," *Proceedings of the 10th Ozone Congress* (Monte Carlo, Monaco, 1991).

34. Hofman, J.A.M.H., Noji, H.M., and Schippers, J.C. " Removal of Pesticides and Other Organic Micropollutants with Membrane Filtration," *Proceedings of the IWSA Specialized Conference AQUATECH92* (Amsterdam, NL, 1992).

35. Eisenberg, T.N. and Middlebrooks, E.J. *Reverse Osmosis Treatment of Drinking Water* (Stoneham, MA.: Butterworth Publishers, 1986).

36. Aptel, P. and Vial, D. "New Membrane Plants for Water Treatment," *Proceedings of the B.E.T.I.I.E., 92* (Toulouse, France, 1992).

37. Anderson, M., Gottler, R.A., and Bellen, G.E. "Point-of-Use Treatment Technology to Control Organic and Inorganic Contaminants," in *Experiences with Groundwater Contamination,* Proceedings of the A.W.W.A. Conference (June 10, 1984).

38. Goodrich, J.A., Lykins, B.W., Adams, J.Q., and Clark, R.M., "Safe Drinking Water for the Little Guy: Options and Alternatives," E.P.A., Trans. Tech., EPA/600/M-*Innovations in the Water and Wastewater Fields* 90/015 (September 1990).

39. Glaze, W.H., Koga, M., Cancilla, D., Wang, K., McGuire, M.J., Liang, S., Davis, M.K., Tate, C.H., and Aieta, E.M. "Evaluation of Ozonation By-Products from Two California Surface Waters," *J. A.W.W.A.* 81(8): 66-73 (1989).

40. Jacangelo, J.G., Patania, N.L., Reagan, K.M., Aieta, M.E., Krasner, S.W., and McGuire, M.J. "Ozonation: Assessing Its Role in the Formation and Control of Disinfection By-Products," *J.A.W.W.A.* 81(8): 74-83 (1989).

41. C.T.R. *Piano regionale di risanamento delle acque*, Regione Lombardia Settore Ambiente ed Ecologia (Milan, Italy, 1992).

42. Personal communication.

CHAPTER 7

Development of Novel Active Ingredients

David I. Gustafson

CONTENTS

Earlier sections of this book have described the nature of the risks posed by pesticides in groundwater. This chapter sets out some methods for developing novel active ingredients and formulations with a lower potential for drinking water contamination.

DEFINITION OF TARGET BEHAVIOR

The target is to develop "environmentally friendly" pesticides. This vague term must be defined. In short, an environmentally friendly product is one that: (1) degrades rapidly and completely after its job is done, (2) does not leave residues in food or water, (3) does not move from the target location, (4) is compatible with manufacturing, formulations, packaging, and application methods that minimize waste and exposure, and (5) does not perturb the treated ecosystem in any undesirable way. Each of these desirable features is described in further detail below.

RAPID DEGRADATION

All pesticides must remain active for a finite time in order to have utility against the target pest. However, all pests (weeds, fungi, and insects) pose a threat to the crop only for a certain time, and it would be most desirable for the pesticide to be present only for the time period during which the pest constitutes a significant economic threat. When an active series of chemical analogs has been discovered, metabolic "handles" can be placed within the chemical structure in order to make them more palatable to soil microbes. Certain atoms and functionalities such as extensive halogenation generally retard degradation in soil and water, and these types of substitutions should be minimized to the extent possible, while still maintaining activity against the target pest.

LACK OF RESIDUES IN FOOD AND WATER

Besides exhibiting rapid degradation after its job is done, a desirable property of the pesticide is to show no detectable residues in either food or water as a result of its normal use. The properties of pesticides that lead to drinking water contamination have already been explored in-depth (see Chapter 3), but their propensity to occur in food has not yet been addressed. For most herbicides of low toxicity, the residues present in food are very unlikely to pose a significant risk. However, most insecticides have at least some toxicity to humans and livestock, and it is important to determine whether toxic levels will occur in such matrices as a result of either normal or unintentionally high application rates.

In order to answer this question, pesticide manufacturers conduct extensive field studies at recommended and exaggerated rates of application to measure concentrations in the portions of treated crops used for food by animals and humans. Residues in the processed forms of these commodities, for instance peanut butter from peanuts, are also measured. In general, food residues are inevitable when a chemical is applied to a crop. However, low use rate, short persistence, extensive plant metabolism, and/or greater time between application and harvest may lead to undetectable residues in the edible portions of the plant. All of these tactics, as well as changes in agricultural practice, represent ways to minimize pesticide residues in food.

LACK OF MOVEMENT FROM TARGET LOCATION

By staying only at the target location, the identity of which is determined by the target pest, the pesticide is the most efficient it can be and severely reduces its potential for causing undesirable effects. Besides enhancing efficiencies, such pesticide behavior reduces the potential for drinking water contamination and other undesirable off-site effects. As will be seen later in this chapter, low mobility is accomplished through strong binding to soil, often accomplished by making the molecule more lipophilic. Having high lipophilicity confers lower soil mobility, but it is not a panacea. Very hydrophobic agricultural chemicals, most notoriously DDT, have been found to accumulate in the food chain. The potential for this phenomenon to occur is usually quantified by the bioconcentration factor (BCF). The BCF is highest for those chemicals that are very resistant to biochemical attack and possess high lipophilicity, generally measured by the base-10 logarithm of the octanol/water partition coefficient, logP.[1]

Besides runoff or leaching, another possible route of off-site movement is through volatilization. Depending on the mode of action of the compound, such movement can have highly undesirable consequences, such as the bleaching of nearby (and not-so-nearby) lawns caused by clomazone in recent years. Extensive volatilization may raise a host of other issues, including contamination of fog and rainwater, or even global climate change for high-sales-volume halogenated compounds. This general subject represents a potential high-visibility environmental topic for the 1990s and beyond.

COMPATIBILITY WITH NEW DELIVERY METHODS

New farming and application technologies are generally more compatible with certain classes of pesticides, primarily those that are foliarly applied after the crop has emerged. Besides avoiding broadcast applications directly to soil as much as possible, the delivery methods involve novel formulations and systems for delivering pesticides from the manufacturer, to the dealer, and finally to the user. Pesticides compatible with this new philosophy of active ingredient delivery should be advanced over the older products that made extensive use of small, wasteful containers and delivery methods onto bare soil — maximizing the potential for contaminating drinking water.

NO UNDESIRABLE PERTURBATION OF ECOSYSTEM

Besides the potential for drinking water contamination, there are other potential environmental risks posed by pesticides. Most of these are minimized by ensuring that the nontarget toxicity of the chemical is as low as possible. Everyone is likely to be familiar with the alleged ability of DDT to thin the egg shells of game birds and thereby accelerate their decline. In addition to such indirect damages, certain pesticides pose a direct and imminent hazard to various nontarget species. Bird kills with pesticides formulated as granular products have been documented for a variety of materials and have led to recent regulatory action against one active ingredient in particular, carbofuran.

Other undesirable effects on the environment include fish kills. A series of fish kills in southern Louisiana during the summer of 1991 caused a statewide ban on the use of a commercially important insecticide, azinphos-methyl, used to treat their vast fields of sugar cane.[2] By late summer, a total of 750,000 fish had died: striped mullet, largemouth bass, yellow bass, freshwater drum, spotted gar, ladyfish, shad, buffalo fish, mosquito fish, carp, southern flounder, blue catfish, and others. Besides the fish, there were allegations reported in *Pesticide & Toxic Chemical News* that aerial applications of the pesticide had killed "a number of birds, turtles, and alligators at 13 sites" in the state.

While not truly an environmental fate issue, lack of nontarget toxicity is a property that may ameliorate other less favorable environmental traits. In order for risk from a chemical to occur, both exposure and toxicity must be present. The dual importance of the two terms in this equation is often lost on the public, who may not tolerate either, even when the other factor is absent.

In completing this section defining environmental friendliness, it is worth noting one pesticide that is almost certainly environmentally friendly: glyphosate.[3] The compound is rapidly mineralized in soil, shows no potential for runoff or leaching due to a high degree of soil binding, has minimal effects on soil microflora, and is virtually nontoxic to all animal life. The relatively strong binding of the compound to soil appears to be through the phosphonic acid moiety of the molecule because phosphate competes effectively for sorption sites. Since it is virtually inactivated once it strikes soil, glyphosate must generally be applied directly to foliage, but this mechanism fits in perfectly with the new, directed application techniques that are now being developed. As environmental issues continue to guide trends in agriculture, pesticides like glyphosate will play an ever-increasing role.

PHYSICAL PROPERTIES DETERMINING BEHAVIOR

The physical-chemical properties of a pesticide responsible for its potential to leach into groundwater or runoff into surface water are its persistence and mobility. In many ways, mobility is the easier of the two to assess. Persistence is a compendium of many different properties, is much more sensitive to environmental factors, and is likely to be more variable.

PESTICIDE MOBILITY IN SOIL

The intrinsic mobility of a pesticide in soil is inversely related to its degree of sorption to (or within) soil surfaces. A pesticide having no molecular interaction with soil material would travel with the water front and inevitably reach groundwater unless it degraded rather rapidly. The effect of the sorptive properties on pesticide runoff has been demonstrated by experiments showing that more strongly sorbed chemicals remain nearer the soil surface than more mobile materials, leading to greater overall losses in runoff. However, the form taken by such strongly sorbed chemicals is as sediment rather than as freely dissolved chemical, making long-range transport difficult, if not impossible.

Since water is the vector by which pesticides are moved through or over soil, via leaching and runoff water, respectively, the definitive characteristic of the pesticide molecule conferring mobility is its inherent propensity to remain dissolved in soil water as opposed to being sorbed on or in solid soil particles. This propensity is usually described by a partition coefficient, K_D (L/kg), which is equal to the ratio, at equilibrium, of the concentration of the pesticide in soil (µg/kg) to its concentration in water (µg/L).

The measurement of the soil/water partition coefficient is generally conducted through a simple slurry experiment. A known quantity of pesticide (usually radiolabeled to facilitate analysis) is added to a slurry of known water and soil content. After shaking the slurry for a set time period (24 to 48 hours) the slurry is centrifuged, and the quantity of pesticide remaining in the aqueous supernatant is determined. The soil/water partition coefficient is easily calculated as the ratio of final concentrations in the soil (µg/kg) and water (µg/L) phases. There is nothing magical about waiting this period of time, but this is the conventional approach.

After this sorption step, the water is decanted off, clean water is added back into the slurry, and another equilibration step is executed. In this manner, the degree of reversibility in the sorption process is measured. Invariably, the partition coefficients determined in these desorption experiments are much higher than the values obtained in the original sorption experiment. This phenomenon, generally known as hysteresis in the sorption/desorption process, is said to exert major influences on pesticide behavior in field situations. These hysteretic effects of sorption on the mobility of pesticides are such that major quantitative differences in leaching patterns can be found when comparing results under saturated- and unsaturated-flow conditions.

At sufficiently low concentrations in soil, such as those typically encountered deep within the soil profile, the sorption process becomes linear, that is, the ratio of sorbed to dissolved concentrations assumes a constant value. However, the K_D ratio for a pesticide is not constant from one soil to another. The variance in K_D is associated with many soil properties, such as soil pH, particle size distribution, and the type of clay present, but, by far, the most important parameter is usually the soil organic matter content, at least for surface soils containing 1% or more organic matter. The situation is more complex for charged molecules and those that are able to interact chemically with the clays present in most soils. Nevertheless, most researchers currently choose to rank the relative intrinsic soil mobility of pesticides by K_{OC}, the soil/water partition coefficient divided by soil organic carbon content. As organic matter increases, a proportionate rise in K_D is generally observed. This common trend has led to the definition of K_{OC} and K_{OM}, the soil water partition coefficient, K_D, divided by soil organic carbon, or soil organic matter, respectively. The main utility of this definition is that either K_{OC} or K_{OM} may be used to predict K_D for a new soil on which the K_D of the compound has never been measured. The K_{OC} of a pesticide also serves as a soil-independent measure of the relative mobility of the pesticide in soil.

Summarizing the current understanding of sorption processes, it is widely accepted that the linear, instantaneous equilibrium model based on K_D and K_{OC} represents an excellent first approximation. However, experimental evidence is accumulating that there are slow processes, undoubtedly related to the slow intraparticle diffusion of pesticides through the solid soil matrix, that mandate the inclusion of sorption kinetics (chemical nonequilibrium models) in any accurate description of pesticide transport through field soils. In addition, when structured soils or soils having rapid leaching rates are present, it becomes necessary to use two-region (physical nonequilibrium) models having mobile and immobile soil–water phases.

PERSISTENCE IN SOIL

Besides mobility, the other factor determining whether a pesticide will exhibit a significant potential for contaminating drinking water is its persistence. For many years, this persistence has been described simply by the "half-life" of the compound. This terminology is a bit unfortunate for several reasons. Unlike unstable radio isotopes such as ^{14}C, which really do decay under strictly linear first-order kinetics, pesticides do not exhibit true exponential decay. In large part, this is due to the host of different processes responsible for the dissipation of pesticides in soil: volatilization, photolysis, hydrolysis, biological degradation, and oxidation, to name just a few. Each of these individual dissipative processes is, in turn, affected by a number of soil properties and environmental factors, such as temperature, pH, and moisture. Since soils are so variable, it follows that the degradation rate throughout the soil will also vary, sometimes by orders of magnitude. As described elsewhere,[4] this spatial variability leads to nonlinear dissipation of pesticides in soil, despite the theoretical fact that linear first-order dissipation kinetics should be observed within a sufficiently small aliquot of the soil matrix having uniform properties.

In contrast to mobility, persistence cannot be accurately predicted from chemical structure alone. Near the soil surface, the rate of field dissipation may be influenced by volatilization, photolysis, and runoff. Once it has passed below the surface, whether by incorporation or movement with rainwater, the rate of dissipation is more closely tied to the susceptibility to chemical and biological transformation reactions.

Hydrolysis and photochemical conversion of many pesticides can complement biotransformation. For example, triazines undergo chemical hydrolysis. Although photochemical conversion cannot occur below the soil surface, this is an important dissipative route for many compounds, such as foliarly applied insecticides.

Most studies suggest that the rate of degradation of many pesticides in groundwater is quite slow. However, the rates of degradation in surface water tend to parallel the rate observed in surface soils. The presence of sediment in estuarine water has been shown to increase dramatically the rates of degradation of most pesticides. Other research has shown that pesticides are able to volatilize very rapidly from natural waters on which they have been inadvertently sprayed.

Summarizing, these data suggest that the simple linear, first-order model for pesticide dissipation so widely utilized is not particularly accurate. Effects of spatially and temporally varying soil moisture, soil temperature, and other environmental factors invariably lead to nonlinear and otherwise quite unpredictable dissipation patterns under field conditions. There is considerable evidence available to suggest that some of these influences can be predicted if the rate of dissipation is well quantified in the laboratory over a range of temperature and moisture conditions. Nevertheless, the situation with respect to predicting rates of dissipation and therefore persistence of pesticides in soil is not nearly as well advanced as that for predicting pesticide mobility.

WATER SOLUBILITY

Water solubility *per se* is not generally an indicator of water contamination potential. A brief review of the environmental concentrations typically encountered demonstrates why water solubility itself is not very predictive of water contamination potential. Typical concentrations of pesticides in soil water near the surface are on the order of 1 mg/L and less, well below the solubility limit for atrazine and other chemicals known to be contaminants of drinking water. As degradation and dissipation occur, soil water concentrations deeper below the land surface are even lower, further removing water solubility from the realm of having any relevance to environmental behavior. Thus, the relatively low water solubilities on the order of even 10 mg/L are not in any way limiting to the mobility of pesticides through soil and down into groundwater.

Having said this, it should be pointed out that there are two ways in which water solubility does play some role in determining the potential for drinking water contamination. Chemicals that are extremely insoluble, having solubilities less than about 1 mg/L, are sufficiently insoluble such that initially applied concentrations are much greater than what the soil water can dissolve. This will obviously decrease the potential for movement either down through the groundwater or laterally into surface water, simply by limiting the quantity available.

The other way in which water solubility exerts some influence on the potential for drinking water contamination is through the correlation between water solubility and the propensity for interaction or sorption to soil. While there is a strong correlation between water solubility and mobility in soil, the correlation is by no means exact, and the relationship does not hold equally well for different classes of chemicals. For neutral, organic chemicals exhibiting ideal behavior, the assumption is generally quite valid, but if there are ionizable groups present or any functional groups on the molecule, such correlations lose their predictive capability. In the case of atrazine, as with many other triazines such as simazine, the water solubility is unusually depressed due to the crystal energies within the solid form of the pure chemical.

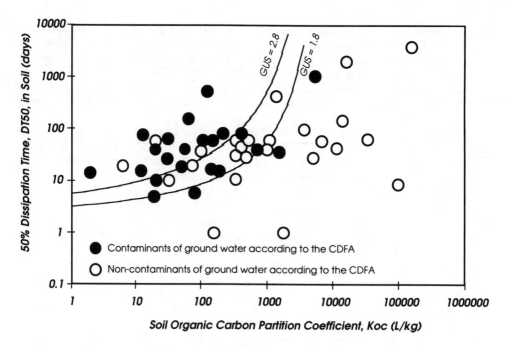

Figure 1   Illustration of the influence of pesticide physical properties (mobility and persistence in soil) on the propensity for occurrence at detectable levels in groundwater.

The fact that the correlation between water solubility and mobility in soil is not perfect should not come as a big surprise. The processes involved in the equilibrium between dissolved and crystalline pesticide are very different from the processes involved in determining whether the pesticide is dissolved in soil water or sorbed to soil. Factors such as crystal energy and other intermolecular parameters are critical in determining water solubility, whereas the fraction of chemical dissolved in soil water is generally determined by other factors such as the degree of interaction with specific chemical functionalities of materials making up soil, such as organic matter and clays.

METHOD OF USE

The rate, timing, and method of application can be important factors in determining the amount of pesticide leached. Compounds applied to the foliage when a crop is actively transpiring are less likely to leach than those that are soil-incorporated before planting (all else being equal). For pesticides used in areas receiving irrigation, the timing and amounts of irrigation relative to the timing of pesticide application may prove critical. In general, those pesticides applied less often, at lower rates, and later in the growing season will be less likely to contaminate drinking water supplies as a result of leaching or runoff processes.

OPTIMAL RANGE OF PHYSICAL PROPERTIES

The method used here to indicate the desired range of physical properties is to simply plot the physical properties of several pesticides with known drinking water contamination potential on a graph, using different symbols for those known to be contaminants. Two figures are given here, with the first (Figure 1) showing pesticides that have been looked for in groundwater and the second (Figure 2) showing pesticides whose occurrence in surface water has been checked. The physical properties forming the horizontal and vertical axes of these two plots are, respectively, the soil/water partition coefficient based on organic carbon, K_{OC}, and the 50% disappearance time, DT_{50}. The physical properties of the pesticides are taken from a recent data compilation.[5] The status of each pesticide with respect to well water contamination is derived from a list published by the

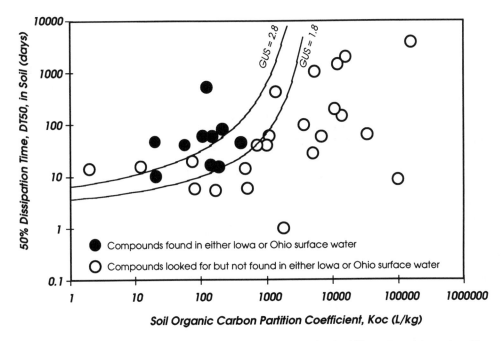

Figure 2   Illustration of the influence of pesticide physical properties (mobility and persistence in soil) on the propensity for occurrence at detectable levels in surface water.

California Department of Food and Agriculture,[6] and the surface water contaminants are taken from statewide surveys in Iowa[7] and Ohio[8].

As shown in both graphs, those chemicals that possess both high mobility and persistence are more likely to occur in drinking water, whether it is derived from surface or ground sources. The two contours included in each of the figures are defined by values of 2.8 and 1.8 for the GUS index, whose value is calculated as log $(DT_{50})* (4\text{-log}(K_{oc}))$.[9] This index, although originally developed to describe specifically the potential for groundwater contamination,[9] also appears to delineate the pesticides likely to be contaminants of surface water (see Figure 2). Values of GUS in excess of 2.8 indicate a high potential for occurrence in drinking water, while values below 1.8 indicate a very low potential.

This graphical approach gives a simple and intuitive assessment of a pesticide's threat to drinking water, once its persistence and mobility properties have been obtained. Armed with such a graphical procedure, the developer of new pesticides can attempt to design molecules with properties in the optimal range, which is accomplished by either having very little persistence or low mobility (K_{OC} > 1000 L/kg) or both. Some methods for predicting these properties are given below.

FUTURE PROSPECTS FOR IMPROVING ENVIRONMENTAL BEHAVIOR

NOVEL FORMULATIONS TO REDUCE CONTAMINATION POTENTIAL

To some extent, even those pesticides with undesirable properties can be made more acceptable through novel formulations. One topic that has occupied much research attention over the past several years is the development of controlled release formulations to enhance efficacy or improve environmental behavior. These research efforts have resulted in the availability of microencapsulated pesticide formulations, which meter the release of active ingredient. One major advantage of such systems is that pesticides that might otherwise degrade too rapidly to be effective against the target pest can be successfully applied as a controlled-release formulation. Such materials with short half-lives have obvious environmental benefits but would not otherwise be economically viable if applied in a conventional formulation. Controlled-release formulations have been shown to reduce the leaching of alachlor, EPTC, and metolachlor under greenhouse conditions.[10]

Nitrogen, although not a pesticide, is the agricultural chemical that has been most widely associated with groundwater contamination at levels over its MCL. Slow-release forms of this essential element will certainly become more widely used in the future as one of the means of reducing the occurrence of above-guideline residues of nitrate in drinking water. In the longer term, there will likely be biotechnology-based improvements to soil microbes that effect soil tilth, but, for the next several decades, there will be a continuing and growing need for better methods of delivering chemical tools to the soil and crop.

COMPUTER PREDICTION OF PHYSICAL PROPERTIES

Prediction of pesticide mobility is much easier than predicting its persistence. Karickhoff discussed the importance of organic matter in the sorption process of pesticides interactions with soils.[11] He showed that the process is dominated by hydrophobic interactions. Although he states that *a priori* estimation techniques are not yet available, reasonable estimates based upon chemical class and sorbent composition can be made. Relatively accurate predictions of K_{OC} can be made based on water solubility, octanol–water partition coefficient, molecular weight, and reversed-phase HPLC retention time.[12] The relationship between aqueous solubility and octanol–water partition coefficients has also been described by Mackay, Bobra, and Shiu.[13] One highly nonlinear regression model was recently proposed[14] based on molecular surface, volume, weight, and charge densities on nitrogen and oxygen atoms of the molecule.

Most of the methods for predicting sorptive behavior assume that the pesticide is an uncharged, neutral organic chemical. If ionizable groups are present, then it becomes important to account for the possibility of dissociation and the resulting effects on sorption. In general, molecules carrying a net negative charge will not sorb appreciably to soil, thus any molecule having a pKa less than 8 will have a very low sorption coefficient. Methods for predicting pKa based on chemical structure have been summarized in a recent textbook.[15]

Certain workers have attempted to predict biodegradation kinetics from a conventional structure–activity relationship.[16] More recently, a very useful functional-group method for predicting the biodegradation potential of pesticides has been reported by the Syracuse Research Corporation.[17] Others have tried the use of molecular connectivity indices and other graph theories to predict both persistence and mobility in soil.[18-23] In the case of predicting sorption coefficients, these apparently sophisticated approaches reduce to a simple correlation of molecular weight and K_{OC}, upon further inspection.[24]

Besides biodegradation processes, volatility plays a role in determining the dissipation rate of pesticides in the field. Field volatility is a function of vapor pressure, Henry's law constant, and sorption to soil. Vapor pressures of many nonpolar organic compounds can be estimated by capillary gas chromatography.[25] This approach was used to measure the vapor pressures of several esters of 2,4-D.[26] Henry's law constants for relatively simple pesticides may be estimated using bond contribution methods similar to those used for estimating partition coefficients.[27]

Together with the simulation modeling tools described in Chapter 3 these methods for predicting physical properties of pesticides present the very real possibility of designing and testing, all on a computer, whether a hypothetical chemical would be likely to contaminate drinking water when used as a pesticide. This promise, when coupled with the rigorous implementation of better application methods, suggests that future contamination of drinking water supplies by pesticides, which is already quite rare, could be entirely eliminated.

REFERENCES

1. Isnard, P. and S. Lambert. "Estimating Bioconcentration Factors from Octanol–Water Partition Coefficient and Water Solubility," *Chemosphere* 17:21-34 (1988).
2. Food Chemical News. "Azonphos-Methyl Linked to Kill of Half Million Fish in Louisiana," *Pestic. Toxic Chem. News*, August 14, 1991, p. 22.

3. Rueppel, M.L., B.B. Brightwell, J. Schaefer, and J.T. Marvel. "Metabolism and Degradation of Glyphosate in Soil and Water," *J.Agric. Food Chem.* 25:517-528 (1977).

4. Gustafson, D.I., and L.R. Holden. "Nonlinear Pesticide Dissipation in Soil: a New Model Based on Spatial Variability," *Environ. Sci. Technol.* 24:1032-1038 (1990).

5. Wauchope, R.D., T.M. Buttler, A.G. Hornsby, P.W.M. Augustijn Beckers, and J.P. Burt. "The SCS/ARS/CES Pesticide Properties Database for Environmental Decision-Making," *Rev. Environ. Contam. Toxicol.* 123:1-164 (1992).

6. Johnson, B. "Setting Revised Specific Numerical Values," October 1989, State of California, CDFA, Sacramento, CA, Eh 89-13 (1989), 12 pp.

7. Iowa Department of Natural Resources. Pesticide and Synthetic Organic Compound Survey, Report to the Iowa General Assembly on the Results of the Water System Monitoring Required by House File 2303, (Iowa DNR: Des Moines) 19 pp.

8. Baker, D.B. "Pesticide Concentrations and Loading in Selected Lake Tributaries — 1982," Final Report, U.S. EPA Grant Erie No. R005708-01, (1983) (EPA:Washington DC), 61 pp.

9. Gustafson, D.I. "Groundwater Ubiquity Score: A Simple Method for Assessing Pesticide Leachability," *Environ. Toxic. Chem.* 8:339-357 (1989).

10. Koncal, J.J., S.F. Gorske, and T.A. Fretz. "Leaching of EPTC, Alachlor, and Metolachlor Through a Nursery Medium as Influenced by Herbicide Formulations," *Horstscience* 16:757-758 (1981).

11. Karickhoff, S.W. "Organic Pollutants Sorption in Aquatic Systems," *J. Hydrol. Eng.* 110: 707-735 (1984).

12. Kanazawa, J. "Relationship Between the Soil Sorption Constants for Pesticides and Their Physicochemical Properties," *Environ. Toxicol. Chem.* 8:477-484 (1989).

13. Mackay, D., A. Bobra, and W.Y. Shiu. "Relationship Between Aqueous Solubility and Octanol–Water Partition Coefficients," *Chemosphere* 9:701-711 (1980).

14. Bodor, D.D., Z. Gabanyi, and C.K. Wong. "A New Method for the Estimation of Partition Coefficient," *J. Am. Chem. Soc.* 111:3783-3786 (1989).

15. Perrin, D.D., B. Dempsey, and E.P. Serjeant. *pKa Prediction for Organic Acids and Bases*, (New York: Chapman and Hall, 1981), 143 pp.

16. Desai, S.M., R. Govind, and H.H. Tabak. "Development of Quantitative Structure–Activity Relationship for Predicting Biodegradation Kinetics," *Environ. Toxicol. Chem.* 9:473-477 (1990).

17. Boethling, R.S., P.H. Howard, W. Meylan, W. Stiteler, J. Beauman, and N. Tirado. "Group Contribution Method for Predicting Probability and Rate of Aerobic Biodegradation," *Environ. Sci. Technol.* 28:459-465 (1994).

18. Sabljic, A., H. Gusten, J. Schonherr, and M. Riederer. "Modeling Uptake of Airborne Organic Chemicals. 1. Plant Cuticle/Water Partitioning and Molecular Connectivity," *Environ. Sci. Technol.* 24:1321-1326 (1990).

19. Niemi, G.J., G.D. Vieth, R.R. Regal, and D.D. Vaishnav. "Structural Features Associated with Degradable and Persistent Chemicals," *Environ. Toxicol. Chem.* 6:515-527 (1987).

20. Hall, L.H. and L.B. Kier. "Estimation of Environmental and Toxicological, Properties: Approach and Methodology," *Environ. Toxicol. Chem.* 8:19-24 (1989).

21. Gerstl, Z. and C.S. Helling. Evaluation of Molecular Connectivity as a Predictive Method for the Adsorption of Pesticides by Soils," *J. Environ. Sci. Health* B22:55-69 (1987).

22. Sabljic, A. "On the Prediction of Soil Sorption Coefficients of Organic Pollutants from Molecular Structure: Application of Molecular Topology Model," *Environ. Sci. Technol.* 21:358-366 (1987).

23. Karickhoff, S.W., V.K. McDaniel, C. Melton, A.N. Vellino, D.E. Nute, and L.A. Ca. "Predicting Chemical Reactivity by Computer," *Environ. Toxicol. Chem.* 10:1405-1416 (1991).

24. Shea, P.J. "Role of Humidified Organic Matter in Herbicide Adsorption," *Weed Technol.* 3:190-197 (1989).

25. Bidleman, T.F. "Estimation of Vapor Pressures for Nonpolar Organic Compounds by Capillary Gas Chromatography," *Anal. Chem.* 56:2490-2496 (1990).

26. Hamilton, D.J. "Gas Chromatographic Measurement of Volatility of Herbicide Esters," *J. Chromatogr.* 195:75-83 (1980).

27. Meylan, W.M. and P.H. Howard. "Bond Contribution Method for Estimating Henry's Law Constants," *Environ. Toxicol. Chem.* 10:1283-1293 (1991).

CHAPTER **8**

Innovative Strategies in Weed Control

Giuseppe Zanin, Maurizio Sattin, and Antonio Berti

CONTENTS

INTRODUCTION

Weed control is a necessity for efficient and profitable agriculture, and herbicides will continue to be a key component of most weed control programs in the future, because no adequate alternative technology currently exists. However, the negative secondary effects of chemical weed control, such as water and food contamination, development of resistance, etc., suggest a lowering of our dependence on herbicides. This can be achieved by setting up control programs, not exclusively with the use of tactical interventions, but based on a strategy incorporating knowledge of the weed–crop–environment system. The challenge is to develop a new weed management system, defined as Integrated Weed Management (IWM), which involves the integration of all knowledge plus the use of alternative weed control measures.

IWM has been defined as the application of different weed control solutions, including agronomic, mechanical, biological, genetic and chemical measures.[1] None of the single control options can be expected to provide an, at least in the long term, acceptable level of weed control on its own, but, if the different components of IWM are systematically implemented, significant advances can be achieved in weed control.[2]

IWM became an accepted and frequently used term in weed science in the early 1970s.[3] It is a part of Integrated Pest Management (IPM), a term that was first used by entomologists during the late 1950s and early 1960s,[4] and has since expanded to include all plant protection disciplines and is now the most promising approach for pest control in agriculture. IWM is made up of two distinct sectors: Weed Population Management (WPM) and Weed Control (WC).

WEED POPULATION MANAGEMENT

WPM is the part of IWM with which weed density and their competitivity can be reduced and weed spectrum shifting can also be avoided. These are essential conditions for the use of a "softer" and more environmentally sound chemical weed control. In practice, the low density and the floral equilibrium of weed communities would lead to a lower reliance on chemical treatments and to a wider choice of herbicides conferring greater flexibility on the weed control. WPM is carried out by means of crop manipulation and weed manipulation.

CROP MANIPULATION

Almost invariably, crop practices influence the competitive ability of both the crop and weed community; crop–weed interactions must be altered in such a way that the balance favors crop growth at the expense of the weeds. This concept is well summarized by an old saying that "the best weed killer is a good crop." The factors that can be manipulated in a cropping system include crop varieties, planting pattern, resource pool management, and clean crop seed.

Crop Varieties

Different varieties of the same crop differ in their ability to compete with weeds. The traits that give competitive advantages are rapid emergence, high root and shoot growth rates especially at the beginning of the growing cycle, a high rate of leaf expansion, and a dense canopy with horizontal leaves and large number of branches.[3] The influence of canopy structure on competitive ability has been demonstrated by the comparative behavior of wheat varieties with different leaf habits. In a weedy situation, the floppy-leaved varieties showed better weed control and gave a good yield, whereas those with an erect-leaf habit had lower yields.[5] In the same crop, the newest short-strawed cultivars were less competitive than the taller traditional cultivars. In sugar beet, the survival of late emerging weeds was lowered by 73% in cultivars with more horizontally arranged leaves. The use of cultivars with an early closing canopy may be promising in low-dosage systems. With these systems, cocktails of herbicides are applied in postemergence at 7 to 10 day intervals at reduced rates to control newly emerged weeds. Sowing more competitive cultivars would probably reduce the number of herbicide applications (normally between 2 and 4).[6] Unfortunately, selection criteria for the more recent cultivars are based primarily on yield in a weed-free environment, while many authors propose that the ability to compete with weeds should be considered in selection programs.

In some cases, tolerance to weeds is tied to not well identified varietal characteristics, often associated with the production of root exudates containing toxins. This type of interference, called allelopathy, is common in natural ecosystems, and some authors have suggested that allelopathy might be exploited for weed control purposes in a variety of agricultural settings. Numerous laboratory experiments support this hypothesis. One accession of the world collection of cucumber (*Cucumis sativus*) inhibited proso millet (*Panicum miliaceum*) and white mustard (*Brassica hirta*) growth by 87% and 25 other accessions inhibited growth by 50% or more.[7] Two lines of soybean out of 141 evaluated inhibited the growth of *Helminthia (= Picris) echioides* but not of *Alopecurus myosuroides*.[8] Fay and Duke[9] isolated 4 out of 3000 accessions of *Avena* spp. germplasm that were able to produce three times more scopoletin (6-metoxy-7-hydroxy coumarin), an allelopathic agent, than a standard commercial oat cultivar.

In the field, the effects are less certain; for example, the allelopathic effect of cucumber was suppressed during periods of high rainfall.[10] These results showed that allelopathic activity could be obtained under specific pedoclimatic conditions, but the results were not considered consistent enough to warrant a plant breeding effort.[11]

Planting Pattern

Resource capture is a very local phenomenon in weeds and crops. The intensity of competition among neighboring plants tends to increase as the distance between them decreases. Spatial arrangement is therefore very important. Planting pattern can be influenced by row spacing, seeding rate, and spatial arrangement.

Crops grown in narrower rows start to compete with weeds at an earlier stage than those in wider rows because of more rapid canopy closure and, probably, better root distribution.[12] Optimization of these variables must be achieved considering the different possible situations because the effect of planting pattern can be modified by other factors such as crop cultivar, weed density, soil moisture, etc. Evidence shows that these factors can be successfully manipulated to provide a competitive advantage for the crop.

Resource Pool Management

Crop and weeds generally compete for the resource pool so, for example, changes in the level and timing of fertilization and irrigation can alter the competitive balance in the agroecosystem. In theory, the reduction of available water and nutrients caused by weeds should be reversible with adequate irrigation and fertilization; however, it must be taken into account that there is a marked difference in the capacity to absorb water and nutrients among different plant species. The increase in nutrient and water availability can sometimes be exploited preferentially by the weeds, limiting the efficacy of the fertilization or irrigation in reversing yield suppression. In a study of competition between winter wheat and Italian ryegrass (*Lolium multiflorum*) that involved different levels of weed density and nitrogen fertilization, Appleby et al.[13] demonstrated that ryegrass was better able to respond to increased nitrogen availability than wheat. The conclusion was that, in the absence of effective weed control, increased nitrogen fertility was of questionable value. Application time can also modify the response to fertilization. For example, in direct-seeded rice, high levels of nitrogen applied after heading of barnyard grass reduced competition, but, when applied at the vegetative stage, it enhanced the weed competitivity.[14]

Nutrient placement is another important variable that can be usefully manipulated. Nitrogen can be banded close to the crop row and phosphorus incorporated into the soil beneath the row, thus enhancing the crop's accessibility to the nutrient. The manipulation of nutrients (level, timing, and modality of application) is a useful but tricky tool for weed management, requiring a good knowledge of the weeds, crop, crop practices, and pedoclimatic environment.

Water is a very important competition factor, and different weed species and crops vary widely in their responses to low soil water availability.[15,16] Wiese and Vandiver[17] showed that competitivity of corn, sorghum, and eight weed species varied with water availability. With high water availability, weeds typical of humid regions or normally present in irrigated crops were more competitive. Conversely, weeds that are adapted to arid regions were more competitive when water was limited. This is generally ascribable to their higher water use efficiency (grams of dry matter produced per gram of water lost). Things are more complicated where water availability fluctuates.

Common cocklebur (*Xanthium strumarium*) exploits the water reserves held in a large volume of soil, thus gaining a competitive advantage over soybean.[18] In other cases, water stress can shift competition to favor the crop: water stress reduced anoda (*Anoda cristata*) and velvetleaf (*Abutilon theophrasti*) biomass, thus giving a competitive advantage to cotton.[19]

Clean Crop Seed

The use of weed-free seed is one of the oldest and most important preventive measures in weed control. As in medicine, "an ounce of prevention is worth a pound of cure." There are two main consequences of crop seed contamination, introduction of alien weeds and spread of weeds within a certain area. The intensive exchange of agricultural products, especially seeds, between different countries or even continents, has caused direct introduction of new weeds. Most governments have introduced rigorous laws trying to reduce the invasion of new weeds (and pests in general), but sometimes not-certified seeds are used, especially in less developed agricultures and generally for certain crops such as soybean and wheat. Crop seed contamination is important especially for the "mimetic weeds". These have adapted their growth and time of maturity in relation to the crop, so dispersal and reseeding are accomplished during crop harvesting and replanting. The classic example is *Camelina sativa* in flax and *Echinochloa oryzoides* and *E. phyllopogon* in rice cultivated with permanent flooding techniques.

WEED MANIPULATION

This can be achieved by seed and bud bank manipulation and alteration of the weed community composition.

Seed and Bud Bank Manipulation

The soil seed and bud bank is obviously a major source of weed infestation. The seed and bud content in arable fields is enormous: the seed bank can reach 1 billion seeds per hectare. In numerous surveys carried out in various European countries, the median value is between 50 and 100 million seeds per hectare.[20] There can also be large numbers of vegetative propagules. Fail[21] reported that in 1 hectare of arable soil there were about 320 km of couchgrass rhizomes (*Agropyron repens*), which is to say 6 million buds.

The stock of the seed and bud bank is regulated by the balance between inputs (seed rain, bud production, and immigration), outputs (emergences, mortality, and predation) and longevity of propagules (seeds, rhizomes, bulbs, tubers, stolons) (Figure 1).

Seed and bud bank manipulation means carrying out procedures that lower the inputs and increase the outputs.

Input Lowering — The best method of lowering seed rain is good weed control; however, once a weed has escaped the various treatments, there are no direct methods for reducing seed and bud production. Once again the only cure is to favor competitivity of the crop, the result of which is twofold: lower seed production per weed plant and lower seed viability.[22,23]

The input can also be reduced by limiting the introduction of seeds and buds from neighboring areas. Prevention must involve the entire agricultural biotope and not just the crop fields. If the weeds growing in noncropped areas are not controlled, they disseminate abundantly, and the seeds can be transported into the fields by means of wind, irrigation, animals, or farm machinery. Mowing vegetation in noncropped areas along roadsides, ditches, canal banks, and in fallow fields to prevent weed seed production has great significance in the philosophy of IWM.

Irrigation is another significant source of contamination. Kelly and Burns[24] calculated that surface irrigation waters distribute, during the season, between 10,400 and 94,500 seeds per hectare. These numbers, in relative terms, represent less than 0.2% of a seed bank of 50 million per hectare; the danger therefore is not so much in the number of seeds introduced but in the possible introduction of alien species. Adequate screening of weed seeds from irrigation water and the

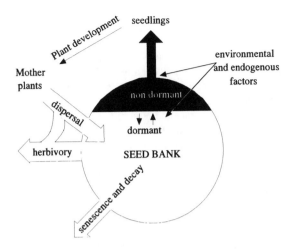

Figure 1 Diagrammatic flow chart for the dynamics of seed population in soils. (From Egley, G. H. and S. O. Duke, *Weed Physiology,* Boca Raton, FL, CRC Press, 1985, pp. 28–64, with permission.)

application of irrigation water from wells and distribution through pipes is ideal for preventive weed control.

Weed seeds can also reach fields by means of other crop operations, particularly harvesting with combine harvesters and the spreading of manure or liquid manure. Cleaning the combine harvester is a very important operation that must be done before it enters a farm and also when it is moved from an infested field to a clean one within a farm.

Farm manure, spread as a fertilizer, can be an important source of seeds for the seed banks of arable soils. Seeds of many weed species can pass through the digestive tracts of livestock and remain viable.[25] The feces are mixed with straw, which may also contain seeds, to form manure that is left to rot for several months. High temperatures inside the manure heap cause many seeds to lose their viability, but some can remain viable because of a greater intrinsic resistance to heat or because they are in the outer part of the pile where temperatures are lower.[26] The use of well-matured manure is therefore a valid option.

Output Increasing — This goal can be reached by speeding up the decline of the seed and bud bank and by reducing seed longevity. A seed bank consists of dormant and nondormant seeds. Seeds that are alive but not germinating are dormant; dormancy is a mechanism for surviving unfavorable periods such as prolonged drought or severe cold. Seeds in the soil may cycle between dormancy and nondormancy depending on the soil environmental conditions. On average only 3 to 6% of seeds present in the soil produce seedlings in the course of a year.[27] While there are many possibilities for treating emerging or germinating plants, there is, at least at present, very little for dormant seeds. According to Harper,[28] seed dormancy is a major barrier to advances in weed control. It is, in fact, the cause of the building up of a large soil seed bank and the reinfestation of arable land. It is therefore important to reduce the size of the seed bank by increasing the seed germination rate, by death or predation of seeds.

Soil cultivation is an excellent means of encouraging weed seed germination: it aerates the soil, favoring the loss of soil bound volatile germination inhibitors and brings seeds that have lost their dormancy near the surface. Each flush of weeds following tillage is then killed by subsequent tillage operations or chemical treatments, which results in a decrease of viable seeds in the upper soil layers. With frequent cultivation and no seed return, an annual reduction of seed number of about 50% can be achieved.[29] The mean half-life of seeds is about 1 year; i.e., half of the seeds disappear by germination, death or predation in each 12-month period. The depletion of the seed bank is therefore relatively slow; after 7 years, 1.0% of seeds are still present, that is 1 million per hectare if the seed bank size was of 100 million. Furthermore a single year of dissemination is enough to regain the original seed bank size. There is an old saying that "a year's seeding means seven years weeding." The difficulty of managing the seed bank and the constant need to carry out the different interventions is therefore obvious.

The most promising approach for reducing a seed bank is to use chemicals that break dormancy even in buried seeds. Many chemicals, such as plant growth regulators (e.g., ethylene), nitrogen compounds (e.g., nitrate, thiourea), plant products (e.g., strigol), and some herbicides at low rates (e.g., carbamates) stimulate the germination of weed seeds.[30] Ethylene injected directly into the soil or applied as an ethylene-generating compound such as ethephon, stimulates the germination of a wide spectrum of species. The best results have been obtained against witchweed (*Striga* spp.), an obligate parasitic weed that attacks cereals and leguminous crops.[31] After germination, the seeds must attach themselves to the root of the host plant, otherwise they die. This requirement makes the weed vulnerable, as stimulation of germination in the absence of host roots would lead to seedling death. In North and South Carolina (U.S.) the infestation of witchweed has been noticeably reduced by a combined program of ethylene plus herbicide.

Nitrates enhance germination in many weed seeds, probably making some weeds more responsive to brief light exposure when the soil is disturbed.[32] In the field, the results vary depending on weeds, however, in some species, interesting results have been obtained, e.g., with *Avena fatua*,[33] redroot pigweed (*Amaranthus retroflexus*), and velvetleaf (*Abutilon theophrasti*).[34] The effectiveness of nitrates as germination stimulators is confirmed by the observation that emergence flushes of *Capsella bursapastoris* and *Senecio vulgaris* occur at the highest concentration of nitrates in the soil.[35] The potential success of nitrates as a weed germination stimulant in the field is in the interaction with other stimuli, particularly light, alternating temperatures, and ethylene. However, nitrate fertilization could be a useful tool for IWM programs.

Some compound inhibitors of respiration can also stimulate weed seed germination. Sodium azide stimulates wild oat (*Avena fatua*)[36] and tumble pigweed (*Amaranthus albus*)[34] germination. Because sodium azide is rapidly lost from the soil, conditions must be favorable for rapid seed germination at the time of application. Development of satisfactory ways to prolong the soil life of azide, such as controlled release systems, may improve its effectiveness.[32]

There is also a problem of instability with another chemical stimulant of seed germination, called strigol, an active *Striga* germination inducer.[37] Strigol was isolated from cotton root exudates and the compound dl-strigol was then synthesized.[38] Strigol and strigol analogues cause a depletion of *Striga* seed reserves in the soil by triggering germination in the absence of host plants. Evidence shows that breaking seed dormancy in the soil is possible. Nevertheless, more knowledge is required to take decisive steps forward.

Another way of encouraging seed bank decline is an increase in predation, which can be obtained by not incorporating the seeds into the soil. Using conservation tillage, a large percentage of seeds remain on the soil surface to be eaten by birds, rodents, and insects.[39] The most vulnerable seeds are large and have high energy value.[40]

Perennial species can survive unfavorable conditions by means of vegetative propagules as well as seeds. Apical dominance is the mechanism by which only the apical buds will develop while the others remain dormant. The latter can then persist in the soil and therefore, like dormant seeds, have a survival value. The lateral buds are usually induced to grow if the dominant shoots are injured or killed or if the propagules are fragmented by cultivation. The magnitude of apical dominance varies among species.

Tillage promotes bud growth by cutting the propagules into short pieces, reducing the effect of apical dominance and thus maximizing the number of shoots produced, which can then be destroyed with a specific weed control treatment.[41] Conversely, improper tillage can cause the spread of perennial weeds across fields. To avoid this, the tillage must be integrated into a weed control strategy. For example, a rotary harrowing of the soil after wheat harvest, coupled with irrigation and nitrogen fertilization, is a system that notably improves the efficacy of a subsequent treatment with glyphosate, not only because a greater number of buds germinate and grow more uniformly but also because increased nitrogen and soil moisture can significantly enhance the translocation and effectiveness of herbicides.[42] Ethylene can also break apical dominance in rhizomes, but only if a leafy parent shoot is attached. If there is no parent shoot, then dominance appears to be enhanced.[43]

Rather than allowing propagules to be formed and then manipulating the regrowth pattern after cultivation, it would perhaps be more effective to prevent the formation of rhizomes altogether.

This could be done by reducing the gibberellin content in the mother plant or the translocation of this substance to the lateral shoots or alternatively by applying synthetic compounds with cytokinin-like activity to alter the balance between the two hormones and prevent horizontal growth.[30] The use of growth regulators in the parent plant might well be effective in preventing rhizome formation, but the necessity of using it at an advanced stage of the growth cycle limits the practical interest of this system.

The reduction of seed longevity is another valid means for reducing the seed bank. It has been shown that common lambsquarter (*Chenopodium album*) seeds collected from nitrate-treated plants were less dormant.[44] Since nondormant seeds often remain viable for a shorter time than dormant seeds, more seeds might die before germination. Similar results have not been obtained with other species [e.g., *Setaria glauca* (= *pumila*)],[45] but the need for a reduction in dormancy of the seed bank requires additional research.

Alteration of Weed Community Composition

The competitive ability of different weeds is very diverse, but, also within the same species, competitivity varies depending on the timing of emergence; the case of *Alopecurus myosuroides* is typical. If it emerges in autumn, it forms a more vigorous plant with more tillers than a plant of the same species emerging in spring.[46] In the latter, competitivity is clearly lower. The existence of competitive hierarchies between weeds suggests the potential benefit of managing species composition in weed communities.[47] The most efficient method for managing the botanical composition of weed flora is crop rotation; alternating the crops differs the type, timing, and rhythm of the mechanical, chemical, and agronomic disturbances. No species gains a permanent advantage over the others, and the flora stays in equilibrium in the shelter of compensation phenomena and the development of resistance.[12] Soils under rotation usually present not only a smaller seed bank than those under monoculture, but also a greater species richness. From the agronomic viewpoint, the better floral situation is the balanced presence of all periodicity types of species, so that only a part of the seed bank is able to infest the actual crop. Moreover, a well-balanced flora allows a wider range of control solutions and makes interventions less urgent.

On the contrary, the continuous repetition of the same crops and, therefore, of the same disturbances heavily influences seed bank composition. For example, in the Po Valley of northeastern Italy the monoculture of corn has selected the spring and summer species, which now represent 75% of the seed bank.[20] With monoculture, specialist weeds are also favored. For example, wild oat germination and growth are regulated by the biochemical characteristics of soil cropped with wheat. Germinating wheat seeds can stimulate wild oats to emerge at the same time as the crop. This would appear to be an adaptive strategy to ensure germination at the moment most favorable to the growth of the weed.[48]

The botanical composition of weed flora can also be manipulated by varying the timing of crop sowing. In northern Italy, postponing the sowing of winter wheat from mid-October to the beginning of November changes the composition of the weed community. The presence of indifferent and autumn species, the most dangerous for the crop, is reduced while winter and spring species is increased. These species, emerging later than the crop, can develop only limited competition.

Containment of weed flora is also obtained with the "false seed bed" technique, or "weed strike". This means preparing the seed bed some 15 to 20 days before drilling. With normal weather conditions, most of the weed seeds in the upper soil horizon emerge, and seedlings can be eliminated by harrowing or with a herbicide treatment before sowing. With this technique, the number of seeds ready to germinate in the superficial layer of the soil is lowered; the crop can then emerge and grows in conditions of scarce infestation, acquiring an advantage that is often enough to allow it to suppress late emerging weeds without the need for other chemical treatments.

There are numerous possibilities for manipulating weed flora, but they often produce contrasting effects on different species. For this reason, sensitivity and attention to detail are required for organizing different interventions in a "weed control system".

WEED CONTROL

The goal of control practices is to reduce weed populations by chemical or nonchemical methods. Nonchemical methods can be divided into physical, biological, and ecological methods.

NONCHEMICAL WEED CONTROL

Physical Methods

In developing countries where labor is relatively cheap, hoeing and weeding by hand are still widely practiced. The labor requirement is very high for a crop weeded entirely by hand-hoeing; 200 to 400 human-hours per hectare are required for each hand weeding, and at least two weedings, are necessary.[49] Hand weeding is best adapted to small garden areas and to the removal of weed plants growing near to or within crop rows, where they are difficult to reach with mechanical tools. Hand removal, although seemingly an anachronism in this day and age, can pay off in preventing the establishment of pernicious weeds. For example, in the Po Valley of northern Italy, more careful farmers hand remove plants of *Abutilon theophrasti* when they find them in their fields for the first time.[50]

Mechanical tillage categories consist of primary, secondary, and cultivating tillage. The goals of these are different. Primary tillage moves and breaks up soil to reduce its strength and buries or mixes plant materials and fertilizers in the tilled layer. It is more aggressive, deeper, and leaves a rougher surface than secondary tillage. Secondary tillage provides additional pulverization, levels and firms the soil, closes air pockets, and kills weeds. The final secondary tillage operation is seed bed preparation. Cultivating tillage is a shallow postplanting tillage whose main purpose is to assist the crop by loosening the soil and/or eradicating the weeds. All mechanical tillage operations have some effect on the weeds, depending on the implement, timing, and type of weeds present.

The effect of primary tillages on weeds is mostly indirect: many weed seeds buried by the plow die, but those that remain dormant are a source of future weed infestation when they are once again turned up by the plow. In this way, the plow helps to perpetuate weed infestation.[51] The timing of plowing may be important. With perennial weeds, the best time is when food reserves are at their lowest.

Secondary tillages have a more direct effect against the weeds. In the context of weed control, secondary tillages are carried out to stimulate seed emergence and to eliminate emerged or emerging plants. With repeated shallow tillages, weed seed germination is stimulated by exposure to light, warmer temperatures, and a higher oxygen concentration. These mechanical interventions bring weed seeds to the soil surface, where they are also more susceptible to decay and predation. Each flush of weeds following tillage is killed by the subsequent tillage operation. This method is used in the "false seed bed technique" and in "stubble cleaning". Secondary tillage also includes the fallow season repeated tillage used to control perennial weeds. The chief purpose of cultivating perennial weeds is to kill them through starvation by attempting to prevent food manufacture and speed up the use of already stored reserves by encouraging their growth. Cutting off the tops of the vegetative propagules induces bud growth beneath by removing apical dominance. For best results, cultivation for perennial weed control must be properly timed. It is generally more effective in spring and summer than in autumn, probably because the regrowth of the plants tends to be faster during these seasons. Another positive effect of secondary tillages is that of bringing the vegetative propagules to the surface, which may then be damaged by high or low temperatures.

The main reason for cultivation of crops is weed control. With rotary hoe cultivators and other implements, it is possible to weed large areas of field crops rapidly and economically. These methods are nonselective and can thus impose restrictions on the planting arrangement of the crop. To allow the working of the machines, crops are sown in more or less widely spaced rows. The main disadvantage of cultivation is the difficulty or inability to control weeds near or within the crop rows. These must be eliminated by hand-hoeing or by the application of herbicides in narrow bands directly above the crop row. The combination of mechanical and chemical methods can give excellent weed control at low cost.

Finger weeders are becoming more widespread and can be used on thickly sown or broadcast crops, such as winter cereals. From trials carried out in Italy and Germany, the control of a complex weed community varied between 40 and 70%.[52,53] There are, however, marked differences between different groups of weeds; control of broad-leaved weeds is high (70 to 80%), while it is insufficient (10%) on grass-weeds, which are more deeply rooted and emerge from deeper layers of the soil. Efficacy can be improved by repeating the intervention in the opposite direction some days later. In both cases, some damage can occur to crop plants, but increasing the sowing rate and sowing depth can slightly compensate for such losses.

The length of the available season for cultivating is limited by the height of the crop and by the potential for damage to crop roots growing near the soil surface. In any event, late season cultivations are probably not necessary if competitive crop cultivars are used.

One significant problem with mechanical cultivation has been how to get a tractor onto the field to cultivate when the soil is wet. The combination of herbicide banding at sowing and mechanical cultivation significantly reduces the risks of farming. Chemical treatment, in fact, provides the control of the weeds along the rows, thus allowing a greater elasticity in the carrying out of the mechanical interventions.

Fire and Flame Weeding

The use of fire to destroy infesting plants is a long-known practice. Putting virgin ground under cultivation still includes previous burning, and stubble burning is commonly done after harvesting winter cereals. The main aim of the latter is the elimination of the straw, but a quota of the seeds present on the surface are also burnt. The seeds only slightly under the soil are not, however, devitalized. The burning of naturally matured weeds should be considered an expedient for destroying debris rather than a weed control measure.

Flame weeding is a technique for controlling weeds with the use of fire and of heat produced by the combustion of L.P.G. (liquefied petroleum gas). Flame weeding involves heating but not burning the weeds. The high temperatures generated by flame weed control equipment do not kill the plants through actual combustion but rather injure cells in leaves and stems. Heat kills living cells by coagulating protoplasm and inactivating enzymes.[54] The thermal death point for most plant cells is between 45 and 55°C for prolonged exposures.[55] Higher temperatures are necessary to kill cells in a shorter time. Another important variable is the stage of plant development; much shorter exposure to heat is obviously sufficient to devitalize seedlings compared to what is necessary for more developed plants. With currently available equipment, it is considered that a 1 second exposure should be enough to produce temperatures higher than 100°C throughout the aerial tissue of the plant.

Flame weeding can be used immediately before crop emergence or in its presence.[56] In the former, the treatment is carried out along the crop row for a width of 10 to 15 cm. It is a method particularly well adapted for crops with long germination (14 to 18 days), such as carrot, onion, leek, and, in certain cases, sugarbeet. Early seedbed preparation encourages a large proportion of weed seeds to germinate before crop emergence and allows weeds to be flamed without harming the crop. Weeds between rows can be removed by subsequent cultivation. A rapid hand weeding might be necessary to eliminate the few weeds within the crop row that escape flaming and cultivation. Flame weeding can also be used in the presence of the crop, working between rows or also directly on the crop when it has acquired a certain resistance to heat. When flame weeding is carried out between rows, the result is analogous to that obtainable with hoeing. However, flaming as a method of row weeding is mainly of interest when mechanical weeding cannot be done. Examples of this are on very strong soils, in severely drought-affected fields, or when the crop has a very shallow or sensitive root system, such as lettuce or cucumber.[57] Treatment of the entire surface, or selective flame weeding, appears more interesting. Some crops, such as maize and onion, are suitable for selective postemergence flaming, but it must be at a well-defined stage. For example, in corn, within-row flaming at postemergence is effective when it is done at the "match" stage (maize 3 to 4 cm tall).[58]

Flaming should be used to control small weeds and broad-leaved plants, as grasses are usually less sensitive, but, in practice, the efficacy of flaming also depends on the amount of weed emergence after treatment and on weather conditions. Flaming is very effective in hot and dry conditions and with a tilled soil surface, so reducing heat turbulence which may cause a reduction in the efficiency of the treatment.

The disadvantages of this method are that it is at least as energy intensive as the current mechanical and chemical control methods, timing is critical, and the driving speed is slow. The advantages are that there are no residual effects, no soil disturbance, and no induction of further emergence.

Flooding

Flooding kills plants by excluding air from their environment, Therefore, it is effective only when the plants are covered, and the roots completely surrounded with water. This weed control measure is often used for rice crops. With deep flooding, it is possible to control barnyard grass in paddy fields without suppressing the growth of the rice crop. According to Noda,[59] the depth and time of flooding in rice very much govern the level of barnyard grass infestation, especially in association with water temperatures. The importance of the high temperatures was pointed out also by McWhorter[60] on Johnsongrass (*Sorghum halepense*). Timely and thorough drainage of the water from the field also helps to control many aquatic weeds, including algae and ducksalad.

Solarization

Soil solarization is a method of heating moist soil by covering it with plastic film or mulch to trap solar radiation and magnify temperatures. This method destroys weed seeds in the soil in warmer climates.[61] In mulched soils, the temperature increase is mainly due to cutting heat loss by evaporation and to heat convection during the day. The dry seeds are resistant to temperatures as high as 120°C, while hydrated ones are killed at 50°C.[62] This is because less energy is required to damage the peptide chain configuration of proteins in the presence of water.

The response to solarization varies among weed species. Many annual species are sensitive, and most perennials are resistant. The latter, having large vegetative propagules, can emerge from deep soil layers, thus practically escaping the lethal temperatures in the upper layer. Other resistant species, such as some legumes and velvetleaf, have mostly impermeable seeds that are likely to be unaffected by high temperatures. It is worth noting that solarization is effective against imbibed but dormant seeds.[63]

Solarization is interesting because it is a fully integrated method of weed control, based on easily manipulated physical factors. However, its use is restricted to regions with intense sunlight and to limited areas such as nurseries and greenhouses and some vegetable crops. Solarization is also effective against a wide range of soil pathogens; therefore, when estimating its economic value, it must be considered an alternative not only to chemical weed control, but also to the use of other pesticide treatments and nitrogen fertilization. Plastic-mulched soil contains higher levels of soluble mineral nutrients released from organic material and heat-killed soil biota.[64]

The cost of solarization falls within the medium price range of soil disinfestation treatments. As technology advances, e.g., development of finer and stronger films, the use of photodegradable or biodegradable films or more efficient film-laying machinery, the overall costs should decrease.[65]

Biological Control

Weed control can be achieved with biological agents such as fungi, insects, etc. Two main approaches have been developed: classical biological or inoculative control and inundative or mass-exposure biological control.[66] Classical biological control relies on the introduction of a biological agent from the center of origin of the weed and its self-perpetuation on its host to reduce and keep populations at noneconomic levels. This classical approach is most economical and successful when used to reduce populations of introduced weeds that dominate large areas of uncultivated sites such as forests, pasture, rangeland, or aquatic habitats. Interesting results have been obtained with the leaf-eating insect *Chrysolina quadrigemina* (Coleoptera: Crysomelidae), which controlled *Hypericum perforatum,* and the boring insect *Cactoblastis cactorum,* which controlled *Opuntia triachantha*. Table 1 gives the ten most successful cases of biological weed control up to 1980.[67]

Table 1 The Top Ten Most Successful Cases of Biological Control Up to 1980

Rating	Plant	Insect	Place and Date
1	*Opuntia vulgaris* Miller	*Dactylopius ceylonicus* Green	India, 1795
2	*O. vulgaris*	*D. ceylonicus*	Sri Lanka, 1795
3	*O. dillenii* (Kergawler) Haworth	*D. opuntiae* Cockerell	India, 1926
4	*Carduus nutans* L.	*Rhinocyllus conicus* Froelich	Canada, 1968
5	*Hypericum perforatum* L.	*Chrysolina quadrigemina* Suffrian	USA, 1946
6	*O. vulgaris*	*D. ceylonicus*	South Africa, 1913
7	*O. elatior* (Miller)	*D. opuntiae*	Indonesia, 1935
8	*Opuntia* spp.	*D. opuntiae*	Madagascar, 1923
9	*O. tuna* (L.) Miller	*D. opuntiae*	Madagascar, 1928
10	*O. triacantha* (Willdenow) Sweet	*Cactoblastis cactorum* Bergroth	Nevis, West Indies, 1957

From Crawley, M. J., *Proceedings of the VII International Symposium on Biological Control of Weeds*, Rome, Italy, Istituto Sperimentale Patologia Vegetale, 1989, pp. 17–26.

The damage caused can be increased by improving weed acceptability to the insect. There are many possibilities, e.g., the application of nitrogen to chlorotic *Opuntia* spp. plants increased their attractiveness to *Cactoblastis cactorum*, which then more readily attacked and destroyed them.[68]

Herbicides may be used to retard the growth of a weed to synchronize it with the population dynamics of the insects; 2,4 D used on alligatorweed (*Alternanthera philoxeroides*) allowed the imported beetle *Agasicles hygrophila* to catch up and put the weed under greater stress. Cutting or pruning plants can also alter their susceptibility to insect attack, an approach that has also been successful with phytopathogenic fungi.[68]

The first example of the use of classical biological control with fungi was the introduction of the rust pathogen *Puccinia chondrillina* to control *Chondrilla juncea* in Australia at the beginning of the century, where it became dominant in wheat fields. A strain of this rust discovered in Italy had been shown to be specific and efficacious in the Mediterranean region. After its introduction in Australia, it rapidly acclimatized and spread to reduce greatly the narrow-leaved form of *Chondrilla juncea*. Research is now under way to discover other strains of rust effective against the large- and intermediate-leaved forms of the weed.[66] Other successful examples of biological control are the introduction of *Cercosporella ageratinae* from Jamaica into Hawaii to control *Ageratina riparia* (*Eupatorium riparium*), a weed of pastures, forests, and rangelands, and that of *Phragmidium violaceum*, a European rust pathogen of blackberries (*Rubus* spp.), which was introduced in Chile in 1973 to control *Rubus constrictus* and *R. ulmifolius*.[69]

Weeds can also be controlled by other biological agents. The Chinese grass carp (*Ctenopharyngodon idella*), commonly called the whistle amur, can be used to destroy aquatic vegetation in ditches and canals. This species, native to rivers in China and the Amur region of Siberia, is a voracious consumer of aquatic weeds; an adult fish can consume daily a quantity of weeds greater than its body weight. Adult grass carp feed on some 35 or more species of aquatic plants, are totally herbivorous, and suffer no competition from other fish.[70]

The use of livestock is another possibility. Goats prefer brush to herbaceous species, so they can be extremely effective in eliminating brush from pastures and are virtually not competitive to cattle or sheep for available forage when sufficient brush is present. In Mediterranean regions, geese are used to eliminate *Orobanche* spp. in fields of tobacco.

Self-perpetuation is the key to success in classical biological control.[71] The approach eliminates profit incentives for industry, so public sector support is necessary for discovery, development, and deployment.

The inundative biological control approach, also called bioherbicide tactic, involves the use of indigenous pathogens that have coevolved with their hosts. Viable fungus spores, formulated either as a wettable powder or as a suspension, are diluted and applied each season using the standard spray equipment for chemical herbicides. These products, called mycoherbicides, are developed from pathogens that normally incite disease at an endemic level in specific weed populations. The innate deficiency of the pathogen to disseminate normally inhibits epidemic development. When the pathogen is applied as inundative inoculum, it infects and kills the weeds within 3 to 5 weeks and, after the death of the host plant, is reduced to background level by natural constraints.[72]

Consequently, there is little or no residual weed control from season to season after the use of a mycoherbicide, particularly if it is an aerial pathogen.

Two mycoherbicides are at present registered for commercial use in North America. These are Devine and Collego, used respectively against *Morrenia odorata* in citrus and orchards and *Aeschynomene virginica* in rice and soybean. Devine is a liquid concentrate of chlamidospores of a pathotype of *Phytophthora palmivora*, native to Florida, while Collego is based on *Colleotrichum gloeosporioides* f. sp. *aeschynomene*, an anthracnose-inciting pathogen. The control obtained with these mycoherbicides is similar to that of chemical products.[73] Two other mycoherbicides, Casst and BioMal are undergoing evaluation for possible registration in the near future for controlling *Cassia obtusifolia* and *Malva pusilla,* respectively.

There are several similarities between inundative biological control and chemical control, but the former is usually more specific. Although often claimed as an advantage of biological control, specificity can be commercially and agronomically undesirable if the pathogen is active against a single weed species and consequently has limited market potential. However, studies are being carried out to evaluate the possibility of producing broad-spectrum mycoherbicides. This is based on the observation that subspecies or special forms of the same fungus are effective against different weeds. Molecular genetic techniques now available have great potential for improving strains for biological herbicide development.[74] The actual costs involved in applying mycoherbicides are similar to those for chemical herbicides, but development costs are much lower.[75]

Pathogen fungi of weeds are particularly interesting because of the possibility of high host specificity and because they are usually much less expensive to produce than insects.[76]

However, the inundative tactic has also been attempted with insects. For example, Frick and Chandler[77] demonstrated that properly timed, periodic, inundative releases of laboratory-reared *Bactra verutana* larvae and adults suppressed *Cyperus rotundus* infesting cotton in Mississippi. *Bactra* is not naturally effective in suppressing this weed because the population remains at a low level until late in the season. Studies on the feasibility of rearing and releasing weed-eating insects as a means of enhancing their action is still in its infancy.[68]

Biologic control with plant pathogens offers a potentially feasible alternative and supplement to chemical weed control without negative nontarget effects. Biologic control must however be integrated with other control methods because (1) the agents do not provide the desired level of control; (2) biological control is inherently narrow-spectrum, and the target weed may be only a component of a mixed infestation; (3) an intensive crop production system may be detrimental to the biological self-perpetuation of the control agent; and (4) economic considerations may not justify similar control strategies. The combination of biological control with other methods and integration into modern crop production systems are therefore essential.

Cover Crop

Soils cultivated with annual crops remain bare or with little ground cover for considerable periods of the year, providing opportunities for weed establishment and growth.[11] Sowing a cover crop during this period is a way to suppress weeds.

The cover crop limits the development of the weeds during its vegetative cycle and, after its destruction, inhibits the germination and growth of many weed species through the leaching of phytotoxic substances from plant residues. It has also been demonstrated that, in some cases, soil microbial activity can convert allelochemicals released from the cover crop into derivatives more toxic to weeds. This biomagnification may be very significant in helping to explain allelopathic weed suppression under field conditions.[78,79] In addition, the physical effect of the mulch and the lack of soil tillage helps to suppress weeds in no-till cropping systems.

The sowing of the cover crop may take place not only after the crop has been harvested, directly on the stubble or with limited tillage, but also within the crop. In this case, the time of sowing the cover crop must be carefully chosen. The cover crop is then killed before planting the following cash crop, normally with a foliar applied herbicide or, more rarely, by mowing. Sometimes, cover crops such as sorghum or spring oats, which are killed by cold in winter, are used. The cash crop is then sown without tillage through the cover crop that lies as mulch on the soil surface. Cover crops of wheat, barley, oats, rye, grain sorghum, and sudangrass have been used effectively to suppress weeds, primarily annual broad-leaved species. Several winter annual legume crops have also proved effective. Trials carried out by Worsham[80] show that the allelopathic suppression effect is usually adequate only for the first few weeks after the killing of the cover crop, which may eliminate the need for preemergence soil-applied herbicides when planting the cash crop. Late season postemergence treatments are still required. This is potentially a great advantage, as preemergence herbicides generally have a much higher groundwater contamination potential than postemergence ones, according to Weber's ranking index.[81]

The potential of no-till, allelopathic cover crop systems in conserving soil resources, with diminishing requirements for purchased herbicides, suggests that these systems can enhance agriculture sustainability. Disadvantages and problems may arise when cover crops are used: e.g., establishment costs of the cover crop, difficulties in killing (especially with legumes), an increase of certain insects and diseases, lowering of soil temperature in spring, depletion of soil moisture in spring, and nutrient capture by the cover crop. More information is needed to develop a better understanding of nutrient and water relations in these systems.

CHEMICAL WEED CONTROL

Biorational Approach

Many lines of research have been explored to develop new herbicides with better toxicologic, environmental, and agronomic characteristics to reduce dosages and improve safety for operators, consumers, and the environment. Shifts in weed populations with a predominance of species that are more closely related to the crop that they infest are resulting in a greater need for more selective and efficacious herbicides, therefore new strategies in herbicide development are required. Also, the number of chemicals that must be screened to develop a new herbicide with traditional herbicide synthesis is increasing to levels that will make the method economically unjustified. An alternative strategy is given by the "biorational approach". This is based on the understanding of the biochemistry of natural herbicidal molecules, the site of action, and the biophysical plant physiology of the target plant. The design of herbicides around specific sites of action is an example of the biorational approach, which is being pursued by several companies.[82] Another possibility is to seek new phytotoxic compounds from natural sources. More knowledge about their chemistry and site of action in comparison to synthesized herbicides could give an insight into the potential of target sites as yet unutilized by commercial herbicides, perhaps providing good hints for biorational design around these sites. A comparison of known sites of action of common herbicides and natural products shows that there is little overlap,[83] suggesting that we have only begun to explore potential sites of herbicide action. This approach has been successful with insecticides (e.g., pyrethroids). Phytotoxic natural compounds are primarily products of either plants or microorganisms. Allelochemicals produced by higher plants have limited phytotoxicity, are generally nonspecific, and their potential use is frequently limited by cost.[82] Microbially reproduced phytotoxins offer greater potentiality as herbicides. These compounds have considerable toxicity, show evidence of host and nonhost specificity, and are relatively inexpensive to produce by fermentation. In addition, compared to mycoherbicides, they are easier to store, formulate, and apply without risk to the environment, are active at low concentrations and are independent of environmental factors. Although microbial phytotoxin herbicides can be toxic to livestock or humans (e.g., aflatoxin), the vast majority of natural compounds pose little health hazard and would in all likelihood be environmentally safe. In fact, the environmental half-life of most natural compounds is much shorter than the synthetic ones.[84]

Plant pathologists divide phytotoxins produced by microorganisms into host-specific and non–host-specific toxins. Host-specific toxins affect only the species or, in some cases, the variety of a species infected by the toxin-producing pathogen. They are all produced by fungi, and very few of them have been chemically characterized. Because of their extreme selectivity, research should focus on phytotoxins from pathogens that affect weed species with high economic impact, such as *Abutilon theophrasti* and *Sorghum halepense*. Once a host-specific toxin for a weed is found, structure–activity research could manipulate its selectivity and efficacy to maximize its agricultural usefulness. Research in this sector has not yet produced practical results.[82]

Plant pathologists define non-host-specific toxins as those that are toxic to plant species that are not infected by the producing microorganism in nature. Both fungi and bacteria produce this group of toxins. Although termed nonspecific, for weed scientists, many of these compounds are quite selective, and several new herbicides have been developed from this group of chemically characterized toxins. Bialophos, a product of *Streptomyces viridochromogenes*, was the first microbial herbicide to be produced commercially. It is a nonselective herbicide that kills both grass and broad-leaved weeds and is also effective on perennials. It has low mammalian toxicity and a relatively short half-life. Bialophos in higher plants is readily metabolized into phosphino-thricin (PPT), the active compound. Bialophos was the basis for developing a synthetic analogue of PPT, namely gluphosinate. This compound is a strong inhibitor of the enzyme glutamine-sinthetase, which is involved in the assimilation of ammonia to form essential amino acids, a previously unknown mode of action.[83] Microbial toxins also demonstrate that there are numerous potential sites of action that have not yet been exploited by the herbicide industry. Another non–host-specific toxin that shows potential for selective weed control in maize, soybean, and cruciferous crops is tentoxin, which is produced by the fungus *Alternaria alternata*. Current interest is focused on the production of an analogue, although it seems difficult and costly to synthesize.[85] Many other non–host-specific toxins are also being studied, proving that interest in microbial products as herbicides is growing. As well as environmental advantages, these products offer new compounds, a high efficacy level, and new and better selectivity. The promised diversity of chemical structure, coupled with advances in biotechnology will make microbial phytotoxins a worthwhile area for future herbicide research.

Improvement of Herbicide Technology

Improvements in formulation, design, and use of adjuvants offer ample opportunities to reduce environmental impact and, often, worker exposure. Reduced environmental impact can be obtained through (1) the separation of enantiomers and dosage reductions from the use of adjuvants and (2) the use of controlled release formulations.

Many chemicals used to interact with biological systems show different activity depending on whether the "right-" or the "left-" handed form of a chiral molecule is used. In fact, the biological receptors are themselves formed by chiral molecules and have a distinctive shape.[86] Many, if not most, synthetic herbicides are marketed in racemic form, which is a mixture of the two enantiomers. If only one of them is active, a double quantity must be applied. For instance, the postemergence herbicidal activity of the arylphenoxypropanoates is largely due to the R-enantiomer, while the S-enantiomer has little activity.[87] The active enantiomers of many herbicides belonging to the aryloxycarboxylic acid and aryloxyphenoxypropanoate families are available commercially.

Controlled-release technology can improve safety for the user, crop, and environment. The soil mobility of some herbicides has caused concern because of their potential to contaminate groundwater. By incorporating the herbicide with appropriate polymeric materials, the rate of leaching from the root zone can be reduced, allowing for greater uptake by the plants and less leaching in the case of heavy rainfall.[88] Seaman[89] showed that microencapsulation of EPTC (S-ethyl-dipropyl-carbamothioate), a selective presowing herbicide for corn, reduces volatility and thereby dosage. Furthermore, herbicide incorporation into the soil, necessary for this very volatile chemical, can be delayed, thus allowing a more flexible use of machinery. With this technology, the herbicide is also released slowly, providing weed control for a much longer period than that obtained with classic formulations. The controlled-release formulations present greater safety for the operator as the active ingredient is less available for direct contact with the user. Skin and eye irritations are reduced significantly compared with classical formulations.

Adjuvants are compounds that facilitate herbicide action or modify characteristics of herbicide formulations or spray solutions. They include surfactants, paraffin or vegetable oils, methylated fatty acids, and fertilizers. The use of adjuvants is often recommended to improve the efficacy of herbicides against specific weeds. Ammonium sulfate increased the efficacy of sethoxidim on wild oats; the rate required to reduce dry weight by 50% was reduced from 0.08 kg a.i. ha^{-1} to 0.07 Kg a.i. ha^{-1} in the presence of added ammonium sulfate.[90] Nicosulfuron activity can be enhanced four-fold against *Setaria* spp. and ten-fold against *Echinochloa* spp. with a nonionic surfactant.[91]

It is evident that careful manipulation of formulations and judicious selection of adjuvants can help in maximizing the expression of biological activity resulting in significant dose reductions.[92]

Crops Resistant to Herbicides

Recent advances in the understanding of the biochemical and physiological factors that confer resistance to specific herbicides and of the genetic basis of these factors, coupled to progress in biotechnology, have created the basis for developing herbicide-tolerant crop cultivars (HTCs). HTCs can be developed genetically through classic plant breeding techniques or genetic engineering. The former have been successful in the release of triazine-resistant varieties of rapeseed (*Brassica napus*). Beversdorf et al.[93] were able to transfer the cytoplasmatically inherited resistence to atrazine from bird's rape to rapeseed by means of backcrossing.

HTC could also be obtained by genetic engineering by means of two basic mechanisms: the altered target and detoxification mechanisms.[94] The altered target mechanism is applicable when the biochemical site of action of a specific herbicide has been previously identified. The gene encoding the herbicide target can be isolated and cloned from a microbial, fungal, or plant source and introduced into the crop. An alternative strategy for developing herbicide resistant crops is to identify and isolate genes encoding enzymes involved in herbicide detoxification. The genetically modified organisms are defined transgenic plants.

The release of HTCs could have both positive and negative impacts from the environmental and agronomic points of view. If the technology encourages the use of nonresidual herbicides, nontoxic to mammals and nontarget organisms and not likely to move into the groundwater, concern for the environment could be reduced. The main agronomic advantages are an increased selectivity with subsequent reduction of herbicide injury, use of certain herbicides on normally susceptible crops, and reduced risk of crop damage from carry-over effects. Moreover, the development of crop cultivars resistant to postemergence herbicides could encourage the use of the two fundamental concepts of IWMS, i.e., economic thresholds and critical period.

The use of HTCs could also lead to the appearance of agronomic problems. Spontaneous hybridization between crops and wild plants may create "superweeds" as many engineered traits could greatly enhance weed vigor or provide opportunity for wild plants to become weeds in arable lands.[95] A crop itself could become a weed problem. Crop breeding is of concern mainly when weeds are closely related to the HTC.[96]

This technology causes much concern, not only from the environmental or agronomic points of view, but also the ethical.[97] A technology is neither good nor bad, it is how it is used and exploited that makes it good or bad.[98] Nonetheless, some corn hybrids that are resistant to imidazolinones have recently been registered in the U.S.

CRITERIA FOR HERBICIDE USE IN THE IWM CONTEXT

The use of herbicides within IWM is driven by two fundamental principles: Selection Pressure (SP) applied to the weed flora and the flexibility of control. The higher the SP, the more rapidly are the resistant types selected. To avoid this shifting, which is always dangerous, it is necessary to organize a chemical weed control with a low SP. There are a number of ways to reduce SP without losing effective weed control: choice of herbicides with low persistence, combination of band-treatment with mechanical hoeing, rotation of herbicides, use of herbicide mixtures, and use of postemergence weed control. The latter eliminates only the weeds present and does not generally interfere with those that emerge later in the season. SP is therefore applied only to a certain part of the weed flora, that which effectively causes damage.

Numerous studies have shown that, to limit crop loss to an acceptable level, it is sufficient that the weeds are eliminated during an interval defined as the critical period (CP).[99] The length of this depends upon the species, cultivar, and density of the crop, the weed species and their density, and the environmental conditions.[100] For most summer crops, this period varies roughly from 2 to 8 weeks after the crop emergence.[101]

The CP is a very important concept within IWM, because it allows the establishment not only of the timing of weed control, but also the period during which the weed control must be effective to avoid yield losses due to weeds that may emerge later in the season. Although the duration of

CP can be affected by numerous factors, an early assessment of this interval will provide a guide for future studies in alternative methods of weed control and crop–weed interference.

Flexibility is the second principle on which weed control within IWM is based. Flexible weed control measures are based on the knowledge of actual and potential weed densities and their economic thresholds, rather than on a routine treatment.[102] Within this logic, the treatment must be carried out only when it is of use or, in other words, when the infestation density is higher than the economic threshold (ET). ET is defined as the weed density at which the cost of control equals the value of crop loss if no control action is taken. The ET is a fundamental concept in weed population management, which rejects weed eradication in favor of the regulation of weed populations at an economically optimum level. Prediction of the economic damage that a given infestation can produce is basic for ET. Computer models have been developed to help weed management decision making. Recourse to models is necessary because weeds grow in mixed populations and competitive ability differs from species to species, as does the response to various herbicide treatments. An example of these decision-making models is the program HERB,[103] which is probably the most widely used. A few other models have been tested, but the possibilities of limiting the use of the herbicides are consistent. A study carried out in Colorado[102] showed that flexible strategies required 64% less herbicide to control weeds in corn. Similarly, a study in Germany revealed that between 20 and 55% of fixed weed control measures were uneconomical.[104]

The models used in decision-making generally refer to a single crop year and do not include a cost factor associated with possible increases in the soil seed-bank.[105] Cousens et al. introduced the economic optimum threshold (EOT) to include the impact of seed-bank dynamics on the long-term profitability of weed management decisions. EOT has been implemented in some models (e.g., WEEDSIM),[106] but more knowledge is needed to include evaluation of the effect of sub-threshold weed densities on the seed-bank and, therefore, of the economics in the medium and long-term of weed control treatments. The estimated values for EOT are generally 3 to 8 times lower than those of ET, and, for some very competitive weeds which have persistent seeds, the threshold is practically equal to zero. Sattin et al.,[50] working in the Po Valley of northern Italy, demonstrated that densities of velvetleaf that did not affect corn yield in the current year could produce enough seeds to necessitate weed control in future crops for several years. They concluded that all velvetleaf plants should be controlled, so that no seeds could be produced.

The main aim of the threshold is to optimize the treatment and its economic evaluation to maximize the net return of the control. Currently, in cultivated fields, there are fairly substantial seed-banks with a simplified and specialized flora. Under these conditions, recourse to weed control is unavoidable. With the application of IWM, however, a reequilibrium of the flora and a reduction of the seed-bank can be forecast, and this will probably permit a substantial reduction in the recourse to chemical methods.

In any event, use of thresholds is of great importance, especially if it is considered that, with the new herbicide families, almost all of the principal crops, (i.e., soybean, wheat, corn, and sugarbeet), can be weed-treated entirely at postemergence, therefore allowing evaluation, case by case, of the need for treatments.

Although, in practice, the possibility of limiting the use of chemical weed control by using thresholds is limited, the maieutic value of this concept must be underlined; that is, it forces the farmer into an active, thoughtful, and critical participation in operative choices and therefore increases professionalism.

REFERENCES

1. Shaw, W.C. "Integrated weed management systems technology for pest management," *Weed Sci.* 30 (suppl. 1):2-12 (1982).
2. Swanton, C.J. and S.F. Weise. "Integrated weed management: the rationale and the approach," *Weed Tech.* 5:657-663 (1991).
3. Walker, R.H. and G.A. Buchanan. "Crop manipulation in integrated weed management systems," *Weed Sci.* 30 (suppl. 1):17-24 (1982).
4. Thill, D.C., J.M. Lish, R.H. Callihan, and E.J. Bechinski. "Integrated weed management. A component of integrated pest management: a critical review," *Weed Tech.* 5:648-656 (1991).

5. Tanner, J.W., C.J. Gardener, N.C. Stoskopf, and E. Reinbergs. "Some observations on upright-leaf type small grains," *Can. J. Plant Sci.* 46:690 (1966).

6. Lotz, L.A.P., R.M.W. Groeneveld, and N.A.M.A. de Groot. "Potential for reducing herbicide inputs in sugarbeet by selecting early closing cultivars," *Proceedings of the Brighton Crop Protection Conference — Weeds* (1991), pp.1241-1248.

7. Putnam, A.R. and W.B. Duke. "Biological suppression of weeds: evidence for allelopathy in accessions of cucumber," *Science* 185:370-372 (1974).

8. Massantini, F., F. Caporali, and G. Zellini. "Evidence for allelopathic control of weeds in lines of soybean," *Proceedings of the EWRS Symposium. Methods of Weed Control and their Integration,* (1977), pp. 23-28.

9. Fay, P.K. and W.B. Duke. "An assessment of the allelopathic potential in *Avena* germplasm," *Weed Sci.* 25:224-228 (1977).

10. Lockerman, R.H. and A.R. Putnam. "Evolution of allelopathic cucumbers (*Cucumis sativus*) as an aid to weed control," *Weed Sci.* 27:54-57 (1979).

11. Putnam, A.R. "Allelopathy: problems and opportunities in weed management," in *Weed management in Agroecosystems: Ecological approaches*, M.A. Altieri, and M. Liebman, Eds. (Boca Raton, FL: CRC Press, 1988), pp. 77-88.

12. Liebman, M. and R.R. Janke. "Sustainable weed management practices," in *Sustainable Agriculture in Temperate Zones.* C.A. Francis, C.B. Flora, and L.D. King, Eds. (Chichester, U.K.: John Wiley and Sons, 1990), pp. 111-143.

13. Appleby, A.P., P.D. Olson, and D.R. Colbert. "Winter wheat yield reduction from interference by Italian ryegrass," *Agron. J.* 68:463-466 (1976).

14. Chisaka, H. "Weed damage to crops: yield loss due to weed competition," in *Integrated Control of Weeds*, J.D. Fryer and S. Matsunaka, Eds. (Tokyo, Japan: University of Tokyo Press, 1977), pp. 1-16.

15. Patterson, D.T. and E.P. Flint. "Comparative water relations, photosynthesis and growth of soybean (*Glycine max*) and seven associated weeds," *Weed Sci.* 31:318-323 (1983).

16. Berti, A., G. Zanin, M. Sattin, and L. Toniolo. "Utilisation de l'eau chez *Solanum nigrum* L. et *Amaranthus cruentus* L.," *VIIIème Colloque International sur la Biologie, l'Ecologie et la Systématique des Mauvaises Herbes* (Paris, France: ANPP, 1988), pp. 599-607.

17. Wiese, A.F. and C.W. Vandiver. "Soil moisture effects on competitive ability of weeds," *Weed Sci.*, 18:518-519 (1970).

18. Geddes, R.D., H.D. Scott, and L.R. Oliver. "Growth and water use by common cocklebur (*Xanthium pennsylvanicum*) and soybean (*Glycine max*) under field conditions," *Weed Sci.* 27:206-212 (1979).

19. Patterson, D.T. "Growth and water relations of cotton (*Gossypium hirsutum*), spurred anoda (*Anoda cristata*) and velvetleaf (*Abutilon theophrasti*) during simulated drought and recovery," *Weed Sci.* 36:318-324 (1988).

20. Zanin, G., G. Mosca, and P. Catizone. "A profile of the potential flora in maize fields of the Po Valley," *Weed Res.*, 32:407-418 (1992).

21. Fail, H. "The effect of rotary cultivation on the rhizomatous weeds," *J. Agric. Eng. Res.* 1:3-15 (1954).

22. Harper, J.L. and D. Gajic. "Experimental studies of the mortality and plasticity of a weed," *Weed Res.* 1:91-104 (1961).

23. Palmblad, I.G. "Competition in experimental studies on populations of weeds with emphasis on the regulation of population size," *Ecology* 49:26-34 (1968).

24. Kelly, A.D. and V.F. Burns. Dissemination of weed seeds by irrigation water. *Weed Sci.* 23, 486-493 (1975).

25. Harmon, G.W. and F.D. Keim. "The percentage and viability of weed seeds recovered in the feces of farm animals and their longevity when buried in manure," *J. Am. Soc. Agron.* 26:762-767 (1934).

26. Stoker, G.L., D.C. Tingey, and R.J. Evans. "The effect of different methods of storing chicken manure on the viability of certain weed seeds," *J. Am. Soc. Agron.* 26:390-397 (1934).

27. Roberts, H.A. "Seed banks in soils," *Adv. Appl. Biol.* 6:1-55 (1981).

28. Harper, J.L. "The ecological significance of dormancy," *Proceedings IV International Congress Crop Protection,* Hamburg (1957), pp. 415-420.

29. Roberts, H.A. "Studies on the weeds of vegetable crops. II. Effect of six years of cropping on the weed seeds in the soil," *J. Ecol.* 50:803-813, (1962).

30. Chancellor, R.J. "The manipulation of weed behaviour for control purposes," *Philos. Trans. R. Soc.* London, *Ser.* B 295:103-110 (1981).

31. Eplee, R.E. "Ethylene: a witchweed seed germination stimulant," *Weed Sci.* 23:433-436 (1975).

32. Egley, G.H. "Stimulation of weed seed germination in soil," *Rev. Weed Sci.* 2:67-89 (1986).

33. Sexsmith, J.J. and U.J. Pittman. "Effects of nitrogen fertilizers on germination and stand of wild oats," *Weeds* 11:99-101 (1963).

34. Hurtt, W. and R.B. Taylorson. "Chemical manipulation of weed emergence," *Weed Res.* 26:259-267 (1986).

35. Popay, A.I. and E.H. Roberts. "Ecology of *Capsella bursa-pastoris* (L.) Medik. and *Senecio vulgaris* L. in relation to germination behaviour," *J. Ecol.* 58:123-139 (1970).

36. Fay, P.K. and R.S. Gorecki. "Stimulating germination of dormant wild oat (*Avena fatua*) seed with sodium azide," *Weed Sci.* 26:323-326 (1978).

37. Johnson, A.N., G. Rosebery, and C. Parker. "A novel approach to *Striga* and *Orobanche* control using synthetic germination stimulants," *Weed Res.* 16:223-227 (1976).

38. Cook, C.E., L.P. Whichard, B. Turner, M.E. Wall, and G.H. Egley. "Germination of witchweed (*Striga lutea* Lour.): isolation and properties of a potent stimulant," *Science* 154:1189-1190 (1966).

39. Wilson, R.G. "Biology of weed seeds in soils," in *Weed Management in Agroecosystems: Ecological Approaches*, M.A. Altieri and M. Liebman, Eds., (Boca Raton, FL: CRC Press, 1988), pp. 25-39.

40. Louda, S.M. "Predation in the dynamics of seed regeneration," in *Ecology of Soil Seed Banks*, M.A. Leck, V.T. Parker, and R.L. Simpson, Eds. (San Diego, CA: Academic Press Inc., 1989), pp. 25-51.

41. McIntyre, G.I. "The correlative inhibition of bud growth in perennial weeds: a nutritional perspective," *Rev. Weed Sci.* 5:27-48 (1990).

42. McIntyre, G.I. and A.I. Hsiao. "Influence of nitrogen and humidity on rhizome bud growth and glyphosate translocation in quackgrass (*Agropyron repens*)," *Weed Sci.* 30:655-660 (1982).

43. Chancellor, R.J. "The effect of 2-chloroethylphosphonic acid and chlorflurecolmethyl upon the sprouting of *Agropyron repens* (L.) Beauv. rhizomes," *Proceedings of the 10th British Weed Control Conference* (1970), pp. 254-260.

44. Fawcett, R.S. and F.W. Slife. "Effects of field application of nitrate on weed seed germination and dormancy," *Weed Sci.* 26:594-596 (1978).

45. Schimpf, D.J. and I.G. Palmblad. "Germination response of weed seeds to soil nitrate and ammonium with and without simulated overwintering," *Weed Sci.* 28:190-193 (1980).

46. Naylor, R.E.L. "Aspects of the population dynamics of the weed *Alopecurus myosuroides* Huds. in winter cereal crops," *J. Appl. Ecol.* 9:127-139 (1972).

47. Roush, M.L. and S.R. Radosevich. "Relationships between growth and competitiveness of four annual weeds," *J. Appl. Ecol.* 22:895-905 (1985).

48. Purvis, C.E. "Allelopathy: a new direction in weed control," *Plant Prot. Q.* 5:55-59 (1990).

49. Auld, B.A., K.M. Menz, and C.A. Tisdell. *Weed Control Economics.* (San Diego, CA: Academic Press Inc., 1987), pp. 177.

50. Sattin, M., G. Zanin, and A. Berti. "Case history for weed competition/population ecology: velvetleaf (*Abutilon theophrasti*) in corn (*Zea mays*)," *Weed Tech.* 1:213-219 (1992).

51. Cavers, P.B. and D.L. Benoit. "Seed banks in arable land," in *Ecology of Soil Seed Banks*, M.A. Leck, V.T. Parker, and R.L. Simpson, Eds. (San Diego, CA: Academic Press Inc.,1989), pp. 309-328.

52. Bräutigam, V. "Mechanische Beikrautregulierung im Getreide mit Striegel und netzegge nach verschiedener grundbodenbearbeitung," in *III International Conference on Non-chemical Weed Control,* Linz October 10-12, 1989, Bundesanstalt für Agrarbiologie, Linz, pp. 65-78 (1990).

53. Covarelli, G. and U. Bonciarelli. "Possibilità e limiti della sarchiatura meccanica del frumento," in *Atti S.I.L.M. Il controllo della vegetazione infestante il frumento* (Perugia, Italy: Guerra Guru, 1991), pp. 232-244.

54. Ellwanger, T.C., S.W. Bingham, W.E. Chappell, and S.A. Tolin. "Cytological effects of ultra-high temperatures on corn," *Weed Sci.* 21:299-303 (1973).

55. Muzik, T.J. *Weed Biology and Control*, (McGraw-Hill, New York, 1970), pp. 1-273.

56. Trouilloud, M. "Le deshérbage thermique," *15ème Conférence du Columa*, Paris, France (1992), pp. 283-289.

57. Mattsson, B., C. Nylander, and J. Ascard. "Comparison of seven inter-row weeders," *III International Conference on Non-chemical Weed Control*, Linz October 10-12, 1989, Bundesanstalt für Agrarbiologie, Linz, pp. 91-107 (1990).

58. Geier, B. and H. Vogtmann. "The multiple row brush hoe. A new tool for mechanical weed control," *International Federation of Organic Agriculture Movements (IFOAM) Bulletin* 1:4-6 (1987).

59. Noda, K. "Integrated weed control in rice," in *Integrated Control of Weeds*, J.D. Fryer. and S. Matsunaka, Eds., (Tokyo, Japan: University of Tokyo Press, 1977), pp. 17-44.

60. Mc Whorter, C.G. "Flooding for Johnsongrass control," *Weed Sci.* 20:238-241 (1972).

61. Rubin, B. and A. Benjamin. "Solar heating of the soil: effects on weed control and on soil-incorporated herbicides," *Weed Sci.* 31:819-825 (1983).

62. Rubin, B. and A. Benjamin. "Solar heating of the soil: involvement of environmental factors in the weed control process," *Weed Sci.* 32:138-142, (1984).

63. Horowitz, M., Y. Regev, and G. Herzlinger. "Solarization for weed control," *Weed Sci.* 31:170-179 (1983).

64. Stapleton, J.J., J. Quick, and J.E. DeVay. "Soil solarization: effect on soil properties, crop fertilization and plant growth," *Soil Biol. Biochem.* 17:369-373 (1985).

65. Stapleton, J.J. and J.E. DeVay. "Soil solarisation: a non-chemical approach for management of plant pathogens and pests," *Crop Prot.* 5:190-198 (1986).

66. Charudattan, R. "Pathogens with potential for weed control," in *Microbes and Microbial Products as Herbicides*, R.E. Hoagland, Ed. ACS Symposium Series (Washington, DC: American Chemical Society, 1990), pp. 132-154.

67. Crawley, M.J. "Plant life-history and the success of weed biological control projects," in *Proceedings of the VII International Symposium on Biological Control of Weeds*, E.S. Delfosse, Ed., (Rome, Italy: Istituto Sperimentale Patologia Vegetale, 1989), pp. 17-26.

68. Andres, L.A. "Integrating weed biological control agents into a pest-management program," *Weed Sci.*, 30 (suppl.):25-30 (1982).

69. Emge, R.G. and G.E. Templeton. "Biological control of weeds with plant pathogens," in *Biological Control in Crop Production*, G.C. Papavizas, Ed., (Totowa, NJ: BARC Symposium no. 5, 1981), pp.219-226.

70. Crafts, A.L. *Modern Weed Control*, (Davis, CA: University of California Press, 1975), pp. 1-440.

71. Templeton, G.E., R.J. Smith, Jr., and D.O. Tebeest. "Progress and potential of weed control with mycoherbicides," *Rev. Weed Sci.* 2:3-14 (1986).

72. Charudattan, R. "Assessment of efficacy of mycoherbicide candidates," in *Proceedings of the VII International Symposium on Biological Control of Weeds*, E.S. Delfosse, Ed. (Rome, Italy: Istituto Sperimentale Patologia Vegetale, 1989), pp. 455-464.

73. Smith, R.J., Jr. "Biological control of northern jointvetch (*Aeschynomene virginica*) in rice (*Oryza sativa*) and soybean (*Glycine max*). A researcher's view," *Weed Sci.* 34 (suppl.):17-23, (1986).

74. Templeton, G.E. "Weed control with pathogens. Future needs and directions," in *Microbes and Microbial Products as Herbicides*, R.E. Hoagland, Ed., ACS Symposium series, (Washington, DC: American Chemical Society, 1990), pp. 321-329.

75. Templeton, G.E. "Mycoherbicide research at the University of Arkansas. Past, present and future," *Weed Sci.* 34 (suppl.):35-37 (1986).

76. Charudattan, R. and C.J. DeLoach, Jr. "Management of pathogens and insects for weed control in agroecosystems," in *Weed Management in Agroecosystems: Ecological Approaches*, M.A. Altieri, and M. Liebman, Eds. (Boca Raton, FL: CRC Press, 1988), pp. 245-264.

77. Frick, K.E. and J.M. Chandler. "Augmenting the moth (*Bactra verutana*) in field plots for early-season suppression of purple nutsedge (*Cyperus rotundus*)," *Weed Sci.*, 26:703-710 (1978).

78. Liebl, R.A. and A.D. Worsham. "Inhibition of pitted morning glory (*Ipomoea lacunosa* L.) and certain other weed species by phytotoxic component of wheat (*Triticum aestivum* L.) straw," *J. Chem. Ecol.* 9:1027-1043 (1983).

79. Nair, M.G., C.J. Witenack, and A.R. Putnam. "2,2'-oxo-1-1'-azobenzene: a microbially trans-formed allelochemical from 2,3-benzoxazolinone," *J. Chem. Ecol.* 16:353-364 (1990).

80. Worsham, A.D. "Allelopathic cover crop to reduce herbicide input," *Proceedings of the Southern Weed Science Society* (1991), pp. 58-69.

81. Weber, J.B. "Potential for groundwater contamination from selected herbicides: a herbicide/soil ranking system," *Proc. Southern Weed Science Society,* vol. 44, San Antonio, Texas, (1990) pp. 45-57.

82. Duke, S.O. "Microbially produced phytotoxins as herbicides. A perspective," in *The Science of Allelopathy*, A.R. Putnam and C.S. Tang, Eds. (Chichester, U.K.: John Wiley and Sons Inc., 1986), pp. 287-304.

83. Duke, S.O. and J. Lydon. "Herbicides from natural compounds," *Weed Tech.* 1:122-128 (1987).

84. Duke, S.O. "Naturally occurring chemical compounds as herbicides," *Rev. Weed Sci.* 2:15-44 (1986).

85. Lax, A.R. and H.S. Sheperd. "Tentoxin: a cyclic tetrapeptide having potential herbicidal usage," *Am. Chem. Soc. Symp. Series* 380:24-34 (1988).

86. Chaloner, P.A. "Chirality and biological activity," *Proceedings of the Brighton Crop Protection Conference — Weeds* (1989), pp. 697-706.

87. Cartwright, D. "The synthesis, stability and biological activity of the enantiomers of pyridyloxy-phenoxypropionates," *Proceedings of the Brighton Crop Protection Conference — Weeds* (1989), pp. 707-716.

88. Petersen, B.B. and P.J. Shea. "Microencapsulated alachlor and its behaviour on wheat (*Triticum aestivum*) straw," *Weed Sci.* 37:719-723 (1989).

89. Seaman, D. "Trends in the formulation of pesticides. An overview," *Pest. Sci.* 29:437-449 (1990).

90. Smith, A.M. and W.H. Vanden Born. "Ammonium sulfate increases efficacy of sethoxydim through increased absorption and translocation," *Weed Sci.* 40:351-358 (1992).

91. Green, J.M. "Increasing efficiency with adjuvants and herbicide mixtures," *Proceedings of the First International Weed Control Congress,* Monash University, Melbourne, Australia, J.H. Combellack, K.J. Levick, J. Parsons, and R.G. Richardson, Eds. (Melbourne, Australia: Weed Science Society of Victoria Inc., 1992), pp. 187-192.

92. Stock, D. "Effect of adjuvants on biological activity of foliar-applied pesticides," *Proceedings of the Brighton Crop Protection Conference — Weeds* (1991), pp. 315-322.

93. Beversdorf, W.D., J. Weiss-Lerman, L.R. Erickson, and V. Souza-Machado. "Transfer cytoplas-mically-inherited triazine resistance from bird's rape to cultivated *Brassica campestris* and *Brassica napus,*" *Can. J. Gen. Cytol.* 22: 167-172 (1980).

94. Comai, L., and D.M. Stalker. "Impact of genetic engineering on crop protection," *Crop Prot.* 3(4):399-408 (1984).

95. Darmency, H., E. Lefol, and R. Chadoeuf. "Risk of assessment of the release of herbicide resistant transgenic crops: two plant models," *IXème Colloque International sur la Biologie des Mauvaises Herbes,* Annales ANPP, pp. 513-523 (1992).

96. Giaquinta, R.T. "An industry perspective on herbicide-tolerant crops," *Weed Tech.* 6:653-656.

97. Comstock, G. "Genetically engineered herbicide resistance, part one," *J. Agric Ethics* 2:263-306 (1989).

98. Wyse, D.L. "Future impact of crops with modified herbicide resistance," *Weed Tech.* 6:665-668 (1992).

99. Zimdahl, R.L. "Weed-crop competition. A review," International Plant Protection Center, Oregon State University (1980), pp. 197.

100. Oliver, L.R. "Principles of weed threshold research," *Weed Tech.* 2:398-403 (1988).

101. Coble, H.D. and D.A. Mortensen. "The threshold concept and its application to weed science," *Weed Tech.* 6:191-195 (1992).

102. King, R.P., D.W. Lybecker, E.E Schweizer, and R.L. Zimdahl. " Bioeconomic modeling to simulate weed control strategies for continuous corn (*Zea mays*)," *Weed Sci.* 34, 972-979 (1986).

103. Wilkerson, G.G., S.A. Modena, and H.D. Coble. "HERB: Decision Model for Post-emergence Weed Control in Soybean," *Agron. J.* 83:413-417 (1991).

104. Heitefuss, R., B. Gerowitt, and W. Wahmhoff. "Development and implementation of weed economic thresholds in the F.R. Germany," *Proceedings of the British Crop Protection Conference — Weeds,* (1987), pp. 1025-1034.

105. Cousens, R., C.J. Doyle, B.J. Wilson, and G.W. Cussans. "Modelling the economics of controlling *Avena fatua* in winter wheat," *Pest. Sci.* 17:1-12 (1986).

106. Swinton, S.M. and R.P. King. "WEEDSIM: a bioeconomic model of weed management in corn," Staff Paper Series, University of Minnesota (1990), pp. 1-23.

107. Egley G.H. and S.O. Duke. "Physiology of weed seed dormancy and germination," in *Weed Physiology* (Boca Raton, FL: CRC Press, 1985), pp. 28-64.

CHAPTER 9

Agriculture and Land Use Planning

Alessandro Toccolini and Vincenzo Angileri

CONTENTS

RESOURCE ASSESSMENT AS A KEY ELEMENT IN AGRICULTURAL LAND USE PLANNING

Environmental decay, unchecked town planning, and pollution over vast areas, have pointed with increasing urgency to the need for a new planning philosophy and methodology that will give due consideration — as objectively as possible — to environmental and land resources.

The ability to enhance these values within the scope of the existing planning tools demands that each and every land use plan, regardless of its kind or importance, be preceded by a careful assessment of the existing resources. An awareness of these goals was first expounded at the end of the 1960s by Ian McHarg in his well-known work *Design with Nature,*[1] where an effort was made by the author to develop a methodology that would provide more accurate and effective assessment of environmental values in the planning process.

McHarg acknowledges the importance of analysis as a basic step in land use planning. The analytical process he developed comprises essentially two stages. In the first, a detailed census of all land-related information required for the planning process is carried out, and all the relevant details are recorded and organized into a set of maps so as to facilitate their reading and interpretation. In the second stage, areas with special features are pinpointed through a "successive sifting" process effected by overlapping the cartographic evidence obtained. Each of these areas is classified either as "eligible for specific use" or as "incompatible" with uses other than preservation. These observations have led to the development of a set of thematic maps that emphasize the specific features or compatible uses of each area.

In this methodologic framework, the actual planning process enables the researchers to ascribe the most appropriate use to the various areas, as a function of the established goals. In some of these areas, the status is ascribed virtually automatically, as it depends on the elements that have been identified in the preceding stage; other areas, however, call for effective action on the part of the planners, who are then required to interpret, supervise and, occasionally, "rearrange" the existing territorial layout. This conceptual approach (which is objective in kind) stands sharply in contrast with the subjectivist outlook, in which the planner ascribes specific land uses (zoning) on no other basis than know-how and professional sensitivity.

The significance of this approach, which provided the starting point for a new planning concept, depends on its ability to supplement the land use planning process with a scientific tool that is actually capable of minimizing the environmental impact of improper land uses. In addition, if properly implemented, the process provides a sound basis for safeguarding and, in some cases, enhancing the aesthetic, natural, and anthropic values inherent in a specific territorial unit. From a most strictly formal point of view, McHarg's approach is viewed as a "landscape approach", largely based on qualitative assessments.

The approach outlined, which marked a major breakthrough in land use planning methodologies, has undergone adjustments and changes over the years, both of which were advocated in order to enhance the effectiveness of the process in the light of social change and scientific acquisition, in terms of land investigation tools, as well as in view of the fast development of information-processing technologies. The planning processes have been adapted to present-day realities by developing methodologies based on the so-called "parametric approach", which is also partly identifiable in some of McHarg's works, namely those in which the assessment of territorial resources and the comparison of their respective values is made on the basis of a quantitative, systematic survey of a select number of significant parameters. The importance of this methodological approach — a significant instance of which is undoubtedly represented by the METLAND procecure (discussed in the following section) — is huge, for it offers the advantage of supplementing the planning process with the enormous potential afforded by the new technologies.

The transition from a planning mode that may be referred to as traditional in view of the implementational tools it envisages (albeit innovative in content), such as the one suggested by McHarg, to an innovative planning model, increasingly bound up with computers, has proved necessary for a number of reasons, the most significant being the evolution undergone by land-scape-scanning and information-processing technologies,[2] the widespread adoption of geographical information systems (GIS),[3] the decentralization of administrative functions, the growth of vast metropolitan areas with fierce conflicts between alternative land uses, the increasingly marked democratic involvement of the town dwellers in planning policy options, and the new standards achieved by scientific knowledge in all areas.[4]

AN OPERATIONAL TOOL: THE METLAND PROCEDURE

The METLAND procedure (Metropolitan Landscape Planning Model) is a regional-level landscape planning methodology developed in the 1970s by an interdisciplinary research team at the University of Massachusetts at Amherst (USA),[5,6] headed by Prof. Julius Fabos, as a response to the chaotic expansion of the metropolitan areas and the ensuing indiscriminate reallocation of agricultural land and environmental resources.

The METLAND pursues the following goals:

- to discourage urban growth in areas with valuable environmental and landscape resources;
- to discourage urban growth in areas subjected to periodical risks, whether natural or resulting from human activities;
- to safeguard the ecological potential of the territorial ecosystem;
- to set the focus of the development process on areas that already rely on high-standard and/ or inexpensively mangeable utilities.

The method is based on "parametric" assessments and systematically avails information — processing and data — acquisition technologies by means of remote surveying equipment, so as to ensure constant updating of the information flow. It is based on three essential stages: resource assessment, plan formulation, and plan evaluation (Figure 1).

The **assessment** stage summarizes the results of the following operational steps: landscape assessment, ecological compatibility assessment, and public service resource assessment.

Landscape assessment — where the term "landscape" refers to the whole set of environmental and territorial resources and is, therefore, very broad in scope — is based essentially on three separate components: the first, relating to the so-called "special resources" (physical and renewable, such as water or physical and nonrenewable, such as sand and gravel, or, finally, aesthetic–visual in nature, such as views); the second, concerning the environmental risk components (air and water pollution, noise, floods); the third, connected with suitability for the development process (environmental opportunity for alternative types of development). The three components are then combined into a final resource assessment. Each of the three components represents substantially different values: special resource and environmental risk assessment provides information to the effect of conditioning or restricting development; the one based on the concept of suitability for development, on the other hand, singles out the areas that are best for development. Obviously, when the three separate components refer to different surface units, there are no problems; more commonly, contrasting values emerge for one and the same area, thus generating conflict.

Assessments are expressed in economic terms so as to support the planners and decision makers in solving these problems; the monetary unit, therefore, acts as the common denominator for different values. The combined assessment of the landscape resources, therefore, expresses all the potentialities and/or restrictions.

Ecological compatibility assessment (based on Odum's conceptual approach)[7] features two components. The first relates to the classification of the relevant areas according to their ecological function; all land uses are grouped into five classes, depending on the size of the existing biomass and on the production/breathing ratio. The second component is connected with an assessment of the substratum functions: each homogeneous surface unit derived from this classification is given a biological power value as well as an erosion and denudation power value.

By combining the above assessments, an ecological compatibility calculation can be made of worthiness or suitability among the ecological functions of the agricultural landscape and the physical features of the substratum.

The concept of **public service resource assessment** is similar to that of suitability for development: the areas in which public utilities are better and are present in larger numbers are also the ones that are most suitable for new settlements. Seven utility services — ranging from sewers to public amenities — have been taken into consideration.

Moreover, the methodologic approach takes into account the different levels of development, advancement, and accessibility of the research projects pursued upstream of the individual parameters; in other terms, landscape mapping as produced by the U.S. federal government's "Soil Conservation Services" in 40 years of work and with investments of millions of dollars is definitely more advanced and more accurate than any state-of-the-art studies in groundwaters or visual amenity; in addition, it often occurs that the techniques used for determining some parameters are indeed available but entail soaring costs when used extensively over vast areas. A typical case in point is represented by geotechnical technologies as used for determining groundwater potential. In these cases, it is essential for the planner to have a thorough knowledge of the techniques needed for determining a set of indicators that will fit in with his or her operating scale, at a limited cost and/or at one that is consistent with the established goals.

The second stage in this procedure implies the **development of alternative plans** based on land use resources, existing conditions, and preferences expressed by the local population. The procedure that has been developed in this connection aims at optimizing the relationship between land use and resource protection. The option connected with the "existing conditions" hinges on the assumption that the currently enforced town-planning tools will not be changed. Thus, the aim of such a thoroughly defined plan boils down to making the best use of the existing urban utilities. Finally, where the planning process draws on the preferences expressed by the relevant community, plans are developed in accordance with the demands of the population.

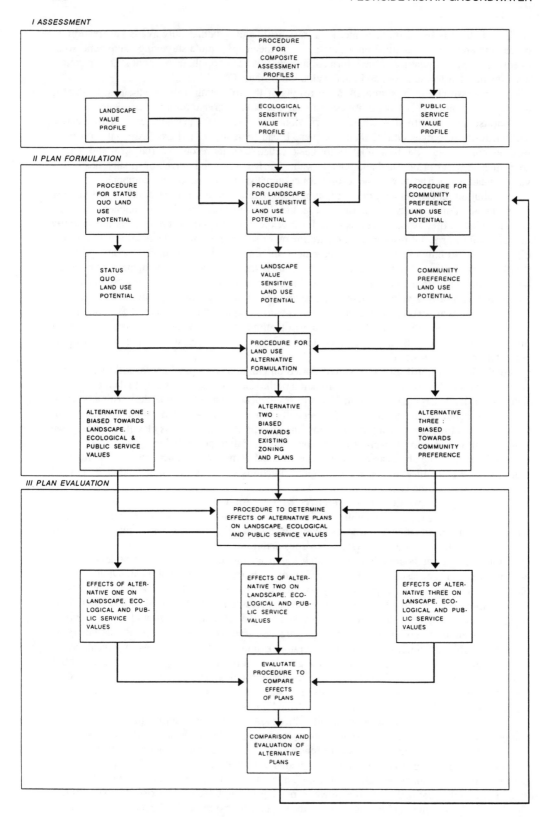

Figure 1   The "METLAND" landscape planning process flow diagram.

The third stage of the process hinges on an assessment of the **effects** of the plans thus developed on environmental and territorial resources, in terms of special resources, ecological compatibility, and public service resource assessment. These effects are also pinpointed by means of economic assessment procedures — as in the first stage described earlier — designed to quantify resources, risks, and development potentials.

Major emphasis is laid, in the model, on the assessment of special resources, which are thus called because of their importance within the planning process: these resources include the following:

- Agricultural productivity;
- Forest productivity;
- Wildlife productivity;
- Groundwater;
- Visual amenity;
- Sand and gravel.

Sometimes their value can be expressed best in qualitative terms, sometimes in quantitative terms, often in both. Drinking water, for instance, must be bacteriologically pure (qualitative assessment) and available in a sufficiently large amount (quantitative assessment).

Agricultural productivity, forest productivity, wildlife productivity, and groundwater are, classically, flow resources; visual amenity is characterized both as a stock and as a flow resource; sand and gravel are stock resources.

It should be stressed that this finite list of resources is in no way exhaustive; however, the resources listed are undoubtedly essential — and all-too-often disregarded — components in any conversion plan from rural to urban land use.

The application of the model leads to a vast amount of cartographic material, in which the "value" of each resource is classified according to a decreasing "value" scale. Each class, after being "weighed", is given a value per surface unit so as to allow for computer-aided overlaying of the various resource assessment maps and for the development of a synoptic resource assessment map.

REVIEW OF A FIELD TEST

The aim of the test was to determine whether the method is universally applicable, with special reference to Italian reality. In this connection, it was decided to apply the method to an area located inside the so-called "Parco Sud Milano", which is a site affected by problems in compatibility between agricultural and recreational land uses as well as by locational difficulties bound up with urban growth. The field test related primarily to the first stage of the process, namely resource assessment, and, in particular, to the analysis of the special resource profile. Moreover, the results obtained for these resources provided a starting point for the definition of a "special resource-based plan", in keeping with the approach followed in the second stage of METLAND, which deals with the development of alternative plans. The draft project is to be viewed as a master plan.

The selected area (including four municipalities and covering roughly 6,000 ha) is located southwest of Milan (Figure 2). The area, on the whole, does not seem to be seriously impaired by urbanization phenomena and is characterized by efficient, up-to-date agricultural land use.

In the resource assessment stage, special emphasis was placed on special resources, in view of their significance within the area under consideration, and specifically on the following:

- Agricultural productivity;
- Groundwater resources;
- Sand and gravel;
- Visual amenity.

Figure 2   Location map of the study area.

The other two special resources envisioned by the method (forest productivity and wildlife productivity) were not taken into consideration, as they were of little — if any — relevance to the study.

AGRICULTURAL PRODUCTIVITY

Land use planning has traditionally overlooked the importance of conservation with respect to the more productive agricultural estates. The problem is readily apparent especially in the metropolitan areas where strong settlement pressure causes a reduction in agricultural soil — which is often top-quality as well.

The assessment technique developed (Figure 3) in METLAND is based on the interaction of two factors:

- the naturally inherent qualities of the soil with respect to farming activities (in this context, the "Land Capability" classification system developed by the U.S. Department of Agriculture has been used);
- the rate of decline in these "qualities", as determined by the present land use (based on MacConnell's land use classification system and on a set of deduction indices).

Three classes of decreasing agricultural productivity were identified in the area under study (Figure 4) by using a computer-aided thematic overlaying of the two plans considered:

- class A was established to include class I and II soils — according to the land capability concept — which are, at present, used for agricultural purposes or other functions that are not in contrast with a prospective shift back to agriculture; in these areas, the resource is present to the highest degree;
- class B ranks class III soils destined to agricultural activities as well as class I and class II soils, the present use of which, though not entirely at odds with a prospective return to agriculture, prevents them from being included in class A;
- class C includes areas of little or no interest in terms of agricultural activities: soils with a poor agricultural potential; soils that, regardless of their agricultural potential, have been irreversibly taken up by other land uses, ruling out any reasonable prospect for a return to agriculture.

Class A includes the greater part of the farmland located within the relevant area; class B ranks land presently used for agricultural purposes or made up of untilled plots that, though external to the urban fabric, belong in class III land capacity; woods and poplar groves that grow on estates that have been ascribed a class II land capacity; class C, finally, comprises woods and poplar groves located on class III land capacity estates as well as all the urbanized areas.

GROUNDWATER RESOURCES

The "METLAND" ascribes a crucially important role to groundwater resource assessment. However, the authors of the method themselves point out that research in groundwater assessment, even in the U.S., is rather sluggish and cannot compare with the soil nature investigations carried out by the Soil Conservation Service. For the area under study, the subject was tackled with reference to two major aspects: the first hinges on a correlation — never mind how loose — existing between land use, on the one hand, and possible groundwater pollution on the other; the second relates to the potentiality — in terms of water supply — of the aquifer. In this connection, a major contribution is represented by the "Hydrogeological Study of Ground Water Balance in the Plain Comprised between the Rivers Adda and Ticino",[8] promoted by the Drinking Water Consortium in charge of water supply to the entire Milan province. This study provided a basis for plotting a map (Figure 5) of the area under consideration, capable of representing the specific rate of flow of the aquifer. By comparing this map against the information provided by the land use map (in terms of the influence of such uses on groundwater quality), a derived map (Figure 6) of the overall supply was developed for the resource in question.

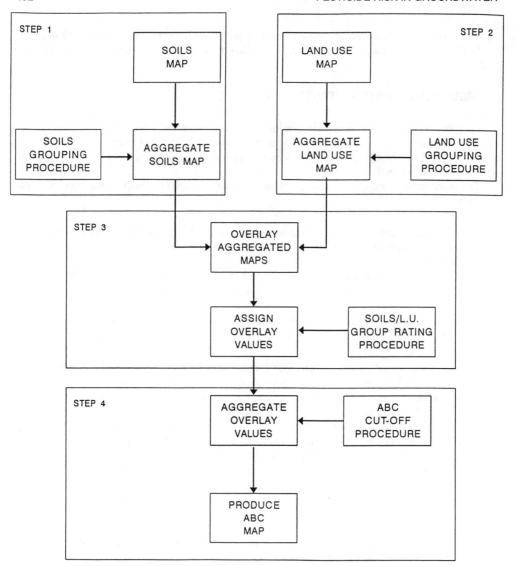

Figure 3   Agricultural productivity assessment procedure.

Class A includes those areas in which water resources attain to the highest degree in terms both of quality and of quantity, or in which only one of the two characteristics attains medium-level standards. According to the survey carried out in the area under study, this class comprises estates having aquifers with a specific rate of flow of 10 to 30 L/s per ebb meter (class A of the quantitative assessment), at present taken up by woodland (class A of the qualitative assessment) or destined to non-polluting mining, such as quarries (class B of the qualitative assessment). This class also includes sites with aquifers with a specific rate of flow of 4 to 10 L/s per ebb meter (class B of the quantitative assessment), at present taken up by woodland.

Class B comprises the remaining land units, not included in class A, with the exception of the urbanized sections. In addition, this class ranks those estates that, though of some interest in view of their sizeable water reserves, qualify as C on the qualitative assessment scale (agricultural land with aquifers having a rate of flow of 4 to 30 L/s per ebb meter), besides those that feature aquifers with specific rates of flow ranging from 4 to 10 L/s per ebb meter, presently used for nonpolluting mining activities, such as quarries (class B of the qualitative assessment scale).

Class C refers to the urbanized areas.

The procedure followed, which is useful for conducting field surveys, is open to further development, as may be justified by more specific, in-depth research work (due to the characteristics of the sites or to particular problems in terms of pollution).

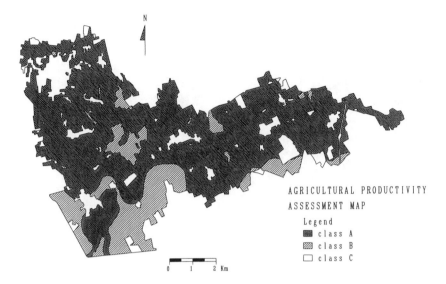

Figure 4   Agricultural productivity assessment map.

Figure 5   First aquifer flow rate map.

SAND AND GRAVEL

Since the sites undergoing major development are also the ones that require the greatest availability of building materials, and, since transport costs are such as to rule out locations far from the relevant construction sites, due to the very nature of these resources, it is clear that a suitably large stock of readily usable materials is essential in metropolitan areas.

In this connection, METLAND provides the basis for sand and gravel assessments according to the diagram shown in Figure 7, which is conceptually similar to the one plotted for agricultural productivity. The final classes (A through C) represent various levels of potential supply in terms of sand and gravel resources (Figure 8).

Class A comprises those soils that are characterized by substantial sand and gravel reserves, namely those areas that should be exploited for mining activities both in the private and in the public interest. The assessment relies on the availability of materials as well as on the ability to

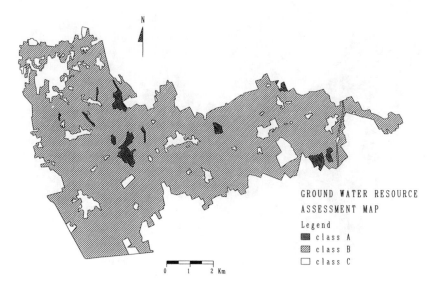

Figure 6   Ground water resources assessment map.

exploit them at reduced costs, depending on the present land use in the area (existing quarries, untilled farmland, etc.).

Class B ranks soils with limited sand and gravel reserves. Should the need arise, in a metropolitan area, to proceed to the exploitation of smaller sand and gravel deposits, these are the areas in which mining should be concentrated. Here, too, the assessment is based both on the area's wealth in natural resources and on the specific land use prevailing in the area. This class, therefore, includes soils with a medium-size stock of easily accessible materials as well as estates that, albeit rich in resources, present some restrictions to their effective exploitation, due to their present land use.

Class C comprises those soils that are of no interest in terms of sand and gravel mining.

VISUAL AMENITY

The assessment procedure envisaged by METLAND in the field of visual amenity — based on the quantitative determination of the "value" of discrete landscape elements, and closely bound up with the concept of inherent natural quality — seemed to be hardly suitable for application within the area under study, because of the differences existing between the American landscape and the Italian one. Italy's rural landscape, in fact, is essentially a "built-up landscape", dominated by anthropic elements that enhance its visual amenity, whereas America's qualitatively more significant landscapes seem to be represented by natural components.

Consequently, an original operational landscape visual analysis technique (the diagram is shown in Figure 9) was developed for assessing this resource, on the basis of previous experience. This technique entails surveying of the landscape units through photo interpretation and groundstation data recording and establishing a visual amenity scale of the various landscape units through interviews with significant samples (experts and laymen) presented with photographs. The respondents were then asked to give an assessment in numerical terms.

The results achieved enabled the researchers to pinpoint four landscape visual amenity classes, defined as follows (Figure 10):

Class A (top quality) — geomorphologically flat landscape, but with a considerable degree of contrast due to the high density of the volumetric masses represented by woodland and poplar groves, enriched by the presence of character elements (poplar and willow rows, artificial canals, springs, poplar groves, and woods); single-crop landscapes combined with poplar groves and woodland.

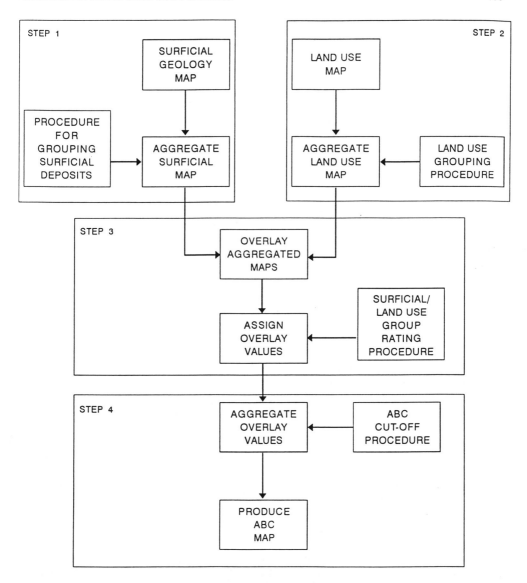

Figure 7 Assessment procedure for sand and gravel supply.

Class B (good quality) — geomorphologically flat landscapes, with some character elements, but, unlike those grouped in Class A, with lesser contrast, due to the limited amount of volumetric masses; agricultural landscapes made up of small-size plots on which a variety of crops are grown.

Class C (fair quality) — geomorphologically flat, wide-mesh landscapes, typically represented by single-cropping utilization or by a limited number of crops; mining activities (quarries).

Class D (poor quality) — urban fringe landscapes, with a set of disorganized character elements developed for different uses (housing, industry, business), further characterized by the presence of urban greenhouses and flower- and vegetable-growing enterprises as well as industrial and business areas; wasteland and untilled land; urbanized areas (included into this class in that they are of little or no significance for conservation from development, as they have already been built up).

Figure 8   Sand and gravel supply assessment map.

DEVELOPMENT OF A PLAN BASED ON LANDSCAPE RESOURCES

As noted earlier, the METLAND method entails the development of plans based on any of the following three alternative elements: landscape resources, existing condition (present mapping), and preferences of the resident population. In addition, the method offers a tool for comparing the different courses of action pursued, and, therefore, the resulting picture is as complete and objective as possible.

In this context, a particularly interesting option, in terms of procedural and operational innovation, seemed to be represented by the development of a landscape resource-based plan: this process enables the researcher to ensure, from the very onset, suitable protection for wildlife and landscape values, without having to assess environmental impact at a later stage. With reference to the area selected for the field study, special emphasis was placed on the assessment of the special resources (agricultural productivity, groundwater resources, sand and gravel, visual amenity) that provided the basis for the development of the plan. From an operational standpoint, the METLAND method hinges on the successive overlaying maps representing individual resources on the basis of specific priority scales. The procedure should be implemented with suitable computer systems so as to ensure effectiveness and flexibility as well as the prospective ability to reuse and reprocess the data at any later stage. The significant "output" of such processing is a synoptic map of the resource assets (Figure 11), which represents the territory's "inherent environmental quality".

Thus, the location is divided into six classes representing just as many levels in terms of resources; the areas with the highest degree and greatest amount of resources are the ones that must be protected against urbanization, which should, therefore, be directed toward the areas that are poorer in terms of resources. The six classes obtained are as follows:

Class "Super A" — this class refers to those areas that qualify as "class A" for at least one of the three special resources — "visual amenity", "sand and gravel", and "groundwater resources". This class, however, does not include areas classed "A" for "agricultural productivity", because the large number of these areas implies less strict protection measures as compared with those enforced in class "Super A".

"Class A" — this class groups areas that qualify as class A for agricultural productivity and as class B for visual amenity. As evidenced by this particular class attribution, the visual amenity of the landscape plays a fundamental role in the development of a prospective plan, since it is the discriminating factor for classifying the areas of greater interest in terms of agricultural productivity;

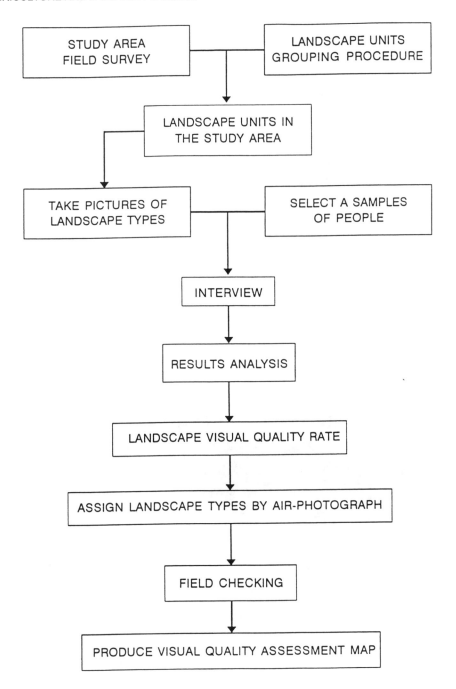

Figure 9 Visual amenity procedure flow diagram.

"Class B" — this class relates to areas that qualify as "B" in terms of both visual amenity and agricultural productivity. As far as the visual aspect is concerned, the agricultural landscape of these areas is similar to that of class A areas but is less suitable for agriculture;

"Class C" — essentially farmland, including top-quality soils, fitted in a poorly planned landscape, or typifying a form of agriculture chiefly based on single-crop or select-crop farming. This class also includes those areas that qualify as "B" in terms of visual amenity and as "C" in terms of agricultural productivity, as exemplified by woods located on less productive land, yet fitted in a

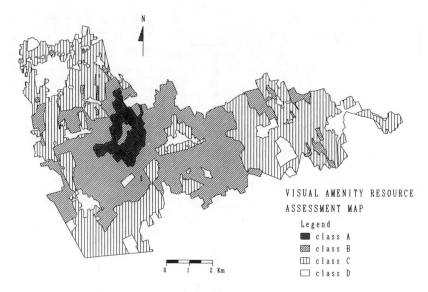

Figure 10   Visual amenity resource assessment map.

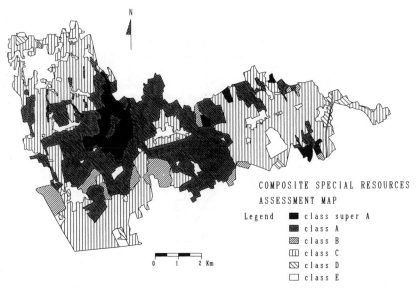

Figure 11   Composite special resources assessment map.

good visual context. The areas that belong in this class should not be subject to exceedingly strict constraints regarding their use, as opposed to the requirements stated for the estates ranked in the upper classes. From a practical point of view, this class is obtained by difference with respect to all the other classes referred to in the process; a great deal of the area under study falls into this class;

"Class D" — this class comprises those areas that, not belonging in the first three classes, qualify as "D" in terms of visual amenity. They are essentially fringe areas with decayed landscapes and should be the first to be used for urbanization purposes and/or redevelopment processes;

"Class E" — this class identifies the urbanized conglomeration, namely the areas ascribed to class C in view of their groundwater resources or — to put the same concept differently — sand and gravel resources.

It will be readily noted that this specification provides a sufficiently clear operational tool for the planning practice.

CONCLUDING REMARKS

This brief outline shows that the adoption of a careful, thorough, transparent land use selection procedure is the very first step towards the effective management of territorial resources. However, while this statement can be shared in principle, it requires a full set of major information-based and management-related processes, if it is to be implemented effectively. In particular, the scope of our present knowledge should be extended by the following:

- in-depth studies concerning the water/land ratio in terms of both surface and ground waters. This effort, however, must lead, in practice, to the identification of study material with a sufficient degree of operational effectiveness, both in terms of costs and of time, while complying with the strict tenets of scientific inquiry;
- an effective system for updating and monitoring changes in land use; here, too, the effort should be directed toward the pursuit of goals that can be attained with the available resources, which are necessarily scarce (e.g., the legend classes should not be too many, so as not to hinder the classification process).

In terms of management, a major effort should be made, on the one hand, to develop sufficiently flexible GIS,[9] capable of conducting an informative dialogue between different softwares and of using material conveyed by different data banks and, on the other hand, to encourage greater integration between GIS and planning methodologies,[10] so as to make use of the potential synergies: this means, beyond and above creating and developing the GIS, responding to the informational requirements of land planning and management methods.

REFERENCES

1. I. McHarg. *Design with Nature* Garden City, NY, Natural History Press, (1969).
2. J. Gy. Fabos and I. Petrasovits. "Computer-Aided Land Use Planning and Management," *Research Bulletin* n. 693, Massachusetts Agricultural Experiment Station, Amherst, MA, (1984).
3. P.A. Borrough. "Principles of Geographical Information Systems for Land Resources Assessment," New York, Oxford University Press (1986).
4. J. Gy. Fabos. "Computerization of Landscape Planning," *Landscape and Urban Planning* 15: 279-289 (1988).
5. J. Gy. Fabos et al. "Composite Landscape Assessment: Assessment Procedures for Special Resources, Hazards and Development Suitability: Part II of the Metropolitan Landscape Planning Model (METLAND)," *Resource Bulletin* n. 637, Massachusetts Agricultural Experiment Station, Amherst, MA, (1977).
6. J. Gy. Fabos et al. "The METLAND Landscape Planning Process: Composite Landscape Assessment, Alternative Plan Formulation and Evaluation: Part 3 of the Metropolitan Landscape Planning Model," *Research Bulletin* n. 653, Massachusetts Agricultural Experiment Station, Amherst, MA, (1978).
7. P. Odum Eugene. "The strategy of ecosystem development," *Science* 18: 262-270 (1989).
8. G.P. Beretta, A. Cavallin, V. Francani, S. Mazzarella, and A. Pagotto. "Primo bilancio idrologico della Pianura Milanese," *Acque sotterranee* 2, 3, 4 (1985).
9. D.F. Marble. "Geographic Information Systems: an Overview," *Geographic Information Systems*, London, U.K., Taylor & Francis, pp.8-17 (1990).
10. E.B. MacDougall, et al. "The Feasibility of Geographic Information System," *Massachusetts Agricultural Experiment Station*, Amherst, MA, (1987).

Section V
Legislation
and
Economy

CHAPTER 10

Regulations and Management

Guido Premazzi and Giuliano Ziglio

CONTENTS

0-87371-439-3/95/$0.00+$.50

INTRODUCTION

In recent decades, pesticides have been developed to be more specific in action and safer for both their users and the environment. Nevertheless, their optimal use and safe disposal remain of crucial importance. Many chemicals that persist in the environment have now been withdrawn from use and their successors are often designed to degrade fairly rapidly in soil or plants. Traces of older pesticides, such as organochlorine insecticides, continue to be found in water, and the newer compounds retain the potential to harm both human health and the environment if they are mishandled.

Contamination of groundwater by pesticides is of particular concern since, unlike surface waters, once contaminated, groundwater may take a long time to regenerate, especially if persistent substances are involved. Groundwaters are valuable resources for drinking water production, but they are limited, vulnerable, and need to be protected from contamination.

Generally, concern over the contamination of drinking water resources has arisen due to increasing public concern, but particularly due to the emergence of the European Economic Community (EEC) Directive on the quality of water intended for human consumption (80/778/ EEC). Monitoring for pesticides in relation to testing compliance with the Directive 80/778 has revealed that some pesticides can exceed the MAC (Maximum Admissible Concentration) in surface and groundwaters and in the derived drinking water. However, only rarely do the concentrations exceed drinking water standards based on the toxicity of the pesticides (i.e., concentrations directly related to health implications). Nevertheless, the finding of pesticides in water resources and the environmental mobility of certain pesticides has been somewhat unexpected and has certainly indicated a lack of understanding about their behavior.

In this chapter the European Community (EC) policy and the EPA strategy towards pesticides pollution in drinking waters and groundwaters are sketched. Some remarks are also made on the future prospects, including the implementation of guidelines for the admission of pesticides on the market, a regular evaluation of the state of the groundwater systems (monitoring) and a potential for a functional groundwater management (modeling).

EUROPEAN COMMUNITY ACTION

The EC has had a policy on the quality of its waters since the early 1970s. This policy, like many developed at this time, was borne out of the 1972 Stockholm Conference on the Environment. Since then, there have been five action programs. They suggest specific proposals for legislation that the Commission intends to put forward over the next few years, and they provide an occasion to discuss some broad ideas in environmental policy and suggest new directions for the future. Rather few national ministries responsible for the environment produce comparable documents. The action programs can be said to provide a policy framework, but, unlike items of Community legislation, they cannot strictly be regarded as constituting Community policy. However, when the Treaty on European Union comes into effect, action programs will be adopted jointly by the Council and the European Parliament under the "codecision" procedure. As a result, they may assume greater legal significance.

The 11 principles of Community environmental policy listed in the first, and repeated in subsequent programs, can be summarized as follows:[1]

1. The principle of prevention: it is better than cure.
2. Environmental effects should be taken into account at the earliest possible stage in decision making.
3. Exploitation of natural resources that causes significant damage to the ecological balance must be avoided. The natural environment can only absorb pollution to a limited extent. It is an asset that may be used, but not abused.
4. Scientific knowledge should be improved to enable action to be taken.
5. The polluter pays principle: the cost of preventing and eliminating nuisances must be borne by the polluter, although some exceptions are allowed.
6. Activities carried out in one Member State should not cause deterioration of the environment in another.
7. The effects of environmental policy in the Member States must take account of the interests of the developing countries.
8. The Community and the Member States should act together in international organizations and in promoting international and worldwide environmental policy.
9. The protection of the environment is a matter for everyone. Education is therefore necessary.
10. The principle of the appropriate level. In each category of pollution, it is necessary to establish the level of action (local, regional, national, Community, international) best suited to the type of pollution and to the geographic zone to be protected.
11. National environmental policies must be coordinated within the Community, without hampering progress at the national level.

The Fourth Environmental Action Program (1987 to 1992) contains a discussion of some general policy orientations, including the consequences of amendment to the Treaty of Rome, the need for improved implementation of Community Directives, integration of the environmental dimension into other Community policies, and the use of economic instruments as an alternative or additional method to legislative regulation for securing environmental protection. Also included is a section containing reflections on pollution control, which, among other things, commits the Commission to participate in the debate taking place in several countries and in other international organizations (e.g., OECD) on what is known as the cross-media or integrated approach to pollution control. This entails an approach that does not regard water pollution control as in a separate category from air pollution control or waste disposal but recognizes that pollutants can move between the environmental media.

The Fifth Environmental Action Program (1993 to 2000) differs from previous programs in a number of significant ways:

1. It seeks to develop a new approach to environmental protection, rather than focusing on a detailed list of specific policy commitments;
2. It stresses the need for the integration of environmental protection into the activities of key actors and economic sectors;
3. It highlights the importance of using a range of policy instruments broader than the traditional legislative "command and control" approach;
4. It emphasises the need for shared responsibility in moving towards sustainable development among governments, public and private industry, and the general public.

By 1987, some 200 items of environmental legislation had been agreed upon under either Article 100 of the Treaty of Rome or under the Article 235. In 1987, important changes to the Treaty of Rome came into effect through the Single European Act, including a new Environment Title (Articles 130r, s and t) that, for the first time, gave an explicit legal underpinning to the Community's environmental policy. A further step in the strengthening of the EC's environmental policy was taken with the signing of the Treaty on European Union in Maastricht, The Netherlands in February 1992. The treaty introduces as one of the EC's basic tasks the promotion of "substainable and noninflationary growth respecting the environment" (Article 2). Environmental policy is to be based, *inter alia,* on the "precautionary principle", and the commitment in the Single European Act that environmental protection should be integrated into other EC policies has been reinforced.

In addition, major changes have been made to the EC's procedures for adopting environmental legislation. Voting by qualified majority in the Council of Ministers now becomes the standard procedure for environmental measures, although the requirement for unanimity is retained for provisions primarily of a fiscal nature, town and country planning, land-use (with the exception of waste management and measures of a general nature), the management of water resources, and measures significantly affecting a Member State's choice between different energy sources and the structure of its energy supply.

FORMS OF EC LEGISLATION

There are a number of different legislative means available for turning the EC environmental policy. The Council of Ministers can adopt the following:

- nonbinding recommendations and resolutions;
- regulations that are binding and directly applicable in all Member States;
- decisions that are directly binding on the persons to whom it is addressed, including Member States, individuals, and legal persons;
- directives that must be implemented by the laws or regulations of the Member States within a designated time limit.

Regulations are usually used for very specific purposes such as trade in products and financial matters; they have not often been used for environmental legislation, except for controls on trade in endangered species. Decisions have been primarily used in environmental legislation to authorize the Community to become a party to international conventions, but also for other purposes, e.g., to set up a system of information exchange on water quality. The directive is the main tool of Community environmental policy. It empowers the Community to define objectives, standards, and procedures but allows the Member States some flexibility in that implementation must take place through national legislation and regulation. In fact, environmental directives have sometimes been similar to regulations by laying down precise limits; controls; or technical, testing or labelling requirements, particularly regarding industrial products.

EC LEGISLATION FOR THE PROTECTION OF WATERS

In 1991, about 30 directives and other regulations were passed to protect water resources. The directives on water pollution may be divided into three fundamental categories. The first includes a group of directives in which quality standards are fixed, which vary depending on the use for which the water is intended. The second category concerns specific sectors or industries, while the third deals with discharges of dangerous substances.[2]

The first group formed a conspicuous part of Community action in the 1970s, mainly concerning the protection of public health. These directives, aimed at containing or reducing water pollution, stimulated the Member States to fix a series of quality standards and to respect them within a certain date and also forced the Member States to draw up suitable protection and clean-up plans. This group included three directives on the quality of water for human consumption, two dealing with fish farming and one concerning bathing water.

- As regards drinking water, two EEC directives, one dating from 1975 and the other from 1980, fix the criteria applicable to surface waters for human consumption. The inspection methods and the frequency of water analyses were in turn harmonized by another directive, adopted in 1979, that defines the methods for measuring the physical, chemical, and microbiologic properties of water. The 1980 Directive lays down quality standards for water intended for human consumption, both directly and by transformation, excepting natural mineral waters or medicinal waters recognized or designated by the Member States. It intends to promote the free circulation of goods in the Community while protecting health and the environment. The Directive EEC 80/778 includes three annexes: Annex 1 fixes maximum admissible concentrations (MAC) and guide levels (GL) for 62 parameters and the minimum concentrations required (MCR) for four parameters in six categories:

a. organoleptic parameters,
b. physical and chemical parameters,
c. substances that are undesirable in excessive amounts,
d. toxic substances,
e. microbiologic parameters,
f. MCR for softened waters.

- The MAC and the GL are not fixed for all the categories. For the parameters and categories (a) and (e) of Annex 1, the Member States must fix values at or below the MAC values, when these are already fixed, or based on the GL values. They can decide not to fix any value for the parameters for which the Directive omits the MAC and GL values. Annex II contains a list of obligatory standard analyses and of obligatory minimum frequencies for these analyses. Annex III lays down indicative analysis methods for the parameters in Annex II. These methods must be applied in as strict a sense as possible. The Member States are bound to fix, for the five classes of parameters, values lower than or equal to the maximum admissible concentrations, following the guide values indicated.

The second group of Directives includes rules to eliminate pollution due to the activities of the titanium dioxide industry and to protect waters against the pollution caused by nitrates from agricultural sources. Proposals, never approved by the Council, were also made for the wood and paper industry.

- The nitrates Directive (EEC 91/676), adopted on December 12, 1991, has the aim of avoiding a concentration of nitrates in surface and groundwaters that would prejudice the legitimate uses of waters and lead to the eutrophication of surface, estuary, coastal, and marine waters. Within a period of 2 years from the notification of the directive, the Member States must designate the areas vulnerable to water pollution caused by nitrogen compounds. Annex 1 reports the criteria for identifying the vulnerable areas. Annex II gives the maximum number of animals per hectare for the spreading of animal organic manure in the vulnerable areas. The Member States must establish maximum quantities of application on the soil for chemical fertilizer and establish rules that discipline the application periods. Annexes III and IV specify the directive lines of good agricultural practice, the control frequencies, and the reference measurement methods.

The third group can be divided into two parts: the first concerns discharges in groundwaters, the second concerns discharges in surface waters.

- Water aquifers are protected by Directive 80/68, which forbids the discharge of the most dangerous products (List I) and defines, subject to authorization, the limits for the discharging of other substances (List II). Direct discharges into groundwaters are not the only discharges involved; indirect pollution caused by infiltrations in the soil is also considered (see next heading).
- Discharges of dangerous substances in surface waters are controlled by the Directive of May 1976 and later daughter Directives. Directive 76/464 undoubtedly represents the EEC's most important contribution to the protection of water resources. It provides for the elimination or reduction of pollution of internal, coastal, and territorial waters caused by particularly dangerous substances, by means of the adoption of "daughter" Directives; it also intends to ensure uniform application at Community level of international conventions. The Member States are bound to take the necessary measures to eliminate the pollution caused by the substances that figure in List I (black list) and to reduce the pollution caused by the substances of List II (gray list). This Directive has given rise to some daughter Directives (e.g., the directive on mercury, cadmium, or hexachlorocyclohexane), without forgetting Directive 86/280, which is intended to accelerate the implementation of the framework directive extending its field of application to pollution from different sources and introducing the concept of BAT (Best Available Technology).

- The directive concerning the treatment of urban wastewaters (EEC 91/271) considers both water from sewerage networks, in which industrial waste may also be emitted, and direct biodegradable discharges from industries of the agriculture–food sector. The aim of the directive is to put an end to the "undisciplined" discharge of municipal waste waters into internal waters, estuaries, and coastal waters. The directive fixes minimum requirements for the treatment of municipal wastewaters and sludge disposal. It proposes to classify the recipient waters into three categories, depending on their different capacity of assimilating the treated wastewaters.

THE GROUNDWATER DIRECTIVE 80/68

Directive 80/68/EEC became the basic legislation in the EC for groundwater protection; it was to be implemented by the Member States by the end of 1981. Its purpose is to prevent the pollution of groundwater by the families and groups of substances given in its annexes. It follows the basic pattern of Directive 76/464/EEC, which remains in force for discharges to all waters except groundwaters. The directive gives the following definitions of groundwater and pollution. Groundwater is all water that is below the surface of the ground in the saturation zone and in direct contact with the ground or subsoil. Pollution is the discharge by humans, directly or indirectly, of substances or energy into groundwater, the results of which are such as to endanger human health or water supplies, harm living resources and the aquatic ecosystem, or interfere with other legitimate uses of water.

As with the dangerous substances Directive (76/464), the groundwater Directive contains two lists of substances that place different legal obligations on the Member States. List I (black list) is again based on a selection of families and groups of substances chosen for their toxicity, persistence, and bioaccumulation. It contains eight groups:

1. Organohalogen compounds and substances that may form such compounds in the aquatic environment;
2. Organophosphorus compounds;
3. Organotin compounds;
4. Substances that possess carcinogenic, mutagenic, or teratogenic properties in or via the aquatic environment;
5. Mercury and its compounds;
6. Cadmium and its compounds;
7. Mineral oils and hydrocarbons;
8. Cyanides.

List II (gray list) contains individual substances or categories of substances that could have a harmful effect on groundwater. The groups are as follows:

1. Metalloids, metals, and their compounds;
2. Biocides and their derivatives not appearing in List I;
3. Substances that have a deleterious effect on the taste and/or odor of groundwater, and compounds liable to cause the formation of such substances in such water and to render it unfit for human consumption;
4. Toxic or persistent organic compounds of silicon and substances that may cause the formation of such compounds in water, excluding those that are biologicly harmless or are rapidly converted in water into harmless substances;
5. Inorganic compounds of phosphorus and elemental phosphorus;
6. Fluorides;
7. Ammonia and nitrites.

As with List I, List II of Directive 80/68/EEC is nearly identical to List II of Directive 76/464/EEC. Directive 80/68/EEC places an obligation on the Member States to prevent the introduction of List I substances into groundwater and prescribes a series of steps in order to achieve this aim. First, Member States must prohibit the direct discharge of List I substances, which is defined as the introduction into groundwater without percolation through the ground or subsoil. Indirect discharges of List I substances can arise as a result of disposal or tipping for disposal. Member States shall subject such activities to prior investigation and, in the light of that investigation, shall:

- either prohibit the activity, or
- authorize it subject to the condition that all the necessary technical precautions necessary to prevent such a discharge are observed.

Other activities might also lead to indirect discharges of List I substances. Member States must take the measures they consider necessary in order to prevent such discharges. The Commission must be notified of such measures so that it might submit proposals to the Council for revision of the directive, should it appear necessary. In practice, the Commission has received little information on these measures, and there has been no proposal to revise the directive. This directive has undeniable merits, but it can now be seen that its substance-by-substance approach leads to some fundamental difficulties. In practice, it is difficult to apply this approach to industrial sectors or to deal effectively with pollution from diffuse sources. Furthermore, the directive does not address the quantitative problems of groundwater protection and consequently does also not cover the interaction between such problems and pollution.

It is clear that a revision is necessary in order to provide adequate quantitative and qualitative protection for this fundamentally important resource.[3] In a resolution of 2/25/1992, the Council of the EC called upon the Commission to draft a proposal to revise this directive by incorporating it into a general freshwater management policy, including freshwater protection.

STATE OF APPLICATION OF THE EEC DIRECTIVE 80/778: THE SPECIFIC CASE OF PESTICIDES

The date for formal implementation of the EEC directive into national legislations was July 17, 1982. Member States were to ensure that, by August 1985, drinking water met values for parameters listed in Annex I. However, the deadline for full compliance (both formal and practical) was January 1, 1986 in the case of Greece and Spain, and January 1, 1989 in the case of Portugal. As of December 1990, correct formal compliance had not yet been fully achieved by a number of Member States.[4] In the context of the directive, the term "pesticides and related products" includes a whole range of substances including herbicides, insecticides, acaricides, nematicides, and fungicides, as well as PCBs and PCTs.

This parameter covers several hundred different chemical substances of widely differing toxicity, persistence and degradability, many of which were introduced after the directive was drafted. The 0.1 µg/L MAC value was set at the limit of detection for chlorinated pesticides. It may have been seen as a surrogate for a zero standard or as a point from which no deterioration should be allowed. Unlike the MACs for other toxic parameters in the directive, it was not set on the basis of toxicology. MACs were originally intended to be based on an 85% compliance rather than 100%.

With regard to the compliance with the pesticide parameter in Member States, derived by comparing national legislations, the present situation is summarized in Table 1. The situation existing in other Countries outside the Community is also indicated. The position of Member States on EEC limits are favorable, even if the U.K. has formally requested in the past a revision of this parameter, and France fixes toxicologic limits for three products. About use restrictions for pesticides, it must be noted that Germany, Italy, and the Netherlands have adopted more or less stringent limitations in some water catchment areas. As regards non–Member States, toxicologic limits for each compound are fixed with the exception of Switzerland, which has adopted the EEC limits (Table 2).

Table 1   Compliance of Parameter No. 55 of the Directive 80/778 in Member States at
           October 1989

Country	Compliance	Monitoring	Use Restrictions	Comments
Belgium	Yes (1988)	Programmed	No	EEC limits
Denmark	Yes (1988)	—	No	EEC limits
France	Yes (1989)	Running	No	EEC limits, Toxicologic limits for three products
Germany	Yes (1986)	Running	Yes	EEC limits
Greece	Yes (1986)	—	No	EEC limits
Ireland	Yes (1988)	Running by local authorities	No	EEC limits
Italy	Yes (1986)	Running by local authorities	Yes	EEC limits
Luxembourg	Yes (1985)	—	No	EEC limits
Portugal	—	—	—	EEC limits
Spain	—	—	—	EEC limits
The Netherlands	Yes (1984)	Running	Yes	EEC limits
U.K.	Yes (1989)	Running	No	EEC limits

From Premazzi G., Chiaudani G., and Ziglio G., EUR Report 12427, October 1989.

Table 2   Present Position About Pesticides in Some Nonmember States

Country	To Be in Force	Monitoring	Use Restrictions	Comments
Australia	1987	Done	No	Toxicologic limits for each product
Canada	1987	Done	No	Toxicologic limits for each product
Norway	1987	—	Yes	Limits for total pesticides
Sweden	1988	Running	50% Limitation in 1990	—
Switzerland	1985	No	No	EEC limits
U.S.	1988	No	No	Toxicologic limits for each product

From Premazzi G., Chiaudani G., and Ziglio G., "Scientific Assessment of EC Standards for Drinking Water Quality," EUR Report 12427, October 1989.

GERMANY

The most comprehensive compilation of the occurrence of active ingredients of pesticides in drinking water in Germany is published about every 3 months by the Federal Association of the German Gas and Water Industry (BGW). Over 40 active ingredients have been identified up to now in the µg/litre range (Table 3). Atrazine seems to be a problem in many federal Länder; it has been found in Bavaria in 33% of all results, and, in Baden–Württemberg, already 1 of 16 wells examined exhibits contents far above the limit of 0.1 µg/L; 1 out of 7 wells is contaminated at levels between 0.05 and 0.1 µg/L. The extensive Drinking Water Monitoring Program conducted during 1984 until 1986 in several Länder has shown the occurrence of pesticide residues at levels above the EC limit (Table 4). From a total of 16 active substances under the most stringent W1 class, only 7 are accessible for detection in this range (W1). However, many substances may have escaped detection because, even for those classified as W1 and W2, the analytical methods for the µg/L range are not always available (Table 5). From a total of 16 active substances under the most stringent W1 class, only 7 are accessible for detection in this range (W1). Similar figures exist for the less severe W2 substances, of which about 60% cannot be detected due to methodologic deficiencies (Table 6). W1 class includes pesticides with the most stringent restricted use; the W2 class includes less severe substances with restricted use. At present, in Germany, 35 substances are banned, 10 are partially banned, a further 23 are banned in all protection zones, and another 50 must not be used in inner zones of water catchment areas.

ITALY

Surveys carried out in 1986 to 1987 regarding the occurrence of atrazine, simazine, bentazon, and molinate in Italian regions, where rice and maize are produced (i.e., Lombardia, Veneto, Emilia-Romagna, Marche, Piemonte, and Friuli Venezia Giulia) revealed that:

Table 3   Pesticides Found in Groundwater (GW), Bank Filtrate (U), Wells (B), Springs (Q), or Drinking Water (TW) in the Former West Germany

Restricted Use in Water Catchment Areas	Use Not Restricted in Water Catchment Areas
Atrazine (GW,TW,B,U)	Azinophos-ethyl (U)
Bentazon (B)	Chlorphenvinphos (GW)
Bromacil (B)	Chloridazon (B)
Clopyralid (GW)	Chlortoluron (U,GW,TW)
1,2-dichlorpropan (B)	2,4-D (U,B)
1,3-dichlorpropen (GW)	Desethylatrizine (GW)
Dimethoat (GW,U)	Desethylsimazine (GW)
Dinosebacetat (B)	Disulfoton (U)
	Diuron (U)
Lindan (U,GW)	Isoproturon (B)
Metazachlor (U,GW,B)	Linuron (GW)
Methylisothiocyanate (B)	MCPA (GW)
Propazine (GW)	MCPP (Q,GW,B)
Pyridat (B)	Methabenzthiazuron (GW)
Simazine (GW,B,U,Q)	Terbutylazine (GW)
2,4,5-T (GW)	Metobromuron (U,GW)
	Metolachlor (U,GW,B)
	Metoxuron (U)
Not allowed at all	Monuron (U)
	Parathion-equivalents (U)
Aldrin (GW)	Pendimethalin (B)
Heptachlor (GW)	Prometryn (GW)
Hexachlorobenzene (U)	Vinclozolin (B)

From Premazzi G., Chiaudani G., and Ziglio G., "Scientific Assessment of EC Standards for Drinking Water Quality," EUR Report 12427, October 1989.

Table 4   Results of a Raw Water Monitoring Program

Analytical Findings Regions	No. of Sampled Well	Above the EC Limit		Notes
		No. of Wells	No. of Active Ingredients Detected	
Baden-Württemberg	65	6	2	Atrazine, bentazon
Bavaria	21	5	4	Atrazine, bentazon, chloridazone, simazine
Hesse	10	0	0	Simazine
North Rhine-Westphalia	45	3	1	Bentazon
Lower Saxony	38	2	1	Pyridate
Rhineland-Palatinate	5	2	2	Bentazon, CMPP
Slesvig-Holstein	22	4	1	Dichloropropane

From Premazzi G., Chiaudani G., and Ziglio G., "Scientific Assessment of EC Standards for Drinking Water Quality," EUR Report 12427, October 1989.

- among 318 sites examined, in most of the cases (93.4%) a single contaminant is present, namely:
  - atrazine (88%),
  - simazine (1%), and
  - bentazon (4.4%).
- In particular, atrazine is present in 279 sites with the following distribution:
  - in 49% at a level less than or equal to 0.1 µg/L;
  - only in 8% is it present at levels from 0.3 to 0.6 µg/L, with a few isolated peaks on the order 0.7 to 0.8 µg/L.

Table 5   Pesticides with Restricted Use in Water
                Catchment Areas

Active Ingredient W1 Ban	Detection Method in the µg/L Range	
	Available	Not Available
Aldicarb	+	
Amitrol	+	
Bromacil	+	
Chlorambene		−
DNOC		−
Dazomet		−
Dikegulac		−
Dichlorpropene	+	
Metam-Sodium		−
Methylisothiocyanate	+	
Sodium Chlorate		−
Oxamyl		−
Picloram	+	
Tebuthiron		−
Terbacil	+	
Thiofanox		−

From Premazzi G., Chiaudani G., and Ziglio G., "Scientific Assessment of EC Standards for Drinking Water Quality," EUR Report
12427, October 1989.

- In 21 sites (6.6%), only two herbicides were present simultaneously, namely;

  - atrazine with simazine: 2.8% (9 sites);
  - atrazine with molinate: 0.3% (1 sites);
  - atrazine with bentazon: 3.4% (11 sites).

At present, the situation is in compliance with the limits of the directive. Italian authorities state that since February 14, 1991, all derogations concerning the parameter 55 are expired. Consequently, since this date, one is no longer allowed to supply and distribute noncomplying waters in Italy.

THE NETHERLANDS

The Soil Protection Act (effective since January 1, 1987) forms the legal framework for soil and groundwater protection in The Netherlands. Two levels of protection are distinguished, i.e., the general protection level and a specific protection level. The general protection level will be realized by rules set by the national government. These rules may consist of effect-oriented (setting of quality standards) as well as source-oriented measures towards pollution sources. The Dutch soil and groundwater protection policy aims at maintaining the multifunctionality of the soil. This implies a quality level of soil and groundwater that allows for the soil to function as a growing place for agricultural crops and as a reservoir for drinking water abstraction, both now and in the future. Diffuse sources of pollution, e.g., from agriculture, will have to meet the following criteria:

- reaching an equilibrium between load and crop extraction as much as possible;
- manmade chemicals must meet requirements on persistency and leachability;
- the concentration of chemical substances in soil and groundwater will have to meet quality standards.

Since 1974, the protection of areas around pumping stations for public water supply was seen as a governmental task in a number of provinces. In 1980, the Dutch waterworks association published guidelines for the dimensioning of groundwater protection areas, which were accepted by the national government in 1984. Protection areas are based on the travel time concept with the outer boundary at 25-year flow distance from the pumping stations. This implies that only a part of the recharge area will receive additional protection based on provincial regulations. As of January 1, 1989, all 12 provinces are obliged to draw up a groundwater protection plan and a protection ordinance under the Soil Protection Act.

Table 6   Pesticides with Restricted Use in Water Catchment
          Areas

Active Ingredient W2 Ban	Detection Method in the µg/L Range	
	Available	Not Available
Afrikal 67		−
Alloxydim-Salt		−
Anizalin		−
Asulam		−
Atrazin	+	
Bac. Thuring	+	
Bendiocarb		−
Benodanil		−
Bentazon	+	
Carbetamide		−
Carbofuran	+	
Chinonamide		−
Chlorthiamide		−
Cyanazin	+	
Dicamba		−
Dichlobenil		−
Dimethoate	+	
Dinoseb(-Acetate)	+	
Dinoterp	+	
DNOC		−
Ethidimuron		−
Ethiofencarb		−
Ethoprophos	+	
Fenaminosulf		−
Hexazinon		−
Isocarbamid		−
Karbutilate		−
Lindan	+	
Maleic Acid Hydrazide		−
Mefluidid		−
Metalaxyl	+	
Metazachlor	+	
Methomyl		−
Methylbromid		−
Monochlorbenzene		−
Napropamid		−
Oxycarboxine		−
Pirimicarb		−
Propoxur	+	
Pyridat	+	
Secbumeton		−
Terbumeton		−
2,4,5-T and Salts	+	
TCA		−
Thiazfluron		−

From Premazzi, G., Chiaudani, G., and Ziglio, G., "Scientific Assessment of EC Standards for Drinking Water Quality," EUR Report 12427, October 1989.

The policy against pesticide pollution requires an authorization at the level of the national government. This holds too for restricted authorization concerning the use of pesticides in water protection areas. In 1987, about 10% of the 314 pesticides (active chemicals) that may be used in agriculture were prohibited in these areas. Moreover, for another 17 pesticides the use in these areas was restricted with respect to time of application and soil composition.

In fact, since 1974, the restrictions in use were advised to the provincial authorities to be inserted in provincial ordinances. However, enforcement of these regulations left much to be desired. Besides, in a number of provinces, these ordinances did not even exist. Since 1982, the Pesticides Act regulates the use of pesticides in protection areas. Since 1983, the prohibitions as well as the restrictions are mentioned in the label attached to the package. A list of pesticides not admitted in groundwater protection areas is given in Table 7.

Table 7   Residues from Pesticides Not Admitted
           in Groundwater Protection Areas in
           The Netherlands.

Compound	Concentration (µg/L)
1,3-Dichloropropene	< 0.1–80
1,2-Dichloropropane (Contamination in 1,3-Dichloropropene)	< 0.1–160
1,2,3-Trichloropropane	0.1–30
1,2,2-Trichloropropane	0.1–1
Aldicarb (including metabolites)	4.5–130
Metolachlor	< 0.04–0.21
Bromacil	0.1–2.0
Bentazon	< 0.1–1.0
Atrazine	< 0.02–0.8
Desethyl-Atrazine	< 0.05–0.3
Desisopropylatrazine	< 0.05–0.28

From Premazzi, G., Chiaudani, G., and Ziglio, G., "Scientific Assessment of EC Standards for Drinking Water Quality," EUR Report 12427, October 1989.

Groundwater quality problems arose especially from the pesticides authorized before 1975. These older pesticides still take a dominant position on the Dutch market. In 1985, the National Institute of Public Health and Environmental Protection started a research program on 14 pesticides in groundwater in four areas with soil types vulnerable to leaching. Though detection limits of the studied pesticides ranged from 0.2 to 10 µg/L, four pesticides showed concentrations above the drinking water standard of 0.1 µg/L, viz: atrazine (max. 0.6 µg/L), dinoseb (max. 9.2 µg/L), 1,3 — dichloropropene (max. 80 µg/L), and aldicarb (max. 130 µg/L). The latter two pesticides are already prohibited in groundwater protection areas under the Pesticides Act, which is justified by these results.

In order to minimize the groundwater pollution potential of pesticides, the following measures will be taken in The Netherlands:

- critical reassessment of the "older" pesticides as to whether authorization for the Dutch market can be continued;
- the number of pesticides that are not permitted for use in groundwater protection areas will be raised;
- the behavior of pesticides in soil types vulnerable to leaching will be given more weight in the authorization procedure;
- periodical testing of application devices will be required.

UNITED KINGDOM

The U.K. has formally requested a revision of this parameter so that limits more closely related to health risks can be used. The U.K. government considers the MACs for total and individual pesticides inappropriate for the following reasons:

- it is not possible to assess whether the MAC for total or individual pesticides has been exceeded, because current analytic methods are unable to detect many pesticides even at concentrations above the MAC;
- no account is taken of the widely differing toxicities of individual pesticides in applying the same value to each pesticide irrespective of its toxicity.

Information from water undertakers during the last years (as from 1984) on pesticides detected in water supplies have led to the following general conclusions.[5]

- triazines, particularly atrazine, have been detected in many supplies, derived from both surface and groundwater sources;
- phenoxyalkanoic acid compounds, particularly mecoprop and MCPA but also 2,4-D and MCPB, have been detected in surface- and groundwater-derived supplies in intensive arable agricultural areas;
- other pesticides such as the persistent organochlorine compounds, have been detected occasionally, but they were probably associated with pollution incidents (e.g., spillage) rather than regular use;
- water undertakers have not attempted to monitor those pesticides that might be present in water supplies at very low concentrations and for which reliable and sufficiently sensitive methods of analysis are not available.

Much of the information on the levels of pesticides in the aquatic environment is derived from analysis of water in public supply or from monitoring of water sources to assess their suitability for abstraction. Recent surveys by the Drinking Water Inspectorate, Friends of the Earth, and the Institution of Environmental Health Officers (IEHO) have revealed many cases where the pesticide limits prescribed in the EC Directive have been exceeded.[6] The IEHO survey in London found that nearly two-thirds of 174 samples of drinking water in a 6-month survey carried out in 1989 to 1990 exceeded the EC limits, mainly because of the presence of the triazine herbicides atrazine and simazine. These findings are in line with those reported by the Drinking Water Inspectorate for roughly the same area. Over England and Wales as a whole, there were marked local variations in the incidence of these herbicides in water sources and supplies. The Drinking Water Inspectorate reported that 33 individual pesticides were detected at concentrations above 0.1 µg/L in 1989 and 34 in 1990, a total of 43 individual pesticides in the 2 years. This amounted to slightly over 2% of the analyses carried out. Atrazine, simazine, and mecoprop were detected most frequently in both years. The Inspectorate stated that the concentrations of pesticides detected were far lower than amounts that are known to be harmful or are likely to damage public health.

Table 8 shows the ranges of pesticide concentrations found in the samples that exceeded the MAC and the number of sources affected. (Reservoirs, blended, and unknown sources are excluded.) More breaches and higher concentrations were recorded for surface water than for groundwater. The breaches occurred mainly in the Anglian, Seven-Trent, Thames, and Wessex regions.

THE EC VIEWPOINT ON PESTICIDES IN DRINKING WATERS

Directive 80/778 has provided the occasion for national assessments of drinking water quality and for reviews of national policies. A particular effect has been to stimulate discussions on difficulties of application (technical/scientific, legal, and regulatory) and on specific parameters (e.g., pesticides, nitrates) and on the adequacy of the Directive. As regards compliance with the Directive, water suppliers in most Member States regularly encounter situations in which they must either cut off the water or else supply water that fails the pesticide parameter of the Directive.

Discontinuing the supply was considered nearly always to present far greater risks than continuing to supply water containing pesticides at the levels typically found. In these circumstances, the possible risk to the health of consumers is of the greatest importance. Assessments of the toxicity of a range of pesticides have been made by several organizations. These have been used to determine the concentrations of certain compounds that, on the basis of available information, are judged likely to have no effect on human health. In some cases these "safe" levels are greater than the 0.1 µg/L MAC of the Directive 80/778.

Opinion in Member States was divided on the objectives of the pesticide standard. A number felt that the limit of 0.1 µg/L was an important component of environmental protection policy and was necessary to maintain pressure on users to reduce the application of pesticides. Another group felt that, on the contrary, the function of the EC Directive 80/778 should only be to protect the health and well-being of consumers, and it should not be used for any other purposes. Other EC

Table 8   Comparison Among Concentrations of Some Pesticides Found in Waters and MAC of the Directive 80/778

Pesticide	Groundwater		Surface Water	
	No. of Sources Exceeding MAC	Range of Concentrations (μg/L)	No. of Sources Exceeding MAC	Range of Concentrations (μg/L)
Atrazine	32	0.11–0.48	32	0.11–1.85
Simazine	14	0.11–0.42	26	0.11–1.97
Propazine	4	0.11–0.27	2	0.25–0.47
Mecoprop	1	0.36	13	0.12–0.98
MCPA	1	0.42	7	0.16–0.84
MCPB	7	0.16–0.41	—	—
2,4-D MCPA	—	—	8	0.17–0.96
2,4-D	3	0.24–0.56	2	0.11–0.56
2,4,5-T	5	0.26–0.50	—	—

From "Freshwater quality. Sixteenth Report," HMSO (1992).

Directives dealt with environmental protection. This group felt that, while the existing MAC should be maintained for new substances of unknown toxicity, for substances for which adequate toxicologic data was available, the limit should be set on the basis of toxicology.

The unsatisfactory state of practical implementation of Directive 80/778/EEC in Member States with reference to pesticides and related products (parameter 55) can be summarized as follows:[7]

1. The Commission has the power to propose legislation, including Directives, to the Council of Ministers, and the duty to ensure that Member States apply Community legislation correctly. It is important to note that it is the Council that legislates. There is a clear separation of powers between the legislative (the Council and the European Parliament) and the executive (the Commission). The Commission must ensure that the provisions of the EEC Treaty and the measures taken by the EEC Institutions are applied (Article 155 of the EEC Treaty). On the other hand, Member States must facilitate the achievement of the Community's, (and therefore, the Commission's tasks (Article 5 of the EEC Treaty). In these circumstances, the Commission's right to verify how the drinking water Directive is applied cannot be exercised by the Commission unless it possesses all the necessary information concerning the application of the Directive by Member States. The Commission therefore has a right to request from Member States any information necessary to demonstrate compliance with the Directive, and Member States have the duty to supply the information requested by the Commission because they have to facilitate the Commission's tasks of ensuring the application of the Directive.
2. That is not to say that the Commission has given no thought to the possibility of proposing changes to the Directive. There are two ways in which this might be done within the framework of the EEC Treaty. The first is the usual procedure of presenting a proposal for a Council Directive, in this case for one to modify Directive 80/778/EEC. This is potentially a cumbersome process and, in order to simplify the process of adapting Directives, the Council was invited, in 1988, to agree to set up a Committee to facilitate adaptation of Directive 80/778/EEC to technical progress. This process was to be used where adaptation could be achieved without altering the level of ambition of the Directive or imposing significant economic burdens on Member States. So far, the Council has not agreed to set up the Committee.
3. The proposal for the Directive was adopted in 1980, after lengthy discussion. It is noteworthy that the only change made to the pesticides parameter was the addition of the reference to PCBs and PCTs. The parametric values of 0.1 μg/L for any pesticide and of 0.5 μg/L for all pesticides together remained as in the Commission's proposal. Given that the Directive was adopted unanimously, after discussion that took place over 5 years, it must be concluded that the Council was satisfied with the definition of the parameter and with the Maximum Admissible Concentrations. Member States were surely satisfied that they could meet those standards; it is difficult to understand why Member States now have a problem with the parameter.

4. The Commission has taken note of the problems and given considerable attention to how they should be addressed. A way forward, which has been suggested by some commentators, would be to invite the Council to change the Directive so that standards are set on the basis of toxicologic evidence for individual parameters. In most cases, the standards would be much greater than the present value of 0.1 µg/L, although a few would be stricter. The Commission has so far rejected this approach. Rather, it prefers the view that pesticides have no place in drinking water. The MAC value should therefore be set near to the limit of detection. To set a higher figure or a range of higher figures would not solve the problem of pesticides in drinking water, it would only conceal it. Some Member States have suggested that the standards of 0.1 µg/L and 0.5 µg/L are unworkable because suitable methods of analysis are not available. The Commission understands that suitable methods of analysis are available for pesticides of interest and maintains, in any case, the analysis of pesticides does not present a fundamental problem.

5. The Commission's policy is well established and clear cut: the quality of drinking water should be maintained to a high standard so as to provide the consumer with a high level of protection. Pesticides are not naturally present in the environment, and they should not be present in water. The Commission does not, therefore, have plans to invite the Council to change the pesticide parameter.

6. However, there is no way to avoid the central issue that Member States are legally obliged to apply the Directive. They must, therefore, take the appropriate measures to ensure that its standards, including those for parameter 55, are respected. The choice of methods is for Member States. The Commission's responsibilities extend far beyond the application of the Directive. It must take a broad view and be aware that controls applied for one purpose might well have consequences elsewhere. This is certainly the case with pesticides. Even so, the Commission is convinced that the best way to prevent pollution by pesticides is at the source.

7. Controls on the use of pesticides so as to protect raw water provide, at first sight, a useful way forward. However, such controls alone cannot provide all the protection needed, for the following reasons:

   • pesticides are already present in the soil and in groundwater;
   • no system of control can totally prevent illicit or incorrect use; and
   • accidental spillages can arise in many ways, not all related to permitted uses.

   Therefore, in most case it will also be necessary to remove pesticides from raw water. Activated carbon has been used successfully to do this, and development work is in progress in a number of Member States.

8. A further important step will be to replace the more persistent pesticides with others that degrade rapidly to harmless products. The benefits of doing this will take some time to show, though, because of the large amounts of other pesticides now in the environment. However, it will have important long-term advantages.

9. There is a potential conflict between controls on the use of pesticides, and the provision of facilities to remove pesticides from raw water. It could be said that controls should remove the need for removal at treatment works and that the existence of means to remove pesticides from water would make controls unnecessary. Neither argument is wholly correct. For a number of reasons, it is desirable to restrict the use of pesticides so as to protect the quality of surface and groundwaters. However, such restrictions cannot give a guarantee that drinking water will always meet the Directive's standards without further treatment.

10. Pesticides have an important part to play in agriculture and land management, but it must be accepted that their benefits can entail serious drawbacks. There is an advantage to a more questioning approach to their use. Thus, in considering what controls to apply, the following questions could be asked:

    • Is it necessary to use a pesticide at all?
    • Is it possible to use a less persistent pesticide?
    • Could the rate of application be reduced?
    • Should methods of agriculture that rely for their success on the systematic use of pesticides be introduced?

11. Some Member States, while accepting the need to comply with the Directive's standards for pesticides, have pointed out that compliance cannot be achieved immediately. Several years will be needed for the necessary work to be carried out, and, during this time, there will be a breach of the Directive. It has been suggested that the Directive should be changed so as to provide time for this work to be completed. The Commission has noted these arguments but finds it difficult to agree with them. Article 20 of the Directive provided for derogations from the general obligation; Article 19 states that all drinking water must comply with the Directive standards within five years after notification. However, no new derogations are possible, because the 5-year period has expired. The conclusion is that Member States in which drinking water does not meet the Directive's standards for pesticides are in breach of their obligations.

12. It has been suggested that if drinking water does not comply with the Directive's standards for pesticides then the only correct course is to cut off the public water supply. This might be appropriate in cases of serious pollution but the supply for a short while of water a little outside the Directive's standards will entail little danger for public health. However, cutting off the public water supply will certainly have serious implications for public health. Article 10 of the Directive allows the MACs to be exceeded in emergencies, provided that there is no unacceptable risk to public health and that there is no other way to maintain the water supply. In extreme cases this article might be used with respect to pesticides. However, it relates to emergencies. It could not be used to cover a case in which pesticide concentrations had been rising steadily over a period of years: it might be applicable where there had been a rapid and unforeseen increase in pesticide concentration. However, in these circumstances the duty of Member States is clear. They must arrange for drinking water to meet the Directive's standards as soon as possible. Until it does they are in breach of the Directive.

13. The Commission is satisfied that the precautionary approach of setting a MAC for pesticides near the analytical limit of detection remains correct. Pesticides have no place in drinking water and Member States must ensure that the Directive's limits are respected. This might entail practical problems, but there is no fundamental difficulty. There is no case for proposing any change to the pesticide parameter.

EEC DIRECTIVE CONCERNING THE PLACING OF PLANT PROTECTION PRODUCTS ON THE MARKET

The Council Directive of July 15, 1991 (91/414/EEC) now regulates the placing of plant protection products on the market. It requires in Article 4, *inter alia,* that Member States shall not authorize a plant protection product unless a number of requirements are satisfied, pursuant to the uniform principles provided for in Annex VI. Moreover, according to Articles 18 and 23 of the Directive, the uniform principles shall be adopted by the Council, acting by a qualified majority on a proposal from the Commission, within 1 year from the date of notification of the Directive.

The uniform principles aim to ensure that all Member States, in making decisions with respect to authorization of plant protection products, will apply the requirements of Article 4 in an equivalent manner and at the high level of protection of human and animal health and of the environment sought by the Directive. The Commission considers that, as the uniform principles are addressed in the first instance to the public authorities in the Member States in charge of plant protection products authorization, these principles should concentrate on the aspects of the evaluation of the data submitted by applicants according to Article 13 of the Directive and of decision making on the basis of these evaluations.[8]

Although the Directive 91/414/EEC does not directly relate to the monitoring of water quality, it does have potentially significant implications for the interpretation of data on the environmental occurrence of pesticides in water. The Directive states that a "plant protection product will only be authorized in the EC if it has no unacceptable influence on the environment, with regard to its fate and distribution and particularly contamination of water including drinking water and groundwater". Furthermore, no product will be authorized unless the nature and quantity of its active substances and its residues, resulting from authorized uses, and which are of toxicologic or environmental significance, can be determined by appropriate methods in general use. In order to assure the harmonization of the product evaluation process a series of uniform principles for the evaluation of data for risk assessment and risk management within the process of authorization are now under development. In the latest draft of the uniform principles, for the evaluation of the influence of pesticides on the water environment the following is proposed.[8]

- No authorization shall be granted if the concentration of the active substance or of relevant metabolites or degradation or reaction products to be expected in drinking water extracted from groundwater after use of the plant protection product under the proposed conditions of use exceeds 0.1 µg/L
- No authorization shall be granted if the concentration of the active substance or of relevant metabolites or degradation or reaction products to be expected after use of the plant protection product under the proposed conditions of use:
  - in drinking water extracted from surface water exceeds 0.1 µg/L or
  - in surface water has an impact deemed unacceptable on nontarget species.

GENERAL PRINCIPLES

With regard to current scientific and technical knowledge, Member States shall.[8]

- assess the performance in terms of efficacy and phytotoxicity of the plant protection product for each use for which authorization is sought and
- identify the hazards arising, assess their significance, and make a judgment as to the likely risks to humans, animals, or the environment.

In the evaluation of applications submitted, in accordance with the terms of Article 4, Member States shall ensure that evaluations carried out have regard for the proposed practical conditions of use. This must include, in particular, the purpose of use; the dose; the manner, frequency, and timing of applications; and the nature and composition of the preparation. Whenever possible, Member States shall also take into account the principles of integrated control. Furthermore, Member States shall have regard for the agricultural, plant health, or environmental (including climatic) conditions in the areas of the envisioned use. In interpreting the results of evaluations, Member States shall take into consideration possible elements of uncertainty in the information obtained during the evaluation, in order to ensure that the chances of failing to detect adverse effects are minimized. The decision-making process shall be examined to identify critical decision points or items of data, for which uncertainties could lead to a false classification of risk.

At first, the evaluation shall be based on the best available data or estimates reflecting the realistic conditions of use of the plant protection product and will result in a realistic worst-case approach. This should be followed by a request evaluation, taking into account potential uncertainties in the critical data and a range of use conditions that are likely to occur, to determine whether it is possible that the initial estimation could have been significantly different.

INFLUENCE ON THE ENVIRONMENT

In the evaluation of the fate and distribution of the plant protection product in the environment, Member States shall have regard for all compartments of the environment, including biota, and in particular for the following compartments:

- Member States shall evaluate the possibility for the plant protection product to reach the soil under the proposed conditions of uses; if this possibility exists they shall evaluate the rate and the route of degradation in the soil, the mobility in the soil, and the change of the total concentration (extractable and nonextractable) of the active substance and of relevant metabolites and degradation and reaction products that could be expected in the soil in the area of envisioned use after use of the plant protection product according to the proposed conditions of use.

- Member States shall evaluate the possibility for the plant protection product to reach the groundwater under the proposed conditions of use; if the possibility exists that the plant protection product can reach the groundwater they shall estimate, using, where appropriate, a suitable calculation model, the concentration of the active substance and of relevant metabolites, degradation and reaction products that could be expected in the groundwater in or from the areas of envisioned use after use of the plant protection product according to the proposed conditions of use.
- Member States shall evaluate the possibility for the plant protection product to reach surface water under the proposed conditions of use; if this possibility exists they shall estimate, using, where appropriate, a suitable calculation model, the short-term and long-term predicted environmental concentration of the active substance and of relevant metabolites and degradation and reaction products after use of the plant protection product according to the proposed conditions of use.
- Member States shall evaluate the possibility for the plant protection product to be dissipated in the air under the proposed conditions of use; if this possibility exists, they shall estimate, using, where appropriate, a suitable calculation model, the concentration of the active substance and of relevant metabolites and degradation and reaction products that could be expected in the air after use of the plant protection product according to the proposed conditions of use.

These evaluations will take into consideration the following information:

- the specific information on the fate and behavior in the soil, water, and in air and the results of the evaluation thereof;
- other relevant information on the active substance such as:
  - molecular weight
  - solubility in water
  - octanol/water partition coefficient
  - vapor pressure
  - volatilization rate
  - hydrolysis rate in relation to pH and identity of breakdown products
  - dissociation constant
  - photochemical degradation rates in air and water and identity of breakdown products;
- all relevant information on the plant protection product, including the information on distribution and dissipation in soil, air, and water;
- possible routes of exposure:
  - drift
  - run-off
  - overspray
  - discharge via drains
  - leaching
  - atmospheric disposition;
- where relevant, other authorized uses of plant protection products in the area of envisioned use containing the same active substance or that give rise to the same residues;
- where relevant, data on dissipation including transformation and sorption in the saturated zone.

EC POLICY ON GROUNDWATERS

Recognizing that the groundwater resources are limited and that the pollution (e.g., nitrate and pesticide leaching, landfill contamination) endanger the sustainable use of groundwater by people and the ecosystem, a program of action was agreed upon at the EC Ministerial Seminar on groundwater (November, 1991). The declaration included agreement on the following:

1. to preserve the quality of uncontaminated groundwater;
2. to prevent further deterioration of contaminated groundwater;
3. to prevent long-term overexploitation; and
4. to replenish groundwater systems, where appropriate, to a sustainable level.

These goals can only be achieved by an integrated approach. Surface water and groundwater should be managed as a whole, paying equal attention to both quantity and quality aspects, while interactions with soil and atmosphere are taken into account, and the water management policy is integrated within a wider environmental framework and with other policies, such as land use planning, agriculture, industry, energy, transport, and tourism.

An action program was proposed, to be prepared by the EC Commission by mid-1993, and including the following measures:[9]

1. The mapping and characterization of groundwater systems and the identification of:
   a. aquifers being contaminated and/or affected by overexploitation;
   b. specific threats and their evaluation;
   c. specific uses and demands.
2. The establishment of adequate monitoring and reporting schemes.
3. The establishment of systems of water resource planning and the definition and implementation of measures; this included plans for rehabilitation of polluted groundwaters and soils and of overexploited aquifers and aquifers affected by sea water intrusion.
4. The introduction of geographic zoning for rights and priorities of uses and socioeconomic activities, aimed at stricter groundwater protection.
5. Introduction of a system of permits, general rules, and codes of practice for water extraction and for those activities that could lead to pollution of groundwater or soil.
6. The establishment of general rules for the disposal of waste.
7. Measures to promote efficient water use and reduction of freshwater consumption.
8. Changes in agricultural practices to prevent pollution, for example, from nitrates as provided for in the nitrate directive, including the progressive reduction of fertilizer application to the level required for good agricultural practice.
9. The progressive replacement of persistent and accumulative pesticides by degradable and less harmful products.
10. The progressive limitation of the use of plant protection products and other pesticides for nonagricultural purposes, to prevent adverse effects on groundwater, and with the objective of ensuring, where technically feasible, that, for already contaminated groundwaters, after a period of sanitation, their concentrations are brought back to at least the values for drinking water.
11. Economic, financial, and fiscal instruments, such as appropriate pricing, possible use of Community funds, levies, and fiscal incentives to support the new policies.
12. The strengthening of public information and public awareness, as well as education and training schemes.
13. Research programs to support improvements in water management and protection of water resources.

The new policies are to be supported by economic and financial measures, research programs, education and training schemes, and steps to improve public awareness. The declaration recognized the need to incorporate the policies into other areas of the Community. The integration aspect reflects the general policy in the field of environment as outlined of the 5th Action Program adopted by the EC. Groundwater protection is a particularly good example to demonstrate the need for action in many sectors and addressing many factors. It has to include training, education measures, information, and the promotion of the behavioral changes. The EC and its Member States will, in the coming years, reinforce the qualitative and quantitative protection of groundwater. The increasing stress on groundwater and soil from more and more intense human activities and increasing land use requires action from now into the 21st century. Taking account of all interactions and integrating water management and protection policy in all fields relevant for water are expected to be the main characteristics of the development in the coming decade. Some actions have already started, but the present Directive will still provide the basic rules throughout the Community for some time to come. However, legislation, on the Community or Member State level, dealing mainly with other subject matters than groundwater protection, will also become more and more important.[10]

U.S. EPA PROPOSED STRATEGY

The EPA's long-term approach for managing the pesticides in groundwater addresses three key issues associated with the environmental goal, the prevention strategy, and the response strategy. The factors and options that the EPA considered are first described, followed by a detailed description of the EPA's proposed policy choices, based on consideration of legislative statutes, long-term policies for pesticides registration, and general strategy of groundwater protection already existing in the U.S.[11]

ENVIRONMENTAL GOAL

EPA's proposed environmental goal will be to manage the use of pesticides in order to protect the groundwater resource.[11] The two key questions that the EPA has addressed in defining its environmental goal for pesticides in groundwater are (1) what waters to protect and (2) what criteria determine protection.

The EPA has addressed these questions within the context of its regulatory statutes* and other EPA's policies** including its basic policies for pesticide registration. In developing the strategic approach for pesticides in groundwater, the EPA considered several aspects of its policy for groundwater protection to determine the extent to which they were appropriate for the pesticide effort.

The following discussion outlines the factors and options that the EPA considered in determining its environmental goal for addressing the pesticides in groundwater concern.

What Waters to Protect

The first issue in determining what waters to protect was whether protection efforts should be focused on assuring the quality of the groundwater resource or the quality of water actually delivered to the tap for drinking. The difference between these two options is that the latter choice could allow for the contamination of groundwater in areas where there is adequate treatment capability for reducing contamination to acceptable levels before it is provided for drinking.

A second key decision was to determine whether to spread its protection efforts widely to protect all groundwaters or to narrow its protection focus on some selected waters that provide more valuable use. The EPA had a number of options, based upon the different degrees of use and value of groundwater including the following:[11]

- Protection for all groundwaters;
- Protection for all groundwater resources that are current or potential drinking water sources or of ecologic importance;
- Protection for only that groundwater currently used as a drinking water source;
- Protection for only that groundwater supplying public drinking water systems;

* Federal Insecticide, Fungicide, and Rodenticide Act (FIFRA)
  Safe Drinking Water Act (SDWA)
  Clean Water Act (CWA)
  Resource Conservation and Recovery Act (RCRA)
  Comprehensive Environmental Response, Compensation and Liability Act (CERCLA or "Superfund")

** In 1984, the EPA issued the Groundwater Protection Strategy, setting out general goals and objectives for the protection of groundwaters.

- Differential groundwater protection; with this option, all groundwaters could be considered for protection, but priorities and stringency of prevention and response efforts would be based on the relative use and value of the groundwater.

What Criteria Determine Protection

The second key part of the EPA's goal was identifying the criteria to be used for determining protection. The options that the EPA considered for the criteria for defining protection included the following:[11]

1. **Pristine** — Protection criteria based on pristine conditions would require prevention efforts that are designed to achieve zero contamination in waters chosen for protection. With this option, response efforts would have to restore protected groundwaters to zero contamination levels. While an unspoiled environment is desirable, there are a number of practical barriers to achieving such a goal. First of all, zero is not scientifically measurable and, therefore, not attainable. Second, efforts to attain pristine groundwater would be enormously expensive and would limit greatly the type of water that could be targeted for such protection.

2. **Maximum Contaminant Level Goal** (MCLG) — The SDWA requires EPA to set a Maximum Contaminant Level Goal (MCLG) for any potential drinking water pollutant at a level that has no known or anticipated adverse effect. The EPA's policy has been to set the MCLG at absolute zero for a chemical proven to be carcinogenic in humans or a probable carcinogen in animal studies. For noncarcinogens, the MCLG is set at a level corresponding to a percentage of the Acceptable Daily Intake (ADI). Where the concern is for a carcinogenic pesticide in groundwater, the use of the MCLG as the protection criterion would have the same problems as the pristine option. In this case, a measurable goal could not be established because a zero level is not scientifically measurable. The MCLG criterion approach does not consider the relative potency of different carcinogens, nor does this approach consider that a noncarcinogen could pose a much greater risk than a weak carcinogen. Furthermore, as with the pristine option, the cost of protecting, monitoring, and restoring groundwater based on a zero level may limit greatly the type of waters that could be targeted for such protection.

3. **Negligible Risk** — The EPA has applied the concept of negligible risk to carcinogens. Substances shown primarily through controlled animal studies to have the capacity to cause cancer are considered to be carcinogens. In assessing the risks of these substances, the EPA assumes that there is no level of exposure that has a zero chance or risk of an adverse effect. However, the risks are assumed to be proportional to the level of exposure. Therefore, at some exposure level, the risk posed by a carcinogen can be considered to be significant or negligible ("worst case" or "upper bound", assumption: cancer risk one in a million or less). The use of negligible risk as the basis for a protection criterion provides a level that is considered very protective of human health. Such a goal could be measurable, but, for some chemicals, the concentration of a chemical at a negligible risk level may not be detectable. Because of its stringency, a standard based on the concept of negligible risk may not be achievable as a clean-up standard for certain pesticide contamination incidents.

4. **Maximum Contaminant Level (MCL) Standard** — Following the development of an MCLG, the SDWA requires the EPA to establish a Maximum Contaminant Level (MCL) that is the enforceable standard used to guard the adequacy of the drinking water provided by public water systems. The SDWA requires the EPA to set the MCL as close as possible to the MCLG, taking into account the cost and feasibility of measuring and reducing the contamination. MCLs are set in a range that is considered protective of human health. MCLs as protection criteria have already been established for public water systems, and, as such, they can provide measurable goals. In some cases, though, MCLs may be difficult to achieve, particularly when they are used to protect private well water or the groundwater resource. As with the previous options, MCLs are not based on a consideration of the benefits of the contaminants to society.

5. **Unreasonable Risk** — The Federal Insecticide, Fungicide, and Rodenticide Act (FIFRA) is a risk–benefit statute and requires the weighing of the human health risks posed by a

pesticide's use against its benefits. In some cases, concentrations posing a nonnegligible risk may be tolerated if the benefits of the pesticide's use are found to be uniquely critical. The key question for EPA was how to develop a goal for protecting groundwater from pesticide contamination that was compatible with the requirements under each of its legislative mandates.

EPA'S PROPOSED POLICY POSITION

The EPA will use a Differential Protection Approach to Protect the Groundwater Resource

With this approach, the EPA will focus protection efforts on groundwaters that are current or potential sources of drinking water or that are vital to fragile ecosystems. Additional measures may be taken to ensure protection of certain "High Priority Groundwaters". Since the natural cleansing processes of groundwaters can be extremely slow, contamination of this resource poses a potential hazard for future generations. Contaminated groundwater could be used as a drinking water source before being tested for contamination. Where contamination is identified, efforts could be made to restore the groundwater for use as a drinking water source, but this option could present insurmountable technological difficulties and exorbitant costs. In practice, alternative water sources or point-of-use treatment, if technology is available, may be the only viable preventive measures for a community. The EPA may have to assume that contaminated groundwater that is a potential drinking water source could eventually result in human exposure. The EPA's protection goal, therefore, includes groundwater resources that are potential as well as current sources of drinking water. As part of its differential protection approach, the EPA will take additional measures to protect "high priority groundwaters" that may serve as irreplaceable sources of drinking water for sizable communities, or may be vital to the continued survival of endangered species or critical ecosystems. For these situations, additional measures, beyond baseline protective efforts, may be required to provide further assurances that these "high priority groundwaters" are not contaminated or that they have priority for corrective actions, if already contaminated.

The EPA will use MCLs, as Defined Under the SDWA, as Reference Points for Helping to Determine Unacceptable Contamination of Groundwaters that are Drinking Water Sources

When no MCL exists, EPA will use interim drinking water protection criteria as its reference points. These will be equivalent to an MCLG for noncarcinogenic pesticides and to a negligible risk level for carcinogenic pesticides. EPA will also use ecologically based protection criteria as reference points for helping to determine unacceptable contamination of groundwaters. Under the SDWA Amendments, the EPA is moving to establish new MCLs for potential drinking water contaminants, including a number of pesticides. Since an MCL is developed for human health reasons, these levels could pose concern for fragile aquatic ecosystems risks to endangered species. Because the EPA's mandate is to protect the environment as well as public health, the EPA also will establish ecologically based protection criteria. For response actions, the above criteria will be used as reference points to help determine where and what measures should be taken to protect human health and the environment should unacceptable levels of contamination be reached. For prevention efforts, the protection criteria, described above, will be used as reference points for helping to define unacceptable levels of contamination and to determine when certain pesticide management actions may be needed to reduce the likelihood of groundwater contamination reaching or exceeding such levels.

When a pesticide is found in groundwater that is a current or potential drinking water source, additional management actions may need to be triggered. The need for, and the stringency of, these management actions will depend on the number of sites where the pesticide is detected, and whether the levels found are above or moving toward a reference point. The EPA's presumed rebuttable will be that the risks posed by pesticide contamination of a local underground source of drinking water, at or above the MCL or an *interim* reference point, will be more significant than the benefits derived from the pesticide's use in the area. Depending on the frequency and scope of such contamination, the EPA will consider the cancellation of the pesticide's use in such an area as an appropriate response. However, a number of factors would need to be considered that could make cancellation inappropriate. For example, the benefits of using the pesticide in an area could be substantial, and there may be management measures, other than cancellation, that could reduce contamination to acceptable levels. Also, the cancellation of a pesticide's use would not be appropriate if the groundwater contamination resulted from unusual circumstances such as an accident. Before taking action to mitigate groundwater contamination threats, the EPA must also consider such action in the context of overall pesticide exposure risks, including risks to applicators and farm workers and risks to the general population through dietary exposures.

PREVENTION POLICY AND PROGRAM

This section discusses the options and factors that the EPA considered in developing its prevention strategy.[11] In designing its program for preventing groundwater contamination by pesticides, the EPA addressed two strategic questions: (1) how to address local variability and (2) what are the appropriate federal–state roles. An important consideration, related to the latter issue, was determining the role of the pesticide user and registrant in groundwater protection. Another key consideration was how to control further contamination of groundwater once a pesticide has been detected in an area.

How to Address Local Variability

Ideally, prevention measures would be tailored so that they provide the optimum level of protection without undue restrictions. Any protective measures above this optimum line for a given level of vulnerability would provide more protection than is actually needed. Protective measures below this optimum line for a given level of vulnerability would provide less protection than is actually needed. Three possible management approaches for addressing the variability in groundwater vulnerability exist:[11]

1. **Uniform approach** — A uniform approach essentially ignores area variability in vulnerability and applies the same prevention measures uniformly to all areas. A worst-case basis could result in a great deal of overprotection for less vulnerable areas, while a moderate-case basis could result in both overprotection of low-vulnerability areas and underprotection of high-vulnerability areas. The obvious drawback of underprotection is the likelihood of groundwater contamination exceeding acceptable levels. The problem with overprotection is that it could result in the loss of valuable pesticide uses in important farming areas of the country — a situation that would not be compatible with the intent of FIFRA.
2. **Uniform baseline prevention measures with local exemptions and additional special measures** — Uniform baseline measures would be established for a pesticide in the same manner as the above option, but area-specific exemptions would be allowed for those places that would otherwise be overprotected. Such an approach could be implemented through a permit system in which exemptions from baseline requirements would be allowed for less vulnerable areas. Additional special measures could be specified on a label and the user made responsible for determining if they are applicable to the user's particular area. This approach has some major limitations. A permit system carries major administrative costs and can place significant burdens on pesticide users. It would be highly impracticable for the EPA to operate a national system of area-specific permits. The user may not have the training to determine if the user's area is more vulnerable than the general case and, therefore, underprotected by the uniform measures.
3. **Differential approach** — A third approach for addressing area variability is to tailor prevention measures to different levels of groundwater vulnerability. The number of vulner-

ability levels is practically limited by: (1) the technical ability to differentiate vulnerability accurately; and (2) the number of different prevention measures that could reasonably be used to provide differential protection. While there is still a likelihood of underprotection and overprotection measures for any given area, the number of such measures would be reduced in comparison to the first two options discussed above.

In conclusion, vulnerability to pesticide contamination may be fairly uniform across some large areas, while in other areas it may vary on a farm-by-farm basis or even on an acre-by-acre basis. Although making determinations of vulnerability at the highest degree of resolution would be the preferred technical basis of management, the sheer number of decisions required by such an approach could be overwhelming. The EPA may be able to manage a differential approach to pesticide use at one degree of resolution, whereas a state may be successful in conducting a management program at a much higher degree of resolution. A state should also be in a better position to make decisions on the use and value of groundwater in a given location.

What are the Appropriate Federal–State Roles?

Who should determine the vulnerability and the use and value of groundwater in an area and who should manage a differential approach? The following three options have been considered by the EPA:[11]

1. **The EPA establishes differential measures; users determine applicability** — The EPA may decide that certain pesticides cannot be used in areas of high use and value and/or high vulnerability. For moderately vulnerable areas, the use of the pesticide could be subject to special restrictions, such as changes in the rate, timing, or method of application. Representative monitoring of groundwater for the pesticide in these moderately vulnerable areas could be required of the registrant to ensure that no threat of unacceptable contamination emerges. For low vulnerability areas a requirement for special measures near drinking water wells might be applicable. The user would primarily be responsible for determining the applicability of differential requirements based on a required assessment of local vulnerability and the user's location within an area of "high priority groundwaters". The user would base a decision on label directions and possibly on supplemental instruction or training. One problem with this option is that most users do not have the scientific background in hydrogeology or environmental fate processes to make accurate field decisions. With this option, directions must be provided that translate technical assessments of groundwater vulnerability into directions that a user can understand and apply.
2. **EPA specifies differential measures for individual countries or state; users determine applicability of EPA-specified subcounty measures** — Since the vulnerability within a county can vary greatly, this approach could unnecessarily apply stringent measures, including local bans, to subcounty areas that are less vulnerable than the average for the county in which they are located. Similarly, such variation could result in interprotection for subcounty areas more vulnerable than a county average. The user would be required to make difficult location-specific judgments on the applicability of prevention measures that would be more stringent than the minimal county-wide requirements because of site-specific vulnerability or the user's location within an area of "high priority groundwater".
3. **State specifies differential measures for subcounty areas based on EPA criteria** — The state would have the dominant role in determining differential prevention measures, but such measures would be based on EPA criteria. The states would identify and map their groundwaters in terms of use, value, and vulnerability and provide pesticide users with explicit directions on where and how pesticides could be applied. The user in the field would have more explicit directions as to where and how the pesticide could be used and would not have to determine local vulnerability. Another advantage to this option is that the state could also closely tie user training and enforcement efforts to its own differential approach. Although the states are in a better position than the EPA to understand local conditions and establish a differential

approach that is highly tailored to these conditions, there are a number of drawbacks to this option. These include the potential lack of uniformity among states and possible political and legislative constraints within particular states, which may jeopardize obtainment of EPA's environmental goal of protecting the groundwater resource.

EPA's Proposed Position

The EPA will employ each of the three basic management approaches, described above, to prevent pesticides from reaching unacceptable levels in groundwater. EPA's specific strategic policies are as follows:[11]

1. The EPA will continue to take uniform action for pesticides causing widespread, national concerns and will establish generic prevention measures to address certain pesticide use and disposal practices that pose groundwater threats independent of area-specific vulnerability. National uniform measures will not be differentiated on the basis of local differences. When a pesticide's use poses a serious, widespread groundwater threat, EPA will take steps at the national level and, if necessary, impose a regional or even a national cancellation of the use of the pesticide. The EPA has proposed regulations for the application of pesticides through irrigation waters, often referred to as chemigation. This practice can directly introduce pesticides into groundwater if precautions, such as anti–back siphoning devices, are not used. The EPA is also developing a rule to restrict the use of potential leaching pesticides to certified applicators. The EPA is also considering the development of national rules to address the potential problem of pesticide applications too near a well, which can result in "run-in" of pesticides into groundwater, a particular problem in areas with uncapped, abandoned wells. Finally, the EPA is considering additional generic rules and guidelines for pesticide disposal and for preventing and handling leaks and spills, all of which can be an important source of a number of pesticide concerns, including groundwater contamination.

2. The EPA will also adopt a new approach of differential management of pesticide use based on differences in groundwater use, value, and vulnerability to an extent that is administratively feasible. County- or state-level measures based on groundwater vulnerability will be employed, including use cancellations. In some cases, the user will have to determine the applicability of differential prevention measures based on interpretation of local field conditions and the user's location within areas of "high priority groundwaters". When a pesticide poses serious but localized risks due to local groundwater vulnerability, an appropriate prevention approach is to remove the threat where it exists and allow the use of the pesticide in other areas to continue without undue restrictions. Prevention measures will include minimum, county-wide requirements that must be followed by all users within a designated county. For some counties, the minimum measure may be a ban on a pesticide's use. For other counties, the pesticide's use will be allowed, but minimum county-wide measures will range from general advisories to extensive requirements involving changes in the rate, timing, or method of application or other agricultural practices that can influence pesticide movement to groundwater. In some cases, the EPA may apply minimum prevention measures, including geographic bans, to an entire state. Such a situation may occur where there is generally uniform groundwater vulnerability to contamination by a pesticide in all usage areas of a state. State-wide measures could also occur where the EPA believes that the States do not have the ability to support a differential approach to the management of pesticides, particularly from enforcement and user education perspectives.

3. The EPA will encourage the development of a strong state role in area-specific management of pesticide use to protect the groundwater resource. State pesticide management plans will be used to strengthen the EPA's foundation for decisions on pesticide use. In some cases, the use of a pesticide in a state will depend on the existence of and adequacy of such a state management plan. Under its management plan, a state will develop and implement highly tailored prevention measures based on local differences in groundwater use, value, and vulnerability. This approach places a major burden on the user to determine the applicability

of more restrictive measures that are based on site-specific conditions that the user must assess. Therefore, the EPA is looking for strong State involvement to determine where a pesticide can, and cannot, be used without groundwater restrictions. This alternative cooperative approach between the EPA and a state will lessen the burden on the user to determine the applicability of site-specific measures, and, in turn, reduce noncompliance and increase the likelihood of the attainment of the EPA's environmental goal. Under a state management plan, the state would identify where a pesticide with groundwater contamination concerns is being used or is likely to be used. The state would also use its knowledge of local groundwater resources and determine where protection measures would be differentiated based on local differences in the use and value as well as the vulnerability of groundwaters. The state would adopt prevention measures that would prevent unacceptable contamination of current and future drinking water sources as well as groundwaters that are ecologically important. As part of its management plan, the state could identify areas with "high priority groundwaters", such as wellhead protection areas, where more stringent measures may be required.

It is EPA's intent that a state generally have the opportunity to select from a broad menu of management measures (Table 9).

Development and implementation of a state management plan will require the active participation of all key state agencies including the Departments of Agriculture, Public Health, Water, Natural Resources, or Environmental Protection. A state's pesticide management plan will be a key component of its overall groundwater protection strategy and should be consistent with its wellhead protection program, if one exists. The state should take a holistic approach to protecting its water resources and realize the important relationship between groundwater protection and surface water quality. The state pesticide management plan should be consistent with any state clean water strategy for protecting surface waters.

The EPA will work with each state to determine if a pesticide's use can be managed locally by the state to prevent or reduce the threat of groundwater contamination. In some situations, the EPA will require a state-specific label or supplemental labeling with the approved conditions of use based on a state management plan. In other cases, the EPA will have to take steps, including possible state-wide cancellations, to control the use of a pesticide that poses a significant groundwater threat, if there is no adequate state management plan that can reasonably be expected to prevent or reduce the threat of unacceptable contamination.

Where groundwater resources are continuous under state boundaries and contamination problems in one state could threaten the quality of waters of another state, it would make sense for adjacent states to coordinate their state plan development. In the case of chemical-specific schemes, certain states may also find multi-state or regional management schemes desirable to obtain or continue EPA's registration of a pesticide that is important to a regional crop.

One of the EPA's key responsibilities under this approach will be to provide as much technical support to the states as possible, including information on the physical/chemical/toxicologic characteristics of pesticides of concern and their behavior in the environment. To meet this responsibility, the EPA will keep abreast of monitoring information to detect new

Table 9   Some Pesticide Management Measures Proposed by EPA

Moratorium areas
Wellhead protection areas
Buffer zones: location, depth, and construction only
  for new wells
Change in practice of applications
  (time, rate, and method)
Advance notice of application
Best management practices
Integrated pest management
Training and certification

From "Agricultural Chemicals in Groundwater Proposed Pesticide Strategy," U.S. EPA, Office of Pesticides and Toxic Substances, December 1987, 53–149.

pesticide contamination concerns as well as to assess the effectiveness of various management approaches.

4. The user's role in preventing groundwater contamination is pivotal; the user's decision-making in the field must be better supported. The user will continue to be in the unique position of directly controlling the use of pesticides in the field. Thus, the user has the responsibility to seek better understanding of groundwater concerns. At a minimum, a user must follow the instructions found on the label of each pesticide product and, when required, be trained and certified in the proper use of the pesticide. The best approach is to provide the user with clear instructions either not to use a pesticide or to use it in a certain manner in highly specified areas. Such areas should be familiar to the user, or a map should be provided that clearly delineates the area. To some degree, a state's management plan should have the capability to provide such specificity. The EPA is attempting to improve and expand training and certification programs so that users may become more aware of groundwater concerns and the measures necessary to protect this vital resource.

5. Registrant responsibilities will need to grow in three areas: (1) technical support for the user in the field; (2) groundwater monitoring to ensure the adequacy of pesticide management plans in protecting groundwater, and (3) the development of safer alternative pesticides.

6. Increased monitoring of pesticides in groundwater is critical to the implementation of this strategy. The EPA will establish an "early-warning" or "yellow light/red light", approach to prevent further area contamination, once detected. The approach will use the MCL or other EPA-specified protection criteria as the point of reference to evaluate and, when necessary, change pesticide management plans. When pesticide contamination is found to be moving toward an MCL or other EPA-designated groundwater protection criteria, the EPA or the state will revise the pesticide's management plan, as necessary, to prevent further contamination. The stringency of new measures will depend on the likelihood of contamination.

RESPONSE POLICY AND PROGRAM

One of the most challenging tasks facing the EPA and the states is developing a strategy for responding to groundwater contamination resulting from normal use of pesticides. The EPA will continue to emphasize the development and enforcement of MCLs to ensure the adequacy of drinking water from public water systems. In the future, the EPA will focus on coordinating enforcement activities so that responsible parties can be identified and required to take the actions necessary to eliminate imminent health threats.

Factors Considered

The type and degree of response must be based on the specifics of each case. The type of response needed bears on all of the three questions: federal/state roles, appropriate authority, and who will be liable.[11]

EPA's Proposed Position

EPA's response strategy will be to address the problem of groundwater contamination on a number of fronts. Specifically, the EPA proposes the following policies:[11]

1. Where a pesticide has reached unacceptable levels in groundwater, strong actions must be taken to stop further contamination. These actions can range from enforcement actions to modification of the way a pesticide is managed, including geographic restrictions on the pesticide's use. A response should include increased monitoring as well as site-specific determinations as to extent, source, and circumstances of the contamination. Modifications in the EPA's or a state's management of the pesticide must occur to minimize the likelihood of

further contamination. When a pesticide level has reached or exceeded an MCL or other reference point as a result of normal agricultural use, a more aggressive stance needs to be taken, including the possibility of prohibiting further use of the pesticide in the affected areas. When a state lacks a management plan to respond to incidents of contamination, the EPA will have to decide whether to allow continued use of a pesticide in an affected area.

2. The EPA will encourage a strong State role in responding to contamination. A state's management plan should consider the development of a valid corrective response scheme. At a minimum, a state needs to take steps to identify and track groundwater contamination in order to determine if current drinking water wells will be affected and to notify users of the potential health risks. By integrating its federally delegated authorities and resources with its own authorities and resources, a state can provide an effective overall scheme for responding to, and, where needed, correcting, public health problems resulting from pesticide contamination of groundwaters. In particular, the response scheme should identify the resources and the appropriate corrective actions needed when contamination is found as a result of the approved use of a pesticide. The EPA is also considering the development of a number of assistance measures to support states indirectly in their corrective response efforts. These measures range from site-specific and general technical assistance to providing public information and education.

3. The EPA will continue to develop and stress enforcement of MCLs. Under the SDWA's emergency powers, the EPA will consider issuing orders requiring responsible parties to provide alternative water supplies when levels of pesticides present an imminent and substantial endangerment to public health. The EPA is accelerating the development of MCLs, particularly those for pesticides considered to be potential drinking water contaminants. In those cases where a public water system draws on a contaminated groundwater supply, the state must require the system to reduce contamination to acceptable levels before allowing public consumption of the water. A system may treat the water to remove the contamination, blend the water with noncontaminated water to reduce levels in delivered water, or close the well and find an alternative water supply. In some cases, it may not be feasible for a drinking water system to comply with such requirements, and states may need to close the system or provide resources to the system to comply with these requirements. Exemptions to meeting the MCL can be allowed by a state during the time it takes for a system to implement necessary corrective measures, but the system must notify its users that it is providing water with contaminants exceeding MCLs.

4. The EPA and the states will place greater emphasis on coordinating FIFRA, SDWA, and CERCLA enforcement activities to identify parties responsible for groundwater contamination as a result of the misuse of pesticides, including illegal disposal or leaks and spills.

5. On a case-by-case basis, the EPA may assist states by undertaking CERCLA fund-financed removal actions to provide alternative drinking water supplies where there is an imminent human health threat.

6. The question of who should pay for long-term corrective actions at sites contaminated by the approved use of a pesticide is a legislative question. The EPA believes that several aspects of the problem must be considered before a decision can be made.

MONITORING OF PESTICIDES

If a particular pesticide is known to be present in groundwater, sampling and analysis for that single pesticide may be appropriate, despite the relatively high costs per sample tested. However, for routine screening of groundwater (or drinking water), existing analytical techniques capable of detecting more than one compound tend to be expensive, time-consuming, limited to certain pesticide classes, and/or lacking in sensitivity.[12]

The development of improved multiresidue analytic screens for pesticides in water would enhance monitoring efficiency.[12,13] The Food and Drug Administration (FDA), for example, uses a multiresidue analytic method capable of detecting up to 125 active ingredients in a single test in its food residue testing programs. Any detection of residues should be confirmed by appropriately sensitive tests, preferably by a separate analytic procedure, before actions based on their presence are taken. Programs to sample groundwater could not be conducted in the absence of formal, established protocols concerning selection of test wells, collection and handling of samples, and development of quality assurance/quality control programs. Therefore, the establishment of scientifically based sampling protocols would not only improve data quality and comparability but also facilitate the training of technicians taking and handling samples.

ERL'S SURVEY

The Environmental Resources Limited (ERL) has undertaken a critical technical review of the monitoring practices for pesticides in water in the EC Member States against the background of present changes in plant protection product authorization procedures within the EC.[14] The objectives of this review were as follows:

- to determine the degree to which monitoring of the occurrence of pesticide compounds in water is undertaken in each of the EC Member States and what the institutional arrangements are for this monitoring;
- to survey the sampling and analytical methods employed in each Member State;
- to evaluate the Analytical Quality Control (AQC) procedures used;
- to determine the factors taken into account when evaluating results;
- to determine if indirect methods of monitoring for pesticides are employed;
- to establish the environmental standards employed for pesticides in water within each Member State.

The key issues arising from the ERL's study can be summarized as follows:

1. Are pesticides in water monitored to an equal extent in Member States?

- there are clear disparities in the degree to which pesticides are monitored in different Member States;
- in some States there are numerous monitoring systems and programs that are collecting data for various uses;
- in other States, however, monitoring is very sporadic and indeed minimal in scope and frequency;
- the reasons for these disparities among Member States are many and varied and relate to a diversity of influencing factors, including the following:
  - location, size and geography of a country;
  - culture and economic background;
  - agricultural practices and the nonagricultural use of pesticides;
  - prime sources of drinking water;
  - activities of the environmental regulations;
  - national legislative requirements;
  - length of time that EC legislation has been applicable.

2. Are different pesticides monitored to varying degrees?

- the rationale behind any choices made varies from Member State to Member State;
- there are obvious priority compounds that could be selected for monitoring such as the EC "Black" and "Gray" lists in particular;
- it is increasingly clear, however, that there are a number of potentially significant compounds that are systematically monitored across the Community;
- use of the major reasons for the variation in the degree to which certain individual pesticide compounds are monitored is related to the case with which they can be measured. Thus, where there is a lack of suitable analytic methods or capabilities, this becomes a limiting factor in the selection of compounds to monitor.

3. Are there disparities in sampling procedures?

- it is clear that, although certain basic principles are being followed, there are clear differences in the application of control procedures over sampling. There is variation in:
  - the vessel cleaning procedures;
  - measures to avoid cross-contamination of samples;
  - storage of either samples or extracts;
  - the length of storage time prior to analysis.

4. Are there disparities in analytical procedures?

- it is clear that there are disparities in the analytic procedures currently being used within Member States. Differences are apparent in:
  - extraction techniques, such as the use of liquid/liquid or liquid/solid methods and the use of various solvents;
  - clean-up methods;
  - instrumentation and method used, e.g., Gas Chromatography-Nitrogen Phosphorous Detector (GC-NPD), Gas Chromatography-Electro Capture Detector (GC-ECD), Gas Chromatography-Mass Spectrometry (GC-MS), High Performance Liquid Chromatography (HPLC).
- Some of these differences may well be historic rather than strategic in nature reflecting often:
  - the origin of a particular laboratory;
  - its ability to generate funds to purchase the necessary technology;
  - the expertise of chemists to develop the appropriate methodology;
  - the fact that pesticides are easier to measure in drinking water than in surface water.
- Differences in analytical procedures between laboratories within individual Member States were cited in some States as resulting in a noncomparable standard of results between laboratories.

5. Are there disparities in Analytical Quality Control (AQC)?

- the disparities in AQC are obvious;
- there are disparities in the use of some particularly important aspects of AQC, namely:
  - external certification of laboratories including regular external audits;
  - frequent interlaboratory comparisons.
- the fact that in some Member States results from different laboratories are not considered to be of a comparable standard is also a result of the inadequate applications of AQC procedures.

6. Are there deficiencies in the interpretation of monitoring results?

- there remain significant gaps in the knowledge of pesticide chemistry, which have influenced the interpretation of concentrations of various compounds in surface and groundwaters;
- interpretation of the results of monitoring programs is perhaps one of the most contentious parts of the process because, in many respects, it is only in comparatively recent and in on-going research programs that we are beginning to understand the environmental factors influencing concentrations of pesticides in water (related to their mobility, transport, and fate);
- key environmental factors that have been frequently omitted in the past but are known to have a considerable influence on pesticide concentrations in water at the time of sampling include the following:
  - weather conditions;
  - soil type and condition (in the case of surface waters);
  - substrate/geologic profile (in the case of groundwaters);
  - land use activity.

SAST* ACTIVITY IN MONITORING

In the recent report on "Monitoring and Modelling for Planning and Management of Sustainable Use of Freshwater Resources,"[15] monitoring practices in EC countries have been reviewed. The main conclusions, which are in general agreement with the ERL's study finding, were as follows:[15]

- Monitoring network design is generally not based on objective criteria and scientifically sound methods, but rather on "rules of thumb," and is often the result of historic, arbitrary developments;
- There is, in general, a lack of available data for decision-making in particular with regard to pollution and contamination assessment and with regard to environmental impact assessment. The actual use of data in decision making is often not exploited to the full potential of the data collected. This may, to a large extent be due to the technical difficulties related to inconvenient access to and transfers of existing data volumes. However, it may also be related to institutional bottlenecks, such as barriers to free and convenient access to data. This may be one of the major constraints for establishment of a reliable database for the EC water policies;
- There are differences among EC Member States in the level of development of monitoring systems, implicating a significant potential and need for technology transfer.

MODELS FOR PREDICTING MOVEMENT AND FATE OF CHEMICALS IN GROUNDWATER AND FOR MANAGEMENT OF SUSTAINABLE USE OF WATER RESOURCES

MONITOR-SAST ACTIVITY IN MODELING

The general principles of environmental models, together with value and limitations of predictive approaches have been described in Section II. Within the SAST Project,[15] needs for research and technologic development have been evaluated related to the use of mathematical models for planning and management of sustainable use of freshwater resources. The scientific and technological status of model application to groundwater problems is summarized in Table 10. Some other statements on this issue were as follows:[15]

1. Groundwater modeling has been carried out on a routine basis for more than a decade. Today a wide variety of modeling systems for groundwater flow, transport, and geochemistry exist. Most of these modeling systems are based on a two-dimensional (2-D) description; however, a few three-dimensional (3-D) modeling systems have been developed recently. For groundwater quality problems 3-D modeling is generally required. Hence, the state-of-the-art is considered to be 2-D modeling for most groundwater resources problems and 3-D techniques for problems related to contaminant transport and geochemical processes.

2. For solute transport modeling (transport and spreading of nonreactive solutes) the main problem relates to obtaining a sufficiently detailed 3-D description of the geology and, in this way, obtaining data on the spatial variability of the hydraulic parameters, which ultimately determine the transport and spreading of contaminants. Hence, there is a considerable research and technology development need in improving the geophysical/geologic field methods and in strengthening the interaction between these field methods and modeling. However, even with improved geophysical/geologic data acquisition methods, considerable uncertainties exist on the exact geologic configuration and the corresponding hydraulic characteristics of the subsurface porous media, resulting in a considerable uncertainty in model predictions.

---

* Strategic Analyses in Science and Technology (SAST) of the EC Directorate for Science, Research and Development (DG XII). SAST activities are part of the Monitor Programme, which aims to identify new directions and priorities for EC research and technologic development policy.

Table 10   Status of Application of Hydrologic Models to Various Problem Types Facing with Groundwater Resources

Field Of Problem	Adequacy of Scientific Basis	Scientifically Well Tested?	Status of Application Validation on Pilot Scheme?	Practical Application	Major Constraints for Practical Application
General:					
1. Assessment of natural groundwater resources	Good	Good	Adequate	Standard-part	Administrative
2. Groundwater pollution					
2.1 Point source (Landfills)	Good	Good	Partially	Standard-part	Technology/Administrative
2.2 Nonpoint (Agriculture)	Fair	Partially	Partially	Very limited	Science/Technology
Models and Monitoring					
3. Design of groundwater monitoring network	Good	Good	Adequate	Very limited	Technology/Administrative
4. Forecasting					
4.1 Groundwater heads/ water table	Very good	Very good	Adequate	Very limited	Data/Administrative
4.2 Groundwater quality	Fair	Poor	Nil	Nil	Science
5. Analysis of historical time series for identification of changes in regime	Good	Good	Very limited	Very limited	Science
6. Prediction of impacts from human intervention					
6.1. Water quantity	Good	Partially	Very limited	Very limited	Science
6.2. Water quality	Poor	Poor	Very limited	Very limited	Science

KEYS TO STATUS OF APPLICATION:

Adequacy of Scientific Basis
- Poor — Large and crucial needs for improvements in scientific basis
- Fair — Considerable needs for improvements in scientific basis
- Good — Some (but not necessarily urgent) needs for improvements in scientific basis
- Very Good — No present significant need for improvements in scientific basis

Scientifically Well Tested?
- Poor — Large needs for fundamental tests of scientific method
- Partially/Fair — Considerable needs for testing (some) of the scientific basis
- Good — Some (but not necessarily urgent) needs for testing of the scientific basis
- Very Good — No present significant need for testing of the scientific basis

Validation on Pilot Schemes?
- Nil — No successful validation on well controlled pilot schemes so far — urgent need for validation on pilot schemes
- Very Limited — A few validation cases — considerable needs for more validation projects on pilot schemes
- Partially — Some cases with successful validation on pilot schemes — some (but not necessarily urgent) needs for further validations
- Adequate — Many good validation — no further present needs

Practical Applications
- Nil — Practically to operational applications
- Very Limited — A few well-proven cases of operational practical applications
- Some Cases — Some cases of well-proven operational practical applications
- Standard-Part — Standard professional tool in some regions of EEC
- Standard — Standard professional tool throughout EEC

Major Constraints for Further Practical Application
- Data — Data availability a major constraint
- Science — Inadequate scientific basis is a major constraint
- Technology — A technology push is required in order to make well-proven methods more widely applicable
- Administrative — Administrative tradition or missing economical motivation is a major constraint

Source: From Commission of the European Communities, SAST Activity. EUR Report 147 25, July 1992.

Therefore, there is a demand for techniques that can provide the decision maker with quantitative information on the uncertainty of model predictions. The most common approach in solute transport modeling is to treat the flow and transport processes independently, i.e., the results from the flow calculations being used as input to the transport calculations without feedback. However, in cases where the feedback mechanisms are important, e.g., high/low density of contaminated water, the flow and transport modeling must be carried out as a fully coupled modeling system.

3. Geochemical modeling covers modeling of a large range of geochemical processes, and, today, no single modeling system is able to accommodate all these processes. The state-of-the-art in geochemical modeling is naturally lagging behind the flow and solute transport descriptions because it is extremely complex in itself and furthermore depends on results from the flow and transport modeling. Thus, most of the modeling experience relates to 1-D and 2-D, which may be sufficient for carefully selected research case studies but definitely not for real world applications, where 3-D is required.

4. The root zone/unsaturated zone modeling can today be carried out by use of a large range of existing modeling systems. Coupled water flow/transport/geochemical modeling is today exclusively based on a 1-D (vertical) description of the flow and transport regime. A few models with 2-D (or even 3-D) descriptions exist for water flow modeling; some of them as fully coupled unsaturated/saturated models. However, for most practical purposes such details are not required and hence seldom used. Therefore, the state-of-the-art can be characterized by 1-D techniques.

5. Integrated catchment modeling comprises an integrated description of surface water, unsaturated zone, and groundwater processes. Lumped, conceptual models are being utilized on a routine basis. These models are adequate for streamflow modeling. Attempts also to apply them to solute transport and water quality modeling have, however, been unsuccessful, because they do not provide a sufficiently accurate and detailed description of the flow pattern, including the subsurface processes, within the catchment.

MODELING AS REGULATORY TOOL

As regards models and decision-making, the US National Research Council addressed two main questions: (1) To what extent can the current state of groundwater models accurately predict complex hydrogeologic and chemical phenomena? and (2) Given the accuracy of these models, is it reasonable to assign liability for specific groundwater contamination incidents to individual parties or make regulatory decisions based on long-term prediction?[16]

Conclusions and recommendation were as follows:[16]

1. Groundwater models are valuable tools that can be used to help understand the movement of water and chemicals in the subsurface. The purpose of the models is to simulate subsurface conditions and to allow prediction of chemical migration. When properly applied, models can supply useful information about flow and transport processes and can assist in the design of remedial programs.

2. The results of a model application are dependent on the quality of the data used as input for the model. Generally, site-specific data are required to develop a model of a site. The model cannot be used as a substitute for data collection. However, model use can help direct a data collection program by identifying areas where additional data are required. Closely linked data collection and model application can provide an adequate representation of site conditions. Incorrect model use frequently occurs when the limitations of the data used to develop the model are not recognized.

3. When properly applied, the results of a groundwater model application can help in making decisions about site conditions. Model results can be used to supplement knowledge of site conditions but cannot be used to replace the decision-making process. The results of the models must be evaluated with other information about site conditions to make decisions about groundwater development and clean up.

4. A generic model may be useful in offering some initial guidance to an investigator. However, only the most naive would rely on the predictions of a generic model in an attempt to understand the details of the movement of groundwater or the behavior of a dissolved pollutant in a specific hydrogeologic environment. It is essential that an investigator gather site-specific information to use as input to the groundwater model of choice and, perhaps, that the model itself be modified and adapted to fit the hydrogeologic conditions at a particular site.

5. Modelers must contend with the practical reality that model results, more than other expressions of professional judgment, have the capacity to appear more certain, more precise, and more authoritative than they really are. Many people who are using these and are relying upon the results of contaminant transport models are not fully aware of the assumptions and idealizations that are incorporated into them or of the limitations of the state of the art. There is a danger that some may infer from the smoothness of the computer graphics or the number of decimal places that appear on the tabulation of the calculations a level of accuracy that far exceeds that of the model. There are inherent inaccuracies in the theoretical equations, the boundary conditions, and other conditions and in the codes. Special care therefore must be taken in the presentation of modeling results. Modelers must understand the legal framework within which their work is used. Similarly, decision-makers, whether they operate in agencies or in courts, must understand the limitations of models.

6. A regulation that requires contaminant transport modeling reflects an implicit decision to require a given level of detail and allow a given level of uncertainty. When regulations require the use of a model, however, they do not imply that the solution to the problem is susceptible to a "black-box" model application. Quite the contrary, the regulations seem to require contaminant transport modeling in the most complicated site-specific problems.

7. In some circumstances, it may be appropriate to specify the use of a particular contaminant transport model. For example, after reviewing site-specific data from a hazardous waste site, an agency or private company may determine that a particular model could be appropriate to apply at the site and such a model may be specified in a consent decree or permit for specific purposes. When a model is used in such circumstances, the consent decree, permit, or other legally enforceable procedure should require actual monitoring to confirm the modeling results and be flexible enough to allow the model to be updated and modified on the basis of new data and recent scientific developments.

8. Models used in regulatory or legal proceedings are required to undergo public comment and review by those whose interest may be affected. The documentation associated with the model therefore must enable any reviewer to do the following:

- understand what was done;
- evaluate the quality of the model, considering issues such as the extent to which the equations describe the actual processes (i.e., model validation) and the steps taken to verify that the code currently solves the governing equations and is fully operational (i.e., code verification);
- evaluate the application of the model to a particular site;
- distinguish between the scientific and policy input.

9. Instead of sanctioning particular models, regulatory agencies should provide detailed, consistent procedures for the proper development and application of models. Detailed specifications of positive aspects need to be developed but should include (1) good documentation of a code's characteristics, capabilities, and use; (2) verification of the program structure and coding, including mass balance results; (3) model validation, including a comparison of model results with independently derived laboratory or field data and possibly other computer codes; and (4) independent scientific and technical review. The guidance must also be written to avoid being misconstrued as providing a list of "approved" models. The mere approving mention of a model in agency guidance may appear to inexperienced and untrained agency personnel as indicating that such models are "approved" or "sanctioned". Agency guidance, therefore, must stress that the descriptions do not sanction the use of any particular model. Instead, the guidance should stress best modeling practice or principles, described previously,

and ensure that only experienced and properly trained personnel are involved in the development and review of such models.

10. The fact that many of the models used in practice have not been validated to a significant extent provides an important source of uncertainty in the predictions that come from the models. Unfortunately, even more uncertainty enters the modeling process from, for example, (1) the inability to describe precisely the natural variability of model parameters from a finite and usually small number of measurement points, (2) the inherent randomness of geologic and hydrogeologic processes over the long term, (3) the inability to measure or otherwise quantify certain critical parameters, and (4) biases or measurement errors that are part of common field methods. When all these sources of uncertainty are properly considered, a single model prediction realistically has to be viewed as one of a relatively large number of possible system responses. Over the past decade, the development of stochastic modeling techniques has been useful in quantitatively establishing the extent to which uncertainty in model input translates to uncertainty in model prediction. What should a decision-maker do now, given existing modeling capabilities? There is obviously no easy or comforting answer to this question. It seems apparent, however, that it would be unwise to rely solely on any single source of information when deciding how to formulate regulations, carry out a clean up, or protect public health. Models should be supplemented by carefully conceived field work, which not only provides data for estimating model inputs but also provides an independent confirmation of conditions in the subsurface environment. Put simply, decision-makers should hedge their bets and distribute their resources, funding different type of modeling efforts and mixing modeling with on-site monitoring. When field data are inconclusive or insufficient, model results may have a significant influence on the predicted impact of a given decision. In this case, the decision-maker should request a quantitative and defensible assessment of the model's accuracy in order to evaluate the risk of making a bad decision. In this regard, environmental management is no different from any other form of management where uncertainty and risk are important. Models are not going to relieve us of the burden of making difficult decisions. They simply provide some additional information to consider. It is unrealistic to expect much more.

CONCLUSIONS AND PROPOSALS

The key issues to be addressed in any strategic approach on preventing and managing groundwater contamination as a consequence of pesticide use can be summarized as follows:

1. criteria (i.e., concentration limits in groundwater/drinking water) to be used for determining protection (or contamination);
2. degree of use and value of groundwater resources to be protected;
3. programs and instruments for preventing groundwater contamination in relation to local variability in resource vulnerability, combined with pesticide use and disposal practices;
4. pesticides (plant protection products) characterization and their procedure of registration;
5. policies for the respective central and local roles in administrative and technical responsibility regarding local pesticide management and monitoring procedures;
6. instruction to be provided to the decision-making users in pesticide application in the field;
7. response policies to groundwater contamination resulting from the use of pesticides and identification of responsible parties.

The EC and U.S. EPA strategies are at a different degree of development. That not withstanding, differences can be appreciated in the criteria for determining protection or contamination of groundwaters and in the preventive measures to be adopted. The EC proposes a generalized 0.1 µg/L for a single pesticide while the EPA is developing MCLs for human health protection integrated with ecologically based protection criteria.

As regards preventive policy, the EC is beginning to authorize, following an established procedure (the so called Uniform Principles), only those pesticides that are not expected to be present, under the proposed conditions of use, in drinking water extracted from groundwater at concentrations lower than 0.1 µg/L. The U.S. EPA proposes an area-specific management of pesticide application, tailored to groundwater use, value, and vulnerability. Overall, the emergence of the problem of pesticide residues in groundwater adds a new dimension to the whole array of public health, environmental protection, pesticide innovation and marketing, and agricultural management.

Variations in approaches may arise from cultural and socioadministrative local conditions, different interests involved, and lack of national/regional leadership in key areas. The most urgent actions that are needed to address groundwater problems arising from field-applied pesticides seem to be:

- Determination of a well-accepted position as regards MACs for pesticides in waters, including relevant metabolites and degradation/reaction products;
- The development of improved, multiresidue analytic tools for screening groundwater for pesticides at the parts-per-billion to parts-per-trillion range;
- The development of specific and systematic monitoring programs;
- The development of improved models of pesticide behavior and fate in soils, unsaturated zones, and aquifers;
- The practical application of the scientifically available, advanced models and the internal comparison of different modeling techniques aiming to an appropriate use of the models and their validation with suitable test data;
- The development of data on environmental fate and pesticide use patterns, in conjunction with local hydrogeologic conditions, at a level of specificity adequate to identify areas especially vulnerable to groundwater contamination;
- The integration of groundwater resource considerations into an array of agricultural management practices and choices, including cropping, tillage, irrigation, pest management, etc.;
- The improvement of the potential for accessing and exchanging of data from different data sources;
- The establishment of a correct approach of functional groundwater management policy;
- The improvement and expansion of training and certification programs for pesticide users in the field;
- The better education of the public concerning pesticide-associated risks in comparison with other risk factors and understanding the role of the media in public perceptions of this risk;
- The evaluation of whether regulations actually result (or will result in the future) in improved health status and benefits or not.

REFERENCES

1. Haigh, N. "Manual of Environmental Policy: the EC and Britain," Longman Industry and Public Service Management, U.K.(1992).
2. Premazzi, G. "Protection and management of water resources in the European Community," Eurocourse "Chimica e Ambiente: legislazione, metodologie e applicazioni", Ispra, June 22-26, 1992.
3. Langer, M. "EC legislation covering the protection of groundwaters," Eurocourse "Technologies for environmental clean-up: soils and groundwaters, Ispra, September 1992.
4. Premazzi, G., G. Chiaudani, and G. Ziglio, "Scientific Assessment of EC Standards for Drinking Water Quality," EUR Report 12427, (October 1989) 200 p.
5. "Standing Committee of Analysts," Department of the Environment (1986).
6. "Freshwater quality. Sixteenth Report," HMSO (1992) 291 p.
7. Mandl, V., "Pesticides: the EC viewpoint," *Water Supply,* 10: 7-11 (1992).
8. Commission Proposal for a Council Directive establishing Annex VI of Directive 91/414/EEC concerning the placing of plant protection products on the market (1992).
9. "Declaration for the Ministerial Seminar on Future Community Groundwater Policy," European Environment Ministers, The Hague, November 26-27, 1991.

10. Van de Ven, F.H., L. Kohsiek, J. da Silva Costa, and I.J. Fried, "EC action program for sustainable use of groundwater," *Eur. Water Pollut. Control*, 2: 9-19 (1992).
11. "Agricultural Chemicals in Groundwater Proposed Pesticide Strategy," U.S. EPA, Office of Pesticides and Toxic Substances, (December 1987), 53-149.
12. Holden, P.W. "Pesticide and Groundwater Quality. Issues and Problems in Four States," Washington D.C.: National Academy Press, 5-13 (1986).
13. Ziglio, G. and G. Premazzi, "Scientific Assessment of EC Standards for Drinking Water Quality, Monitoring, Sampling Frequency and Reference Methods Analysis," EUR Report 13600; 43-47, (1991).
14. "Monitoring pesticides in water and definition of the Uniform Principles" Environmental Resources Limited, Vol.1 European Crop Protection Association (1992).
15. "MONITOR-SAST Activity. Sast Project N° 6," European Communities, Report on Monitoring and Modelling. EUR 147 25; 9-99 (1992).
16. "Groundwater Models: Scientific and Regulatory Applications," National Research Council, Washington, D.C.: National Academic Press, 7-21 (1990).

CHAPTER 11

An Economic Approach

Tore Söderqvist, Lars Bergman, and Per-Olov Johansson

CONTENTS

INTRODUCTION

This chapter offers an economic approach to groundwater contamination. The approach chosen largely follows mainstream environmental economics, a subdiscipline of economics that has grown considerably during the last two decades. The reasons for this growth probably include the increased public attention to environmental problems and the fruitfulness of applying standard economic theory to a new set of issues.

The fact that mainstream environmental economics fits in with the general framework of economics implies shared ethical foundations. It is important to be aware that these foundations are anthropocentric, which means that the basis for any attempt to make an economic valuation of the environment will be the value that humans place on the environment. This implies, for example, that a particular species, say a plant species, only has a value if there is any human who values this species. Economists usually choose to quantify such values by using monetary units as a yardstick.

The general framework of economics is not undisputed. Readers interested in the discussion on whether it is reasonable to put a monetary value on the environment are referred to, e.g., Bojö et al.,[1] Kneese and Schulze,[2] and Sagoff[3] and the references mentioned in these works. Some economists prefer a more pronounced institutional (e.g., see Eggertsson)[4] or ecological perspective than that offered below. The latter perspective gives insights into the reliance of the human society on coexistence with and support by ecological systems and the restrictions for human activity that this reliance implies (e.g., see Common and Perrings,[5] Folke,[6] and Folke et al.[7]).

Another characteristic of economics is that it belongs to the social sciences and thus focuses on the effects of human behavior and interaction (or lack of interaction). Knowledge gained by the natural sciences is, however, of crucial importance to the environmental economist, who should thus aim to establish some kind of interdisciplinary cooperation. Such cooperation can only be really fruitful when there is at least a basic mutual understanding of the involved disciplines. The presence of this chapter in this book may be viewed in that particular perspective.

The chapter is structured as follows. The next section gives a general presentation of groundwater contamination as an economic problem. We then turn to the issue of how to design institutions for solving contamination problems. These theoretical issues are illustrated by some results of earlier economic empirical studies of groundwater contamination. The chapter closes with some final remarks.

GROUNDWATER CONTAMINATION
AS AN ECONOMIC PROBLEM

INTRODUCTION

Human-made contamination of groundwater is often an unintentional effect caused by people's behavior in, for example, agricultural or industrial production, or in their private consumption as households. A fundamental problem, which will be explored further below, is that people whose activities cause the contamination are often not identical to the people affected by the contamination. This means that the activities causing contamination are not automatically restricted by the contamination damage. The contamination is an example of a *negative externality*: the behavior of one group of people has a damaging impact on the well-being of another group of people, but this impact is not taken into account by the former group.

The concept of externality is illustrated by Figure 1. In the figure's society box, we find two private decisions boxes, one including producers and users of pesticides, say firms and farmers. The exchange of pesticides for money takes place at the pesticide market. However, the behavior of firms and farmers are not automatically influenced by the externalities caused by them. The externalities box shows how pesticides are distributed to recipients in the environment (soil, air, surface water, etc.) and eventually reach groundwater, given that we deal with pesticides that are persistent enough. Groundwater may be used as drinking water supplied by public or private waterworks or by household wells, i.e., we now move into the second private decisions box. At least in principle, consumers of drinking water can decide what drinking water they would like to consume and exchange drinking water for money on a market for drinking water.

The boxes for private decisions and externalities are joined by introducing institutions ("the rules of the game in a society"[8]) constraining private decisions in a way that the decisions take the externalities into account, i.e., the externalities are internalized. In the figure, they are illustrated by an institutions box embracing the private decisions and externalities boxes. Ideally, the constraints should be based on an overall analysis, where the *social* benefits and costs of the production and use of a pesticide are estimated. In this cost–benefit analysis, the externalities box is taken into account. A successful analysis presupposes, for example, sufficient natural science knowledge of how and to what extent pesticides are distributed to the groundwater. The analysis may determine the *socially efficient* quantity of the pesticide (which may be zero or above zero), and institutions enforcing the socially desirable situation should then be designed. It has been shown that, theoretically, a simple definition of property rights may be sufficient. In reality, however, a more active intervention of the legal system is necessary by defining liability rules and/or regulations.*

---

* We disregard another and usually very powerful type of institution; namely, restrictions set by ethics, i.e., moral principles.

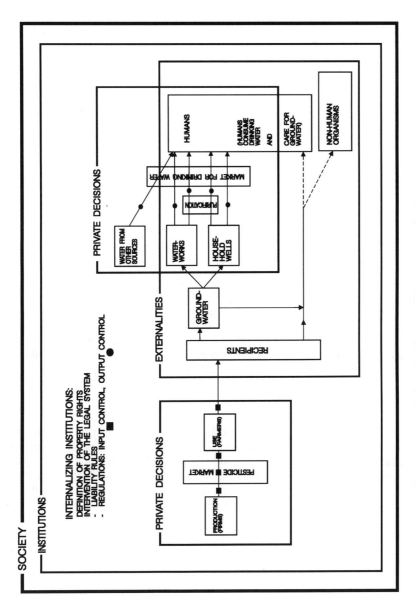

Figure 1 A rough flow chart for a pesticide contaminating the groundwater.

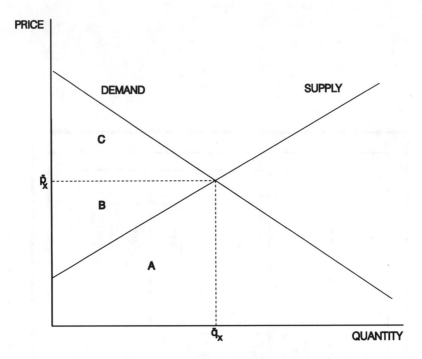

Figure 2   The market for the pesticide x.

A different situation arises if a regulation, say a drinking water standard, is given, and it is discovered that drinking water does not meet the standard. The desirable action to take in order to improve the water quality is then the one that accomplishes the given standard at the lowest costs, i.e., the *cost-effective* action. For example, when dealing with the case of pesticides, the costs of reducing pesticide use may be compared with water purification costs. Both actions could have the potential to accomplish a situation meeting the standard.

A SIMPLE MARKET SITUATION

Many issues raised by Figure 1 will now be dealt with by introducing some basic economic theory. A more penetrating exposition of this theory than the scope of this chapter allows is offered by, e.g., Boadway and Bruce,[9] Johansson,[10,11] and Just et al.[12] Let us begin by studying the first private decisions box. To be concrete, suppose that we deal with a contaminating pesticide x used by farmers. When buying x, farmers are acting as demanders on the market for x. As demanders, they constitute one of two types of agents on the market, which we assume to be competitive. The other type is the suppliers, i.e., the producers of x.

Both demanders and suppliers are likely to have opinions about what quantity of the market's product they are willing to buy and sell, respectively, for a given price. These opinions can be represented mathematically by two functions: a demand function and a supply function. The market for x is described by these two functions in Figure 2. The demand function is represented by a downward-sloping curve and the supply function by an upward-sloping one. It is usually reasonable to think that the demanders (suppliers) would like to buy (sell) large (small) quantities of x when the price is low and small (large) quantities when the price is high, given that other factors that affect the decisions (demanders' income, prices of other goods, etc.) do not change. It is evident that the only combination of quantity and price for which the demanders' and suppliers' opinions are equal is $(\bar{q}_x, \bar{p}_x)$. This combination is called the market equilibrium; $\bar{q}_x$ is the equilibrium quantity, and $\bar{p}_x$ is the equilibrium price. Because of interaction between demanders and suppliers, the prevailing quantity and price of the market are likely to tend to the equilibrium situation.

At this point, some observations regarding the origin of the demand and supply functions should be made.* Economists like to show that the supply function is the logical behavioral outcome of profit-maximizing producers and that the producers' supply functions are, roughly speaking, identical to their *marginal cost* functions. The marginal cost function tells us the cost for producing an additional unit of the product in question. This means, for example, that the marginal cost for a production of $\bar{q}_x$ is $\bar{p}_x$. Summing marginal costs for all quantities gives total variable production costs. Given that fixed costs are zero, the total cost of producing $\bar{q}_x$ is equal to the area A in Figure 2. Profits, or more strictly speaking, the producers' surplus, are total revenues $(\bar{p}_x\bar{q}_x)$ minus area A, i.e., area B.

As regards the demand function, it can be derived in an elegant fashion from individuals' rational behavior, by maximization of their utility subject to their budget restriction. Of course, it is far from self-evident how the behavior of a particular group of demanders, e.g., farmers, should be described. We stick here to standard consumer theory in order to be able to conveniently present some important issues. It could be argued that small family farms fit in with this framework. Larger "agribusiness" farms could very well follow a profit-maximizing behavior. A more elaborate behavioral assumption may in general be more satisfactory, for example maximization of expected utility under risk aversion, where utility depends on profits alone.[13,14]

The relation between the demand function and people's well-being from a particular product is of crucial importance, since it gives a possibility to obtain information on how people value the product. According to the demand curve in Figure 2, the users of x demand $\bar{q}_x$ when the price is $\bar{p}_x$. But it could also be argued that the highest price that the users are willing to pay for the $\bar{q}$th unit of x is $\bar{p}_x$. Economists say that the users' marginal willingness to pay (WTP) for $\bar{q}_x$ is $\bar{p}_x$ and view this willingness as an economic measure of the benefit of consuming the $\bar{q}$th unit of x, i.e., a measure of how people value the $\bar{q}$th unit.

Evidently, the downward-sloping demand curve implies that people's marginal WTP is decreasing in the quantity of x. How could this be explained? People value the consumption of x differently and are thus willing to pay different amounts of money for x. Consider the case when only a single unit of x is available. Among people, there is likely to be someone who is the most keen to consume this unit and is thus willing to pay a lot of money in order to ensure his consumption of x. When two units are available, a person who is slightly less keen to consume x can also get a unit of x. His or her willingness is, however, a bit lower than that of the most keen person. And so on; eventually, we are dealing with persons not particularly interested in x. Their WTP is perhaps lower than $\bar{p}_x$. If so, and, if the price of x is $\bar{p}_x$, the price exceeds their WTP. Consequently, these persons will not buy any x. We can now say that the sum of all WTPs expresses the total value of the consumption of $\bar{q}_x$. As consumers' expenses are $\bar{p}_x\bar{q}_x$, the net value, or the consumers' surplus, is equal to the area C.

The preceding paragraphs indicate that $(\bar{q}_x,\bar{p}_x)$ is not only the reasonable outcome of the interaction between demanders and suppliers. In a special sense, it is also an *efficient* outcome. The easiest way to see this is to recognize the fact that for a quantity such as $\bar{q}_x+1$, the marginal cost to the suppliers is higher than the marginal WTP of the demanders. This means that some supplier's costs of producing this additional unit of x are larger than some demander's benefits of consuming it. Thus, it does not seem to be worthwhile to produce this unit. What about the quantity $\bar{q}_x-1$? For this quantity, the costs of producing an additional unit are lower than the benefits of consuming it. So why not produce it and exchange it for a price that make both the supplier and the demander better off?

The result that a competitive equilibrium is (Pareto) efficient is known as The First Theorem of Welfare Economics. See, e.g., Kreps[15] or Varian.[16] There are, however, some weaknesses with the above method of identifying $\bar{q}_x$ as an efficient quantity. An obvious complication is that pesticide markets are oligopolistic rather than competitive. See, e.g., Friedman[17] on oligopoly theory. We will, however, focus on an underlying presumption that no one except the demanders and suppliers of x is affected by x. If this presumption was correct, there would really be reasons to regard $\bar{q}_x$ as the efficient quantity. In reality, however, things are not so simple. x is a substance contaminating groundwater, which may affect, e.g., the health of people using groundwater as

---

* We disregard problems of aggregation, i.e., how market demand and supply curves are related to the individual consumer's and firm's demand and supply curves, respectively.

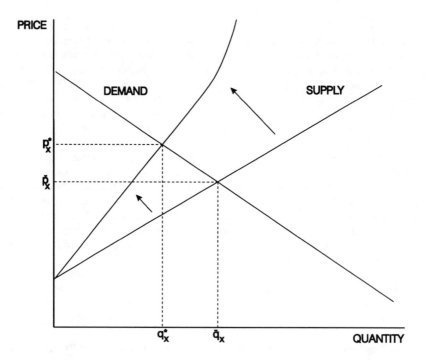

Figure 3   Damage costs are represented by an adjusted supply curve.

drinking water. That is, there are costs caused by the production and use of x that are not taken into account by either the suppliers or the demanders.

The presence of this negative externality could be illustrated by the demand and supply curves. It could be argued that the costs that the producers of x should face are larger than those reflected by the supply curve in Figure 2. The situation where the costs of damage caused by the contaminant x are also considered could be represented by the adjusted supply curve in Figure 3. The figure shows that if the damage costs were taken into account by the market, the new equilibrium would be (q_x^*, p_x^*). As all costs are considered, the equilibrium is now *socially efficient*. The socially efficient quantity is q_x^*, which is lower than the equilibrium quantity when only the producers' private costs are considered $(\bar{q}_x)$. Needless to say, this is just an example. If the damage is severe, the socially efficient quantity may turn out to be zero or close to zero, and in the case of very small damage, this quantity would only be slightly less than $\bar{q}_x$.

It could also be argued that the users of x should face the damage costs, instead of the producers. If a user has to pay an amount reflecting the damage costs caused by his or her use, then his or her demand for x is likely to decrease. This can be illustrated by an adjusted demand curve. Again, the result is a reduction of the equilibrium quantity to a socially efficient level.

A SIMPLE CONTAMINATION SITUATION

We have now seen how a socially efficient use of x can be established by looking at the market for x. Another equally valid perspective is to focus on the actual concentration of x in groundwater. In cases when a contaminant is not a market good, but perhaps a by-product of some industrial production, this perspective is of course the relevant one.

First, let us observe that a reduction of the concentration of x is not likely to be costless. For example, farmers using x may have to turn to other, less efficient pesticides. Secondly, the assumed damage due to the presence of x in groundwater contamination is reduced when the concentration is decreased. These two observations suggest that we must look for a balance between the reduction costs and the damage costs. Figure 4 illustrates the situation and shows us that the socially efficient concentration of x in groundwater is Q_x^*. At this level, the marginal damage costs of Q_x are equal to the marginal costs for reducing Q_x. Any change in Q_x from Q_x^* increases the sum of damage costs

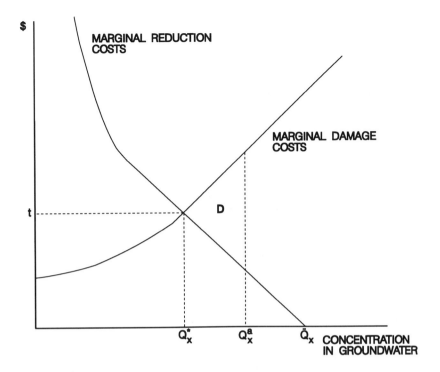

Figure 4 A simple contamination situation.

and reduction costs and is thus not good for society as a whole. If there is no internalization of the externality, the concentration will be $\overline{Q}_x$ (or higher), as neither reduction costs nor damage costs are taken into account by those actually causing the contamination. As an example, suppose that a standard Q_x^a is introduced by the government. This level is closer to Q_x^*, but is not the efficient level. Costs amounting to the area D could be avoided by a more restrictive standard, Q_x^*.

The market and contamination stories may give nice descriptions of the economic problem due to groundwater contamination. However, they also raise new questions. First, the stories suggest that q_x^* and Q_x^* are, in the efficiency sense described above, a socially desirable use of x and concentration of x, respectively. But how could q_x^* be quantified in, e.g., tons of x and Q_x^* in, e.g., mg/L? Secondly, given that we really can determine q_x^* or Q_x^*, how can a situation consistent with these amounts be realized? The latter question is the subject for the section that deals with institutions. The former question will be dealt with immediately.

ESTIMATION OF DAMAGE COSTS

The main obstacle to the estimation of the curves of Figures 3 and 4 is usually lack of data on damage costs. The quantification of these costs requires knowledge gained by the natural sciences of the relation between the use of x and its appearance in groundwater and the damage to humans and the environment caused by different concentrations of x in groundwater. Finally, the economic value of the damage has to be estimated.

A crude estimation method is to follow the so-called human capital approach (e.g., see Cropper and Freeman)[18] and look at lost earnings, costs of medical care, and other costs to society caused by an impaired human health. A valuation with the aid of this approach could in some instances be viewed as a useful lower-bound estimate of the damage costs. However, a more correct economic value is obtained by focusing on what value people place on groundwater quality, so that, e.g., marginal damage costs in Figure 4 can be measured as marginal reduction benefits, i.e., the benefits from improving groundwater quality. The measure that is of principal interest is people's WTP for avoiding damage or attaining an improvement. People's WTP may reflect a concern for their own and others' health, and for the environment. The latter concern may be due to, e.g., care for the existence of a pristine groundwater. Also mere risk aversion may be of importance. Even if risks to human health and to the environment from a certain substance are considered to be extremely small or even nonexistent, some people may be willing to give up much for an environment free from the substance. On the other hand, some people are likely to have other priorities; they may care very little for the environment and are not interested in paying much for a health risk reduction. In principle, people's motives do not matter in an economic analysis, but their valuations do. This issue is related to the fundamental principle of consumer sovereignty in economics, i.e., that each person is the sole judge of his or her welfare.

If we were dealing with the WTP for a good or service that is sold and bought on a market, the estimation of the WTP would be straightforward, at least from a conceptual perspective. Recall Figure 2 and the fact that its demand curve reflected people's marginal WTP for x. However, for nonmarket goods such as groundwater quality and many other "goods" and "services" supplied by the environment, the matter is more complicated. Economists have tackled this problem by designing so-called direct and indirect methods for the estimation of people's WTP for nonmarket goods.[19]

The most used direct method is the contingent valuation method.[20,21] It elicits people's WTP by asking them in an interview or questionnaire setting. Actual payments do not take place. The hypothetical nature of this method contrasts with the indirect methods, which utilize the fact that, often, people's use of a nonmarket good is related to their consumption of a market good.[22] When dealing with groundwater quality, there could indeed be at least one interesting market to study. Suppose that people are informed about the quality of drinking water and have the possibility to choose between different qualities (cf. Figure 1). If an individual worries a lot about, e.g., health risks due to a contaminant, he or she may be willing to pay a comparatively high price for drinking water of good quality, e.g., bottled water, water that has been purified — either by waterworks or by the individual himself — or water from a noncontaminated aquifer. A market for drinking water reflecting a demand for different qualities of drinking water may then reveal how people value these qualities. Some people are, however, likely to place a value on good groundwater quality not only because they care about their own and others' health, but also because of their concern for the environment. The latter part of the value is not likely to be captured when studying market behavior (cf. the second private decisions box in Figure 1, which is only a part of the externalities box). Thus, there are sometimes reasons to believe that estimates of WTP based on indirect methods are biased downwards.

DISCUSSION

In principle, we now know how a socially efficient situation can be determined. The practical problem of extensive data requirements for this determination may be one reason for the desirable use or concentration of a pesticide is often not determined with the aid of an economic analysis. If, e.g., a water quality standard has to be taken as given, an economic analysis may identify how the standard could be met in a cost-effective way, i.e., at the lowest costs. While the importance of an analysis of this kind should be acknowledged, an unrestricted analysis should in general be preferred.

There may be other and more fundamental reasons for constraining the economic analysis. For example, there may be other social goals than efficiency. One common additional goal is some kind of a fair distribution of income. Another type of constraint — but related to the issue of intergenerational fairness — is sometimes imposed by the environment. Irreversibility of damage to essential ecological systems and the limited possibility to substitute human-made capital for natural capital call for some cautionary criterion, e.g., a "safe minimum standard" (SMS) approach.[23-27] An SMS could tell us, e.g., to what limit humans can safely exploit an environmental resource without causing any irreversible damage. Of course, it is difficult to know what damage is irreversible, what ecological systems are essential, and if future substitution possibilities will be less limited. The presence of this uncertainty emphasizes the need for caution and, of course, for increased natural science knowledge.

Note that a cautionary criterion would be added to the degree of caution that is already involved when people place a value on the environment, and thus is involved in the trade-off between benefits and costs in an economic analysis. Again, irreversibility, etc. sometimes imply that this degree of caution is not enough, but the anthropocentric basis of economics and the principle of consumer sovereignty imply that the reasons for digressing from people's valuations must be strong. That is, a cautionary criterion should not be applied carelessly. These intricate issues, which are closely related to the way of interpreting and accomplishing a "sustainable development",[28] call for further cooperation between economists and natural scientists. The appearance of an ecological economics[29] shows that even an integration may be possible.

INSTITUTIONS

LIABILITY RULES VS. REGULATIONS

We now turn to the issue of how to internalize externalities in order to induce private decision-makers to adjust their behavior in a direction that is good for society as a whole. Broadly speaking, an internalization calls for institutions that constrain human behavior.

One very basic form of institution is property rights. Coase[30] showed that the existence of fully defined, enforced, and freely transferable private property rights and no transaction costs implies that resources will find their highest valued use no matter how the property rights are assigned. For example, a socially efficient contamination situation will arise irrespective of whether the contaminators have the right to contaminate or if the people affected by the contamination have the right to an environment free of contamination. Both parties have incentives to reach an economic agreement, and Coase's theorem says that the resulting agreement will be in accordance with social efficiency. However, in a real world, transaction costs are rarely negligible and sometimes property rights for environmental resources cannot be fully defined. In such cases, the legal system with its formal institutions could intervene in order to enforce a socially desirable situation.

What type of legal intervention could then be chosen in our case of groundwater contamination by pesticides? In broad terms, there is a choice between defining *liability rules* or introducing a *regulation*. Tort liability only occurs if damage really appears. A liability suit may then follow, possibly resulting in the injurer having to pay the injured parties the amount of money that compensates for the damage. The threat of a potential tort liability may, however, induce the potential injurer to take the care required to avoid the risk of damage. In contrast to tort liability, a regulation, e.g., a water quality standard, is introduced by the government before any damage has appeared. Administrative or criminal sanctions that will follow if the regulation is not complied with are also specified. These sanctions also apply if noncompliance does not result in any damage.

When should liability rules and regulations, respectively, be the preferred type of intervention? Faure[31] identifies the following criteria for regulation:

1. **Information asymmetry.** In some instances, the government can acquire sufficient information on damage risks at a lower cost than would be incurred by any potential injurers and injured.
2. **Insolvency risk.** The wealth of a potential injurer is sometimes smaller than the damage. If so, liability rules cause underdeterrence, i.e., the potential injurer considers only damage equal to his or her wealth when making decisions about preventive care. This problem is also present in the case of regulations with monetary sanctions.
3. **The threat of a liability suit.** The threat is negligible when, for example, (a) there is a long period of time between the injurer's activity and the appearance of damage; (b) it is difficult to establish a causal relationship between the activity and the damage; and (c) each victim is only slightly affected by the damage. The case of cancer is likely to fit in well with cases (a) and (b), and case (c) often follows from damage to common property, such as many environmental resources.
4. **Administrative costs.** The administrative costs of a liability rules system are likely to be large only when damage occurs and a suit follows. In contrast, regulations are designed at a stage when it is unknown whether damage will appear or not.

Faure concludes that liability rules alone are insufficient in the case of pesticides in groundwater. The potential injurers are farmers and, possibly, producers of pesticides. At least farmers cannot be expected to assess fully the environmental risks of their activities, and their capacity of paying for damage is generally very limited. Above all, however, the threat of a liability suit is likely to be negligible because of cases (a) and (b) of criterion 3 above. Regulations thus seem necessary, but nothing hinders the coexistence of regulations and liability rules. Faure notes that this combination can, in fact, be found in most legal systems.

Broadly speaking, there are two regulatory approaches: input and output control. In our case of pesticides in groundwater, it is quite easy to see the distinction between the two approaches (cf. Figure 1). Input control is directed to emitters or producers of pesticides and could be specified as, e.g., registration of pesticides, production restrictions, or use restrictions. Output control is directed to suppliers of drinking water, e.g., waterworks. Such a control could be a drinking water quality standard that waterworks have a legal responsibility to meet.

There is a large variety of instruments aimed at enforcing the chosen control(s). Traditionally, a *direct regulation* is interpreted in the economic literature as a directive to individual decision-makers requiring them, for example, to restrict the use of a pesticide to a certain quantity (to zero in the case of a ban) in order to enforce an input control. In contrast, *economic instruments* provide economic incentives for decision-makers to come to decisions consistent with an overall objective, e.g., an overall quantity restriction or a quality standard. In the following, we will have a brief look at economic instruments. See, e.g., Baumol and Oates,[32] Bohm and Russell,[33] Opschoor and Vos,[34] and Tietenberg[35,36] for a more detailed presentation of economic instruments than is offered below.

ECONOMIC INSTRUMENTS

A list of economic instruments would include charges (emission charges, product charges, tax differentiation, etc.), permit markets, deposit–refund systems, and subsidies (grants, soft loans, etc.). We will begin by focusing on the two former instruments. In order to illustrate some principles, let us return to the pesticide x and for a moment assume that the curves of Figure 4, which describe the reduction costs and the damage costs, are known. For simplicity, let us also assume that there are only two farmers, say A and B, causing the contamination and that the relation between each farmer's use of x and its appearance in groundwater is known, i.e., the farmers know what pesticide use is consistent with a certain concentration of x in groundwater. We know from Figure 4 that Q_x^* is the efficient contamination of x. Suppose that A and B do not have identical marginal reduction cost curves as shown by Figure 5. If there are no restrictions against contamination, A's contamination is $\overline{Q}_x^A$ and B's is $\overline{Q}_x^B$, and $\overline{Q}_x^A + \overline{Q}_x^B = \overline{Q}_x$. One way of realizing a reduction of the total contamination to Q_x^* would be to introduce a *direct regulation* stipulating that both A and B have to reduce their contamination to $Q_x^*/2$. However, the figure shows that, for this level, A's marginal reduction costs are lower than B's. This means that the total reduction to Q_x^* is not accomplished at the lowest costs. This is evident by the fact that total costs would decrease by forcing A to contaminate somewhat less and allowing B to contaminate somewhat more, but without changing the total contamination. Total reduction costs are minimized only when A's and B's marginal reduction costs are equal. That is, the direct regulation should stipulate Q_x^{A*} for A and Q_x^{B*} for B, where $Q_x^{A*} + Q_x^{B*} = Q_x^*$.

An alternative is to introduce a *uniform emission charge*. Faced with a charge, farmers have to balance their costs of contamination (paying a charge per unit of contamination) against the advantage of reducing contamination (paying reduction costs, but less charges). If the charge is set to t in Figure 4, all reductions from $\overline{Q}_x$ to Q_x^* will be worthwhile, as the reduction costs per additional unit reduced are lower than the charge in this interval. The charge also ensures that individual farmers reduce their contamination to cost effective levels, i.e., Q_x^{A*} and Q_x^{B*}, respectively.

Another alternative is to create a *permit market*. This would mean that farmers are allowed to contaminate Q_x^* in total, and permits summing to Q_x^* are distributed to them. They are then free to trade permits among themselves. If the permit market is well-functioning, the trade will tend to equalize the farmers' marginal reduction costs, i.e., a cost-effective situation is the likely outcome. For example, if $Q_x^*/2$ permits are distributed to A and B, respectively, there are prices of permits that would induce A and B to trade permits. Suppose that A is offering B the chance to buy some

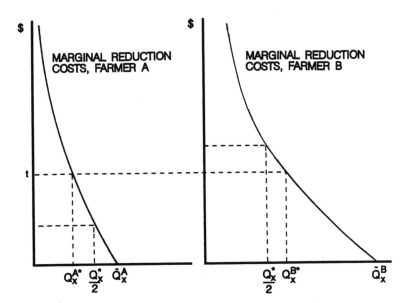

Figure 5   Marginal reduction cost curves for two farmers; the horizontal axes measure their respective contamination.

permits at a price of t. We see from Figure 5 that B will accept this offer and buy $Q_x^{B*} - Q_x^*/2$ permits. Why? Because B is saving money; the price B pays for every additional unit of contamination between $Q_x^*/2$ and Q_x^{B*} is t, but, if B does not buy the permits, a reduction to $Q_x^*/2$ is necessary. These reduction costs are larger than B's expenditures for permits. Also, A is happy to sell; each additional unit of reduced contamination between $Q_x^*/2$ and Q_x^{A*} costs him less than t. Note that the overall standard Q_x^* is not affected by the trade; it serves as a "bubble" within which a market is established.

Apparently, in a situation with enough information, direct regulations as well as the economic instruments described can be designed in a way that ensures cost-effectiveness. However, in reality, we usually know little about damage costs and reduction costs. The choice between instruments then becomes considerably more intricate. In terms of Figure 4, uncertainty about damage costs and reduction costs means that the efficient contamination Q_x^* or the charge consistent with Q_x^*, t, cannot be derived. What is then the best thing to do: restrict the contamination by a direct regulation or introduce a charge that may be far from Q_x^* or t, respectively? Weitzman analyzed this question in a seminal paper.[37] He compared the expected net losses from an inefficient direct regulation and an inefficient charge. Loosely speaking, his results showed that, if the slope of the marginal damage cost curve is larger than the slope of the marginal reduction cost curve, a direct regulation should be preferred. The reason is that a flat marginal reduction cost curve implies a large deviation from the efficient contamination if the charge is wrongly set. If, at the same time, the marginal damage cost curve is steep, considerable damage costs may follow from the deviation. Of course, the opposite result follows from the case when the relation between the slopes of the curves is reversed.

The uncertainty problem analyzed by Weitzman is only one of many practical difficulties that arise when an actual application of an economic instrument or a direct regulation is considered. Bohm and Russell[38] suggested the following dimensions along which potential instruments for achievement may be judged.

1. **Static cost-effectiveness.** Which instrument realizes a given objective, e.g., a water quality standard, at the lowest costs?
2. **Information intensity.** For example, what information is needed for the use of a certain instrument?
3. **Ease of monitoring and enforcement.**
4. **Flexibility in the face of economic change.** For example, how easily can the instrument cope with unexpected changes in the economy in order to maintain the desired objective?

5. **Dynamic incentives.** For example, how does the instrument influence the development of better technology for, e.g., pollution control?
6. **Political considerations.** For example, what effects on the income distribution in an economy does the instrument have?

In general, emission charges and permit systems have the potential to meet especially the first and fifth criteria in a better way than direct regulations. Direct regulations consistent with cost-effectiveness can be found only if information about marginal reduction costs for every source of contamination is available. In the case of permit markets and charges, cost-effectiveness is consistent with individual polluters' profit-maximizing behavior. As regards dynamic incentives, the technological effects of emission charges and permit markets are likely to be more positive than those of direct regulations. Improved technology that lowers reduction costs may be induced by the emission charges that polluters have to pay as long as there is any contamination. In the case of permit markets, polluters have the chance to sell permits if they improve their reduction technology. A direct regulation does not provide any incentives to improve the technology when its requirements have once been met.

While these favorable properties of emission charges and permit markets should be noted, it is difficult to give general recommendations on what instrument to use. The existence of practical complications such as location, costly monitoring, exogenous change in technology, and natural environmental systems preclude any general recommendation.[39] Such practical complications and authorities' traditional preference for "command and control" explain why emission charges and permit markets have hitherto been applied to a relatively small extent. In addition, when emission charges have been applied, their incentive impact has been low (often deliberately).[40] This means that we do not know very much about the actual performance of these instruments. This lack of knowledge and the difficulty to give general recommendations must not, however, hinder the application of emission charges and permit markets to situations in which they seem to be a promising alternative or complement to direct regulations. Such situations include the internalization of externalities because of agricultural activities.[41-43]

As regards deposit–refund systems and subsidies, we limit ourselves to just noting that the former instrument should probably be used to a considerably larger extent than today, and that subsidy schemes are often not compatible with the Polluter-Pays Principle accepted by most OECD countries. Note that when a deposit–refund system is introduced for environmental reasons, it is important that it involves both a "penalty" on the production of the units of the product in question that are causing environmental damage and a "reward" for the return of consumed units.[44,45]

Let us now be a little less general and turn to our case of groundwater contamination by pesticides. We made an assumption above that it is possible to establish a relationship between pesticide use and the concentration of the pesticide in groundwater. In practice, however, such a relationship is very difficult to establish, though not impossible.[46,47] Let us first note that this means that a combination of input and output control is likely to be suitable. An output control would ensure a desirable drinking water quality but has no influence on either the groundwater quality or the origins of contaminants. On the other hand, input control does influence the origins, but it may not be possible to design a set of input controls ensuring a certain water quality; the inputs may be too diffuse, and again, the knowledge of how and to what extent contaminants reach groundwater may be insufficient, etc. The two regulatory approaches thus seem to be complementary, at least if there is reason to protect groundwater and not only drinking water or if purification alone is not believed to be the optimal way of protecting drinking water.

Especially for enforcing input controls, economic instruments, and in particular permit markets, may be promising. They can induce a cost-effective situation at low information costs. In a "bubble" consisting of, e.g., an overall quantity restriction, the farmers' selfish behavior, in fact, works for cost-effectiveness. The cost-effective situation is, however, only likely to be reached if the permit market is well functioning. A necessary condition for well-functioning markets is a large number of agents on the market. At least in the case of a commonly used pesticide, a considerable number of farmers are indeed likely to be agents, and, thus, this condition could be satisfied. When compared to charges, another often attractive property of a permit market is that it works subject to an overall quantity restriction. As was discussed above, charges might be risky to use in a case where damage costs are rising quickly with the level of contamination.

The need for a combination of instruments is emphasized by Mason and Swanson.[48] They studied pesticide manufacturer incentives by a dynamic model and showed that a combination of patents and bans on specific pesticides may, in fact, encourage producers to introduce pesticides whose presence in groundwater will accumulate up to ban levels over the duration of the patent. In the model, the accumulation is assumed to be used by the producer as a strategic variable. This implies that a producer's response to a charge on the use of a pesticide would simply be to increase the accumulation rate of the pesticide. Mason and Swanson suggest instead an accumulation charge combined with a rent on the remaining stock of water quality. In practice, we suppose that an accumulation charge would have to be placed on characteristics of a pesticide that are related to the risk for accumulation.

As will be evident from a case study presented in the last section of this chapter, any restriction of the use of a specific pesticide can be meaningless in terms of groundwater quality, if close substitutes with, e.g., similar leaching properties as the restricted one are available. Irrespective of what instrument is used, it is thus of crucial importance that farmers' and producers' likely responses are taken into account. There may be reasons to, e.g., introduce restrictions also on substitutes in such a way that the substitutes' relative dangerousness is reflected (cf. an accumulation charge). Another type of response is that farmers start cultivating another crop. This could increase the use of a totally different type of pesticide. Note also that the possible presence of residues in pesticide containers may cause environmental damage if they are not disposed properly. This problem could probably be handled by a deposit–refund system. The trivial conclusion is that a comprehensive and long-term view is needed in any application of economic instruments or direct regulations.

WHAT CAN BE LEARNED FROM EARLIER STUDIES?

SOME FINDINGS IN EARLIER EMPIRICAL ECONOMIC STUDIES

In order to illustrate some of the issues presented above, we now turn to a brief review of some empirical economic studies dealing with groundwater contamination. No empirical study seems to have overcome fully the problem of the extensive requirements for data and scientific knowledge and thus succeeded in determining a socially efficient groundwater contamination situation. More successful research is found on valuation issues and cost-effectiveness.

First, some studies show that people really do adjust their economic behavior in response to groundwater contamination. Averting behavior (purchases of bottled water, purification equipment, etc.) is analyzed in, e.g., Abdalla,[49] Harrington et al.,[50] Lichtenberg et al.,[51] Malone and Barrows,[52] and Smith and Desvousges.[53] In most of these studies, households' averting expenditures are estimated and used as a measure of the welfare loss due to groundwater contamination. For example, Abdalla focused on a groundwater contamination episode in a community in Pennsylvania. The groundwater was found to contain high concentrations of perchloroethylene (PCE). No safety standard existed for PCE, but authorities nevertheless sent out information about potential health risks to households. Some households responded by deciding to protect themselves by, e.g., boiling water, installing purification equipment, or buying bottled water. Abdalla estimated the annual welfare loss per household to be about $290 to 440 (1990 U.S. dollars). It seems to be a general fact that individuals influence health risks by their own economic behavior. This endogenous nature of health risks is underscored by Crocker et al.,[54] who devote a large part of their paper to the potential impact of endogenous risk on cost–benefit analyses of groundwater contamination.

Second, there are some studies using the contingent valuation method for deriving estimates of how people value better groundwater quality, e.g., Edwards,[55] Hanley,[56] Mitchell and Carson,[57] and Silvander.[58] Let us use the study by Mitchell and Carson, who focused on drinking water quality, as an illustration. In the beginning of the 1980s, local water companies in southern Illinois had difficulty in meeting the U.S. standard for trihalomethane (THM), a substance considered to pose potential carcinogenic risks to humans. A sample of individuals living in this area were asked to state their WTP for a reduction of the THM level from some hypothetical levels to the standard. The corresponding mortality risk reductions were communicated with the aid of risk ladders. The largest and smallest considered risk reductions were a decrease in annual mortality risk from 9.50 and 0.61 per 100,000 persons, respectively, to 0.57 per 100,000. The smallest risk reduction is comparable with avoiding the risk of being struck and killed by lightning. The estimated average WTP per household and year was found to be about $52.00 and $3.50 (1990 U.S. dollars) for the largest and smallest risk reduction, respectively. Mitchell and Carson underscore that people really seem to understand the extremely small nature of the smallest risk reduction and value such a reduction on symbolic grounds. The results of Mitchell and Carson suggest that it is important not to disregard a groundwater quality project causing seemingly negligible risk reductions. If sufficiently many individuals would benefit from the project, it may very well turn out that total benefits outweigh the costs of the project.

Third, nothing general can be said about which method of solving a groundwater contamination problem is cost-effective. It may be efficient to take preventive actions in order to reduce the probability of contamination, but, in other cases, remedial response may be a better alternative. This is illustrated by Raucher[59,60] (see also Smith and Desvousges),[61] who focused on the benefits and costs of policies designed for the reduction of the risk of groundwater contamination caused by waste disposal sites. The interrelationship between prevention, monitoring, and remedial response is shown by his theoretical framework, which is applied on case studies of three U.S. waste disposal sites affecting aquifers (in Miami, Fort Lauderdale, and Nashua, respectively). Preventive as well as remedial actions for each site, and the costs of these actions, are described and discussed. By comparing costs, he concludes that, for the Miami and Fort Lauderdale sites, remedial response is cost-effective. However, for the Nashua site, response costs generally exceed preventive expenditures. Raucher then estimates costs per statistical cancer case avoided thanks to the cost-effective corrective actions. For each site, these costs are compared with estimates of the value of a statistical life from other economic studies. With the aid of this comparison, it is possible for Raucher to conclude that remedial response is warranted for the Miami site, but not for the Fort Lauderdale site. For the Nashua site, he finds that prevention is not necessarily more efficient than no action at all.

A CASE STUDY ON ATRAZINE

As the economics part of an interdisciplinary research project, Söderqvist[62] studied the case of groundwater contamination by atrazine in the Po Valley in northern Italy. Atrazine is an herbicide that, in Italy, has mainly been used in maize cultivation in the Po Valley since the beginning of the 1960s. One of its advantages in maize cultivation is its high flexibility; it can be used at presowing, preemergence or postemergence of weeds, and at any phase of maize growth. To this general efficacy should be added the advantage of a relatively low price. However, atrazine also has leaching properties that make it likely that it will eventually reach groundwater, especially in areas with permeable soils. This may be worrisome, since groundwater is the main drinking water source in northern Italy. Surveys of drinking water in the 1980s revealed the presence of pesticides and, in particular, of atrazine. The detected concentrations often exceeded 0.1 μg/L and sometimes even 1 μg/L.[63]

A European Communities (EC) Council Directive on drinking water of July 15, 1980, set the maximum admissible concentration for pesticides in drinking water at 0.1 μg/L. In February 1985, the Directive was implemented into Italian law. Faure[64] carefully describes the Italian implementation and enforcement of the directive. Local restrictions against use of supplies of drinking water containing atrazine were introduced in 1986. In March 1990, the Italian Ministry of Health decided to introduce a nationwide ban against the sale and use of atrazine.

Söderqvist first approaches the question of whether the EC standard could be viewed as socially efficient and discuss benefits and costs of less atrazine in groundwater. Actual concentrations of atrazine in Italian groundwater do not seem to cause any health effects or environmental damage, but some less tangible benefits are likely to exist. These benefits could be due to, e.g., the reduction in anxiety caused by the presence of atrazine in drinking water and people's valuation of groundwater as a pristine environmental resource. The available information on these benefits was found to be too limited to permit any cost–benefit analysis. Instead, the EC standard is taken as given, and the costs of two ways of meeting the standard are compared: water purification by granulated active carbon or a ban on the use of atrazine. A representative area in the Veneto region is selected for this cost-effectiveness analysis.

Because of the difficulties of separating out purification costs attributable to atrazine, an interval for these costs is estimated. It is found that, in the short run, the increased costs for farmers due to a ban on atrazine are slightly higher than the lower bound of the purification cost interval. This makes the comparison between the costs inconclusive. Söderqvist then considers whether there are reasons to believe that the costs could be balanced against the value of actually meeting the EC standard, i.e., the question of social efficiency is again approached. A separate survey for eliciting people's WTP is not undertaken (but suggested as future work). Instead, the findings of the study by Mitchell and Carson[57] presented above are used. If it is assumed that each individual living in the studied area in Veneto has a WTP for reducing the concentration of atrazine in drinking water equal to the WTP per person for the smallest risk reduction studied by Mitchell and Carson, the sum of individual WTPs amounts to a number only slightly lower than the costs of a ban. It is emphasized that this comparison is interesting but not conclusive, since the application used by Mitchell and Carson concerned neither atrazine nor Italy, and, in contrast to the case of atrazine, there were actual and quantified health risks, though extremely small.

Finally, let us take a closer look at the design of institutions in the case of atrazine in Italy. The type of control present was an output control — the EC standard. Apparently, Italy found it unreasonable to meet the standard only by water purification. There are probably many reason for this: water purification is not particularly cheap, it takes time to increase the capacity of purification systems, etc. Since there is no clear relationship between atrazine use and its appearance in groundwater, it is not easy to specify what input control would ensure a groundwater quality that meets the EC standard. This is, in fact, an argument for the type of input control chosen in Italy, i.e., a ban. With a ban, one can be sure that the EC standard will be met sooner or later. However, the farmers' response to the ban yielded absurd results: at least in the short run, farmers mostly replaced atrazine by herbicides (mainly terbuthylazine) that are likely eventually to be found in groundwater. That is, the ban on the specific herbicide of atrazine merely accomplished a costly change in the mix of contaminating substances. This is not an example of the comprehensive view when designing restrictions and instruments recommended above. It also implies that the use of the WTP figures of Mitchell and Carson must be assigned another qualification, since the people questioned in their study were considering a permanent reduction of a contaminant. A more satisfactory enforcement strategy would have been to specify input controls that cause a gradual change to a more desirable situation, i.e., a weed control situation consistent with, e.g., a total groundwater quality that will meet relevant standards.

FINAL REMARKS

We would like to close this chapter by drawing the reader's attention to two of the key words of the chapter. The first of these is *information*. One aspect of information that is touched upon is that; even if people's subjective interpretations of information on, e.g., health risks due to pesticides in drinking water are likely to differ; a correct estimate of the benefits of a reduction in these risks would be based on these persons' valuations. However, there is sometimes a need for a pragmatic interpretation of this principle. For example, the case study on atrazine described above has resulted in the initiation of a contingent valuation study on groundwater quality in northern Italy. This study will focus on *total* groundwater quality and not on the probably less robust value of reducing the concentration of individual substances in groundwater. It does not seem reasonable to ask for a WTP for reducing atrazine concentration, and a WTP for terbuthylazine, and a WTP for linuron, etc. In this case, a more pragmatic approach should be adopted, namely, people can tell us how they value, e.g., a safer drinking water but experts have to find out what "safer" is and how it can be accomplished. Another important aspect of information is the uncertainty about irreversible environmental damage, etc. and its significance for the need for a cautionary criterion.

The second keyword is *comprehensive*. The need for a comprehensive view was emphasized in the section describing ways of internalizing negative externalities. These ways mainly imply some kind of intervention by the government. A comprehensive view also includes, however, awareness that the existence of, e.g., lobbying implies that actions by the government are not likely to be impartial (cf. the so-called public-choice view).[65] This may strengthen the importance of the major change in environmental attitudes that seems to be in progress within parts of the business community.[66]

ACKNOWLEDGMENTS

We have benefited from constructive comments by, among others, Michael Faure, Timothy M. Swanson, and Marco Vighi. Söderqvist is grateful to the European Science Foundation and the Swedish National Council for Planning and Coordination of Research for financial support, and to the Beijer International Institute of Ecological Economics, Stockholm, for hospitality.

REFERENCES

1. Bojö, J., K-G. Mäler, and L. Unemo. *Environment and Development: An Economic Approach,* Second Revised Edition (Dordrecht: Kluwer Academic Publishers, 1992).
2. Kneese, A. V. and W. D. Schulze. "Ethics and Environmental Economics", in *Handbook of Natural Resource and Energy Economics*, Vol. I, A. V. Kneese and J. L. Sweeney, Eds. (Amsterdam: Elsevier Science Publishers, 1985), pp. 191-220.
3. Sagoff, M. *The Economy of the Earth* (Cambridge, U.K.: Cambridge University Press, 1988).
4. Eggertsson, T. *Economic Behavior and Institutions* (Cambridge, U.K.: Cambridge University Press, 1990).
5. Common, M. and C. Perrings. "Towards an Ecological Economics of Sustainability," *Ecological Economics* 6:7-34 (1992).
6. Folke, C. "Socio-Economic Dependence on the Life-Supporting Environment", in *Linking the Natural Environment and the Economy: Essays from the Eco-Eco Group*, C. Folke and T. Kåberger, Eds. (Dordrecht: Kluwer Academic Publishers, 1991), pp. 77-94.
7. Folke C., M. Hammer, R. Costanza, and A.-M. Jansson. "Investing in Natural Capital: Why, What, and How?", in *Investing in Natural Capital: The Ecological Economics Approach to Sustainability,* A.-M. Jansson, M. Hammer, C. Folke, and R. Costanza, Eds. (Washington, D.C.: Island Press, 1994), pp. 1-20.
8. North, D. C. *Institutions, Institutional Change and Economic Performance* (Cambridge, U.K.: Cambridge University Press, 1990), p. 3.
9. Boadway, R. W. and N. Bruce. *Welfare Economics* (Oxford, U.K.: Basil Blackwell, 1984).
10. Johansson, P.-O. *An Introduction to Modern Welfare Economics* (Cambridge, U.K.: Cambridge University Press, 1991).
11. Johansson, P.-O. *Cost–Benefit Analysis of Environmental Change* (Cambridge, U.K.: Cambridge University Press, 1993).
12. Just, R. E., D. L. Hueth, and A. Schmitz. *Applied Welfare Economics and Public Policy* (Englewood Cliffs, N.J.: Prentice-Hall, Inc., 1982).
13. Andréasson, I.-M. "Costs of Controls on Farmers' Use of Nitrogen", Ph.D. Thesis, Stockholm School of Economics, Stockholm, Sweden (1988).
14. Andréasson-Gren, I.-M. "Regulating the Farmers' Use of Pesticides in Sweden", Beijer Discussion Paper Series No. 10, Beijer International Institute of Ecological Economics, Stockholm, Sweden (1992).
15. Kreps. D. M. *A Course in Microeconomic Theory* (Hemel Hempstead, U.K.: Harvester Wheatsheaf, 1990).
16. Varian, H. R. *Microeconomic Analysis*, 3rd Edition (New York: W. W. Norton, 1992).
17. Friedman, J. *Oligopoly Theory* (Cambridge, U.K.: Cambridge University Press, 1983).
18. Cropper, M. L. and A. M. Freeman III. "Valuing Environmental Health Effects", in *Measuring the Demand for Environmental Quality*, J. B. Braden and C. D. Kolstad, Eds. (Amsterdam: North-Holland, 1991), pp. 165-211.

19. Mäler, K.-G. "Welfare Economics and the Environment", in *Handbook of Natural Resource and Energy Economics*, Vol. I, A. V. Kneese and J. L. Sweeney, Eds. (Amsterdam: Elsevier Science Publishers, 1985), pp. 3-60.

20. Mitchell, R. C. and R. T. Carson. *Using Surveys to Value Public Goods: The Contingent Valuation Method* (Washington, D.C.: Resources for the Future, 1989).

21. Arrow, K., R. Solow, E. Leamer, P. Portney, R. Randner, and H. Schuman. "Report of the NOAA Panel on Contingent Valuation", *Federal Register*, 58(10):4601-4614 (1993).

22. Freeman, A. M. III. *The Measurement of Environmental and Resource Values: Theory and Methods* (Washington, D.C.: Resources for the Future, 1993).

23. Ciriacy-Wantrup, C. V. *Resource Conservation: Economics and Policy* (Berkeley, CA: University of California Press, 1952).

24. Bishop, R. "Endangered Species and Uncertainty: The Economics of a Safe Minimum Standard", *American Agricultural Economics*, 60:10-13 (1978).

25. Goodland, R. and G. Ledec. "Neoclassical Economics and Principles of Sustainable Development", *Ecol. Modelling*, 38:19-46 (1987).

26. Pearce, D. W., E. Barbier, and A. Markandya. *Sustainable Development: Economics and Environment in the Third World* (London, U.K.: Edward Elgar, 1989).

27. Pearce, D. W. "Cost–Benefit Analysis and PMPs", in *Persistant Pollutants: Economics and Policy*, J. B. Opschoor and D. W. Pearce, Eds. (Dordrecht, The Netherlands: Kluwer Academic Publishers, 1991).

28. World Commission on Environment and Development. *Our Common Future* (London, U.K.: Oxford University Press, 1987).

29. Costanza, R. "What Is Ecological Economics?", *Ecological Economics*, 1(1):1-7 (1989).

30. Coase, R. "The Problem of Social Cost". *J. Law Econ.*, 3:1-44 (1960).

31. Faure, M. "The EC Directive on Drinking Water: Institutional Aspects", in *Environmental Toxicology, Economics and Institutions: The Atrazine Case Study*, L. Bergman and D. M. Pugh, Eds. (Dordrecht: Kluwer Academic Publishers, 1994), pp. 39-87.

32. Baumol, W. J. and W. E. Oates. *The Theory of Environmental Policy*, 2nd Edition (Cambridge, U.K.: Cambridge University Press, 1988).

33. Bohm, P. and C. S. Russell. "Comparative Analysis of Alternative Policy Instruments", in *Handbook of Natural Resource and Energy Economics*, Vol. I, A. V. Kneese and J. L. Sweeney, Eds. (Amsterdam: Elsevier Science Publishers, 1985), pp. 395-460.

34. Opschoor, J. B. and M. B. Vos. *Economic Instruments for Environmental Protection* (Paris: OECD, 1989).

35. Tietenberg, T. *Environmental and Resource Economics*, 3rd Edition (Glenview, IL: Scott, Foresman and Company, 1992).

36. Tietenberg, T. "Economic Instruments for Environmental Regulation", *Oxford Review of Economic Policy*, 6:17-33 (1990).

37. Weitzman, M. L. "Prices vs. Quantities", *Rev. Econ. Studies*, 41:477-491 (1974).

38. Bohm, P. and C. S. Russell. "Comparative Analysis of Alternative Policy Instruments", in *Handbook of Natural Resource and Energy Economics*, Vol. I, A. V. Kneese and J. L. Sweeney, Eds. (Amsterdam: Elsevier Science Publishers, 1985), pp. 395-460.

39. Bohm, P. and C. S. Russell. "Comparative Analysis of Alternative Policy Instruments", in *Handbook of Natural Resource and Energy Economics*, Vol. I, A. V. Kneese and J. L. Sweeney, Eds. (Amsterdam: Elsevier Science Publishers, 1985).

40. Opschoor, J. B. "Economic Instruments for Controlling PMPs", in *Persistent Pollutants: Economics and Policy*, J. B. Opschoor and D. W. Pearce, Eds. (Dordrecht, The Netherlands: Kluwer Academic Publishers, 1991).

41. Dubgaard, A. "Taxation as a Means to Control Pesticide Application", Report No. 35, Statens Jordbrugsøkonomiske Institut, Copenhagen, Denmark, (in Danish).

42. Andréasson, I.-M. "Costs of Controls on Farmers' Use of Nitrogen", Ph.D. Thesis, Stockholm School of Economics, Stockholm, Sweden (1988).

43. Andréasson-Gren, I.-M. "Regulating the Farmers' Use of Pesticides in Sweden", Beijer Discussion Paper Series No. 10, Beijer International Institute of Ecological Economics, Stockholm, Sweden (1992).

44. Mäler, K-G. *Environmental Economics: A Theoretical Inquiry* (Baltimore: The Johns Hopkins University Press, 1974), pp. 235-242.

45. Söderqvist, T. "A Deposit-Refund System for Mercury Batteries". Mimeo, Stockholm School of Economics (1987), (in Swedish).

46. Vighi, M. and A. Di Guardo. "Predictive Approaches for the Evaluation of Pesticide Exposure", this volume, Chapter 3.

47. Giuliano, G. "Groundwater Vulnerability to Pesticides", this volume, Chapter 4.

48. Mason, R. and T. M. Swanson. "Regulating Chemical Waste: Addressing Manufacturer Incentives". Mimeo, Faculty of Economics and Politics, University of Cambridge, U.K., (1993).

49. Abdalla, C. W. "Measuring Economic Losses from Groundwater Contamination: Investigation of Household Avoidance Costs", *Water Res. Bull.*, 26:451-463 (1990).

50. Harrington, W., A. J. Krupnick, and W. O. Spofford, Jr. "The Economic Losses of a Waterborne Disease Outbreak", *J. Urban Econ.*, 25:116-137 (1989).

51. Lichtenberg, E., D. Zilberman, and K. T. Bogen. "Regulating Environmental Health Risks Under Uncertainty: Groundwater Contamination in California", *J. Environ. Econ. Management*, 17:22-34 (1989).

52. Malone, P. and R. Barrows. "Groundwater Pollution's Effects on Residential Property Values, Portage County, Wisconsin", *J. Soil Water Conserv.*, 45:346-348 (1990).

53. Smith, V. K. and W. H. Desvousges. "Averting Behavior: Does It Exist?", *Econ. Lett.*, 20:291-296 (1988).

54. Crocker, T. D., B. A. Forster, and J. F. Shogren. "Valuing Potential Groundwater Protection Benefits", *Water Resources Res.*, 27:1-6 (1991).

55. Edwards, S. F. "Option Prices for Groundwater Protection", *J. Environ. Econ. Management*, 15:475-487 (1988).

56. Hanley, N. "Problems in Valuing Environmental Improvements Resulting from Agricultural Policy Changes: The Case of Nitrate Pollution", AEEA Seminar on Economic Aspects of Environmental Regulations in Agriculture, Thune University of Agriculture, Copenhagen (1989).

57. Mitchell, R. C. and R. T. Carson. "Valuing Drinking Water Risk Reductions Using the Contingent Valuation Method: A Methodological Study of Risks from THM and Giardia", draft report to the U.S. Environmental Protection Agency, Resources for the Future, Washington, D. C. (1986).

58. Silvander, U. "The Willingness to Pay for Angling and Groundwater in Sweden", Dissertation 2, Department of Economics, Swedish University of Agricultural Sciences, Uppsala, Sweden (1991).

59. Raucher, R. L. "A Conceptual Framework for Measuring the Benefit of Groundwater Protection", *Water Resources Res.*, 19:320-326 (1983).

60. Raucher, R. L. "The Benefits and Costs of Policies Related to Groundwater Contamination", *Land Econ.*, 62:33-45 (1986).

61. Smith, V. K. and W. H. Desvousges. "The Valuation of Environmental Risks and Hazardous Waste Policy", *Land Econ.*, 64:211-219 (1988).

62. Söderqvist, T. "The Costs of Meeting a Drinking Water Quality Standard: The Case of Atrazine in Italy", in *Environmental Toxicology, Economics and Institutions: The Atrazine Case Study*, L. Bergman and D. M. Pugh, Eds. (Dordrecht: Kluwer Academic Publishers, 1994), pp. 151-171.

63. Vighi, M. and G. Zanin. "Agronomic and Ecotoxicological Aspects of Herbicide Contamination of Groundwater in Italy", in *Environmental Toxicology, Economics and Institutions: The Atrazine Case Study*, L. Bergman and D. M. Pugh, Eds. (Dordrecht: Kluwer Academic Publishers, 1994), pp. 111-139.

64. Faure, M. "The EC Directive on Drinking Water: Institutional Aspects", in *Environmental Toxicology, Economics and Institutions: The Atrazine Case Study*, L. Bergman and D. M. Pugh, Eds. (Dordrecht: Kluwer Academic Publishers, 1994), pp. 39-87.

65. Buchanan, J. and G. Tullock. *The Calculus of Consent* (Ann Arbor, MI: University of Michigan Press, 1962).

66. Schmidheiny, S. *Changing Course: A Global Business Perspective on Development and the Environment* (Cambridge, MA: MIT Press, 1992).

CHAPTER 12

Conclusions

Marco Vighi and Enzo Funari

CONTENTS

The basic role of groundwater as a primary source of good-quality water supply for human consumption justifies the growing public concern about the possibility of contamination of this irreplaceable resource, which may cause, in some cases, its use for drinking purposes to be severely limited.

In the last few decades the relevance of the problem has led the international scientific community to devote a significant amount of its energies and human and economic resources to the study of the various aspects of the complex concept of groundwater pollution. At the same time, the administrative and political authorities, both at the international and local levels, have developed strategies for the protection and recovery of groundwater resources.

The task of protecting groundwater in order to provide good quality water for drinking purposes has been faced with extremely different approaches. A not negligible part of the scientific and administrative community supports the thesis that, at least in highly developed and heavily crowded areas where human activities unavoidably produce a lower quality of environmental resources and where the demand for water supply is extremely high and growing with the level of welfare, it is economically unacceptable to provide good quality water for all human uses. Thus, one proposal is the setting up of a double pipe system, even for domestic purposes. High-quality water would be supplied for drinking purposes, whereas lower quality water would be used for other purposes (washing and other domestic needs).

In this case, only few high-quality water supplies would be severely protected. For other water sources, less stringent quality standards could be acceptable. Apart from the technical and practical difficulties of such a solution, the proposal is unacceptable from an environmental point of view.

The idea of a two-level water quality supply is based on the principle of the consensus for an unrecoverable degradation of a natural resource, a concept in contrast to the growing awareness of the danger of environmental damage. The correct solution should be a suitable strategy of prevention of environmental damage, and the main aim of applied environmental sciences must be the production of conceptual instruments and technologies for the management of potentially dangerous substances, in order that the protection of environmental resources keep pace with the need for social and economic development.

0-87371-439-3/95/$0.00+$.50
© 1995 by CRC Press, Inc.

With this frame of reference, the management of pesticides, which may represent a significant reason of concern for groundwater pollution, must be faced with a proper strategy, based on sound scientific knowledge and appropriate regulatory measures. In this book, an attempt is made to describe the multidisciplinary information needed for this task.

A synthesis of the more relevant statements and conclusions, deriving from the different aspects treated, can be made to evaluate the present "state of the art" of our scientific knowledge about this problem, both from the agronomic and the environmental point of view, and to point out the current situation in relation to social sciences (economy, politics) and regulations.

EXTENT OF THE PROBLEM

In the first part of this book, examples of experimental data on pesticide occurrence in groundwater were described, obtained both from a literature review of specific investigations and from an extensive, nationwide monitoring program. A critical evaluation of the available data indicated that a reliable picture of the worldwide extent of groundwater contamination from pesticides is practically impossible due to several limitations of the monitoring approach. Nevertheless it has been demonstrated that large-use compounds are likely to occur in groundwater, even at relatively high levels, particularly if they show some agronomic characteristics (application on soil, treatment during wet season, etc.) and some molecular properties (high water solubility, low soil binding capability, high persistence, etc.). All of this indicates that herbicides are products of major concern, with the chemical class of triazines most frequently involved in groundwater contamination. Thus, having established that pesticides can be included among significant contaminants of groundwater, it must be evaluated if this contamination represents a threat to human health.

The impossibility of defining "typical" concentrations for pesticides in groundwater makes it impossible to evaluate the extent of the risk for humans deriving from the use of groundwater as drinking supply. However, even the relatively scattered data reported in Chapter 1 allow some comparison with the WHO Guidelines or with the U.S. EPA Health Advisories described in Chapter 5. For example, of the 32 pesticides reported as nonsporadically occurring in groundwater, a WHO guideline is defined for 20. For 17 of these compounds, at least one set of data indicate a maximum value above the WHO guideline while, for 9 compounds, the geometric mean of maximum reported values is above the guideline (Table 1).

This comparison cannot be assumed as a quantification of the risk from pesticides and, in any case, must not be overvalued. The precise meaning of maximum reported values is not easy to interpret, particularly in a general evaluation of monitoring data. This evaluation can only indicate that, at least under particular conditions (high agricultural use of the territory, high vulnerability of aquifers, etc.), pesticides may impair the use of groundwater as a drinking water supply.

SCIENTIFIC KNOWLEDGE AND NEED FOR RESEARCH

The most unwanted properties of a pesticide, from an environmental point of view, are its toxicity on nontarget living organisms, its persistence, bioaccumulation, and mobility, i.e., its capability to reach environmental compartments or sites far from the site of application.

In this context, the latest generation of active ingredients is very different from the old compounds, such as the first organochlorinated insecticides or other chemicals produced at the beginning of the pesticide era. Compounds of growing selectivity in their mode of action have been produced, to increase their effectiveness against target organisms and to reduce the danger for nontargets. The environmental persistence of new pesticides is generally relatively low, in order to be effective only during the period of pest presence on the crop. In comparison with the half-life of many years of the old organochlorinated insecticides, such as DDT or lindane, the half-life of the new active ingredients is measured in terms of days, weeks or, at the most, a few months.

As for environmental distribution and mobility, the situation is more complex and heterogeneous. From the environmental point of view, the less dangerous compounds should remain where

Table 1   Pesticides Frequently Detected in Groundwater (See Chapter 1) Showing Values above the U.S. EPA and WHO Limits

Molecule	U.S. EPA Lifetime HAs µg/L	WHO Guidelines µg/L	Number of Data Set Showing Maximum Values Above the Limit Related to the Total		Geometric Mean of Maximum Values Above (+) or Below (−) the Limit	
			U.S. EPA	WHO	U.S. EPA	WHO
Herbicides						
Alachlor		20		5/33		−
Atrazine	3	2	28/41	31/41	+	+
Bentazone	20	30	3/9	1/9	−	−
Bromacil	90		3/8		−	
Chlortal Dimethyl (DCPA)	4000		0/7		−	
Cyanazine	1		12/17		+	
2,4-D	70	30	0/6	1/6	−	−
Dicamba	200		1/7		−	
Dinoseb	7		5/9		+	
Isoproturon		9		0/2		−
MCPA	10	2	0/6	4/6	−	−
Mecoprop		10		1/7		−
Metolachlor	100	10	3/20	6/20	−	−
Metribuzin	200		2/17		−	
Molinate		6		4/4		+
Pendimethalin		20		0/2		−
Picloram	500		0/2		−	
Propazine	10		0/2		−	
Simazine	4	2	4/10	8/10	−	−
2,4,5-T	70	9	0/3	1/3	−	−
Trifluralin	5	20	2/9	0/9	−	−
Insecticides						
Aldicarb	1	10	15/15	14/15	+	+
Carbofuran	40	5	2/9	6/9	−	+
Dieldrin		0.03		1/2		+
Fonofos	10		0/3		−	
Lindane	0.2	2	2/3	2/3	+	+
Oxamyl	200		1/5		−	
Nematicides						
DBCP		1		3/4		+
1,2-Dichloropropane		20		6/10		+
1,3-Dichloropropene		20		4/8		+

they have been applied, for example, in the surface layer of soil where they will be more or less readily biodegraded. From the agronomic point of view, the distribution and mobility properties of a pesticide must be set as a function of its agronomic effectiveness and reliability. Some cationic herbicides effective in postemergence on green plant tissue (e.g., paraquat) are strongly bound with the inorganic soil matrix and are not residual and practically immobile because of their very rapid inactivation by irreversible adsorption in contact with the soil. On the contrary, preemergence herbicides must be carried by water up to the soil layer, where they can come in contact with weed seeds, and this can explain their relatively high leachability. Pesticides used as fumigants in stored products or for some soil treatments must be highly volatile, and, as a consequence, they are extremely mobile in air. The high Kow of some insecticides can improve their effectiveness against targets and reduce their mobility in water and air, but this increases their bioaccumulation potential.

In any case, even if the primary objective in pesticide design is, obviously, to improve effectiveness against pests, the recent evolution of these compounds is generally towards more environmentally friendly products. Moreover, because of the large variety of active ingredients now existing on the market, the most suitable compound can be selected in relation to different situations, in order to optimize agronomic effectiveness and to minimize environmental impact.

At this point, a question must be asked: is the present knowledge about the environmental behavior of pesticides and environmental vulnerability developed enough to provide the information

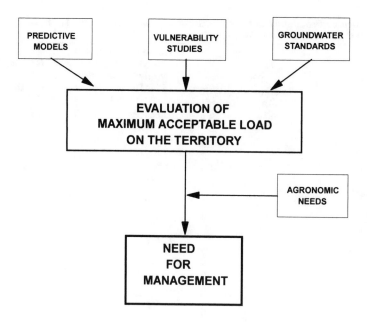

Figure 1   Scheme for the evaluation of the need for intervention in the management of pesticide use.

needed for the management of pesticides? An answer to this question, at least in relation to groundwater risk, has been given in the Section II of this book. With the development of suitable predictive approaches, able to describe the environmental behavior of pesticides and to predict distribution and fate, together with hydrologic studies on the vulnerability of aquifers, the maximum acceptable load of a pesticide on the territory can be evaluated in order to comply with groundwater quality standards. The admissible load should be compared with agronomic requirements, in order to evaluate the need for interventions and management strategies for the protection of groundwater resources (Figure 1).

If this is true in theory, in practice some problems still need to be solved. In particular, the predictive capability of environmental models, highly reliable on the field scale, tends to decrease as the scale of the application increases. Nevertheless, as described in Chapter 3, the development of relatively simple models can be the right way to obtain reliable results on a scale large enough to produce useful indications for a proper planning strategy of pesticide management. The recent results in this field seem to be extremely promising.[1,2]

Even in the field of hydrogeologic studies, the conceptual instruments needed to evaluate groundwater vulnerability and to contribute to the prediction of pesticide occurrence in groundwater are available at a suitable level of reliability. In order to increase the effectiveness of these instruments, our knowledge of the hydrogeologic structure, still lacking in many countries even in the developed world, should be improved.

Thus, if suitable information for proper pesticide management can be provided, the second question is: are the right means to reduce the impact of pesticides and to protect groundwater yet available? Some answers to this question were given in the Section III of the book.

As for agricultural practices, there is no doubt that integrated pest management is the most effective way to reduce the amount of pesticide used. In the case of insecticides, through integrated pest management, the use of chemical compounds can be reduced for some crops by more than 50% without significant effects on productivity. Unfortunately alternative strategies to chemical weed control are not so effective as those for insects. Weed control management is an extremely difficult task, influenced by a number of factors that are difficult to predict. Economic and regulatory policies, at national and international levels, may affect the annual turnover of crops, particularly in the European Economic Community (EEC). Agronomic practices and farmer know-how are rapidly evolving. All of these factors affect changes in weed community and make a long-term strategy of integrated weed management very difficult (Figure 2).

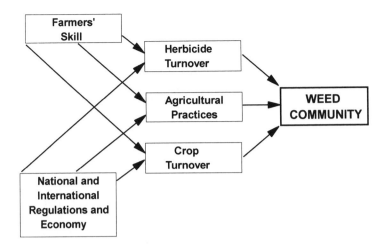

Figure 2   Agronomic and social factors affecting weed communities. (Modified from Zanin et al., *Terra e Vita* 7: 11–13, 1993.)

This statement is confirmed by a recent management project carried our in Denmark. The Danish government planned a strategy to reduce pesticide use by up to 50% (with reference to the period 1981 to 1985) by 1997, with an intermediate objective of 25% by 1990.[3] The intermediate goal was attained for insecticides and fungicides, whereas unsatisfactory results were obtained for herbicides. Consequently, the development and refinement of methods for integrated pest management and generally for innovative pest control strategies that are able to substitute, at least in part, for the use of agrochemical compounds, are presently among the most important fields of research in agronomic sciences.

From a wider perspective of prevention, the protection of environmental resources from potentially dangerous chemicals, particularly the protection of groundwater from pesticides, must be included in a general strategy of land use management. The conceptual instruments to optimize land use planning on a rational and scientific basis are available, and an example has been given in Chapter 9. Obviously, this is a very difficult task from an administrative and political point of view, and actual results on a wide scale can hardly be obtained in the short run. In many important agricultural areas, even in developed countries, there is still a lack of knowledge of some basic information needed for a sound application of land use management models. There is no doubt that this important gap needs to be filled in the near future.

Finally, for emergency situations, various possible treatment options are available and have proven effective for many pesticides. Nevertheless, water treatment technology must be continuously updated as a function of the evolution of pesticide use, and specific methods must be optimized for possible new compounds likely to be found in water intended for human consumption.

SOCIAL SCIENCES AND REGULATIONS: ARE THEY ADEQUATE TO PROTECT GROUNDWATER?

The scientific development of the last few decades has produced enough information and know-how to understand the problem of pesticide risk and to suggest some possible solutions, even if with some uncertainties. However, to ensure the maximum protection of groundwater resources and, in the mean time, to avoid an excessive burden for agricultural activities, economic, social, and political factors should be taken into account when planning the most effective solutions. In Section V of this book the main approaches used to protect groundwater from pesticides in different parts of the world have been described and compared, in relation to the production of drinking water for human consumption. The EEC approach, based on "analytic" and not "toxicologic" standards, is the most restrictive and is subject to several objections and criticisms even among some EEC member countries. From the point of view of environmental protection and human health, the EEC approach, based upon the philosophy of total protection from xenobiotic chemicals of one of the most precious natural resources, is without doubt a good position and must be defended. From the scientific and technological point of view, as mentioned above, the conceptual instruments are available for the prediction of pesticide behavior and minimization of the impact on environmental resources. Though not an easy task, the objective of groundwater free from pesticides can be realistically attained by means of an integrated strategy based on a proper management of pesticides and land use.

The regulatory instruments to plan such a strategy are difficult to produce and, for the time being, are not yet defined. An attempt to produce suitable legal tools for the management of pesticides in order to protect groundwater is at present under discussion in the EEC and will be proposed as an annex (the so called "Uniform Principles") of the Directive 91/414 concerning the placing of plant protection products on the market.[5] Another question is if the EEC quality objective for pesticides in drinking water is acceptable from an economic point of view, taking into account that less stringent standards, based on WHO toxicological guidelines, are still safe for human health and more easily attainable. An economic evaluation is difficult because, while some aspects are relatively easily accounted for, others are not. First of all, the cost of high-level protection must be evaluated not just in terms of direct expenses required for the realization of the interventions needed to attain the objective. A strategy of pesticide control could heavily affect the price of agricultural products, reducing the EEC's competitivity in the international market. This would lead to economic loss, mainly for countries where agriculture plays a significant role in the national budget (France, Italy, Spain, Greece, Portugal). On the other hand it is extremely important to heed the growing public concern for environmental protection and human health.

Even if, from the scientific point of view, there is no reason for not trusting the reliability of toxicologic standards proposed by international organizations (FAO, WHO), it must be taken into account that human welfare is also determined by individual judgment. Thus, the fear (justified or not) of the presence of pesticides in drinking water, even if below toxicologic standards, must be internalized in any economic evaluation of the problem.

Public willingness to pay for a safe environment and, in particular, for pesticide-free drinking water, could counterbalance the high cost of a stringent protection strategy. At this point, a further question must be asked: is this kind of strategy sufficient to avoid or at least reduce anxiety and fear of contamination? Is the public aware that the "near to zero" philosophy of the EEC standards for pesticides means total protection of drinking water resources and that this is the maximum allowable level of water safeness? To maximize the results of water protection strategy, the administrative and political authorities should direct their efforts to improving the level of public information and awareness about the real situation and the meaning of some interventions to protect water.

Finally, the suggestions proposed by agronomic and environmental sciences, and accepted as realistically applicable on the basis of an economic evaluation, must be made operative by means of suitable regulatory interventions and economic incentives, in order to meet the standards for groundwater protection. These interventions must be developed at various levels in order to produce an integrated strategy to control the risk from pesticides.

On one level, a regulatory approach should be implemented that ensures that pesticide manufacturers are aware the current incentives when developing new chemicals i.e., that encourages manufacturers to design pesticides with environmentally friendly and socially desirable characteristics. An example of such an incentive system for manufacturers is that proposed by Mason and Swanson briefly described in Chapter 11. A second level of intervention could deal with orienting the farmer's choice towards the best compound for each condition. At present, this choice is determined mainly by agronomic properties and marketing characteristics (Table 2).

On the other hand, it is impossible to claim that the farmer has more than a rough idea about pesticide risk and that he evaluates environmental hazard as a function of pesticide properties and environmental characteristics. The ban of specific pesticides generally does no more than force farmers to substitute other equally or more unpleasant pesticides. Moreover, a generalized ban could be inappropriate; a pesticide could be extremely dangerous for groundwater in some situations

Table 2 Main Criteria for the Choice of a Maize Herbicide by the Farmers, Listed in Order of Decreasing Importance

1 Personal experience
2 Cost
3 Absence of residues
4 Selectivity
5 Toxicologic class
6 Wide action range
7 Retailer advice
8 Ease of use
9 Product's agronomical persistence
10 Rapidity of action
11 Specific action on dicotyledons/grass weed
12 Company's assistance

From Informark, Market survey for Dupont, personal communication.

(sandy soil, shallow aquifers) and harmless in other conditions (clay soil, deep aquifers). Information on pesticides generally is provided by the seller or the manufacturer. There is the need for a system of information (provided by the public administration) and incentives in order to orient the choice toward the "best" compound, selected as a function of sound knowledge of agronomic and environmental characteristics. An example of this kind of incentive could be a differentiated charge for pesticides that reflects environmental risk. In this case, the price could, in some way, convey information to the customer and the farmer on the social costs of pesticides. The principle of differentiated taxation of pesticides is under discussion in many countries. From the scientific point of view, the major difficulty is in producing simple and reliable instruments to classify pesticides as a function of the risk for man and the environment. An interesting example of this kind of classification for taxation purposes has been recently proposed in Switzerland.[8] A further step of interventions could be to substitute chemical treatments with the implementation of suitable agronomic practices. There is a growing public consensus to pay for more wholesome food, and this is a good incentive for increasing the presence on the market of agricultural products resulting from integrated pest management strategies. On the other hand, it should be remembered that, for large scale projects of integrated pest management, suitable incentives for the farmer and appropriate expert support are needed. Moreover, precise regulation is necessary to control the market and to avoid abuses and fraud for the purchaser. At the fourth level of intervention, the possibilities of planning and organizing land use management in agricultural areas from the administrative and political point of view should be evaluated.

At present, in many developed countries, at least in Europe, there is the tendency to reduce cultivated surfaces with the aim of controlling agricultural surplus. The so called "set aside" approach should be oriented toward the protection of more vulnerable and valuable natural resources, taking into account not only marketing aspects but also environmental protection and landscape ecology. In a long-term perspective, suitable institutional and political solutions should be studied.

Thus, there is the need for an integrated strategy, and it appears evident that there is the opportunity for a close collaboration between agronomic and environmental scientists on one side and social scientists and administrators on the other. The task is very difficult but not impossible, either from the scientific or the economic and political points of view, and the stakes involved are extremely high. A "near to zero" level of pollution for groundwater means the protection of one of the most important natural resources for human life.

We would like to conclude with a quotation from professor Finn Bro-Rasmussen, president of the EEC Scientific Advisory Committee on Toxicity and Ecotoxicity of Chemicals, taken from a statement at the EEC Drinking Water Conference in Brussels, September 1993: "We are dealing with a question which could not be approached or treated solely within the context of toxicological sciences. It is a matter — and I dare say an urgent matter — of protection of a natural resource and preservation of a cultural heritage."

REFERENCES

1. Mackay, D., S. Paterson, and W.Y. Shiu, "Generic Models for Evaluating the Regional Fate of Chemicals," *Chemosphere* 24:695-717 (1992).
2. Barra Rios, R., M. Vighi, and A. DiGuardo, "Prediction of Runoff of Chloridazon and Chlorpyrifos in an Agricultural Wateshed in Chile," *Chemosphere* 30, 485-500, (1995).
3. Haas, H. "Danish Experience in Initiating and Implementing a Policy to Reduce Herbicide Use," *Plant Protection Quarterly* 4: 38-44 (1989)
4. Zanin, G., A. Berti, and M. Sattin. "Razionalizzare il Controllo delle Malerbe," *Terra e Vita* 7: 11-13 (1993).
5. EEC Council Directive 91/414 Concerning the placing of plant protection products on the market. O.J. N. L 230/1 (1991) pp. 1-32.
6. Mason, R. and T. M. Swanson. "Regulating Chemical Waste: Addressing Manufacturer Incentives," Mimeo, Faculty of Economics and Politics, University of Cambridge, U.K., (1993).
7. Informark. Market survey for Dupont. Personal communication. (Milan, 1990).
8. Delucchi, V. and M. Bieri. "Lenkungsabgaben auf Planzenbehandlungsmitteln. Machbarkeitsstudie zur Differenzierung nach Umweltrisikopotential" Bericht zu Handen des Bundesamtes für Umwelt, Wald und Landshaft, Bern (Zurich, 1992), p. 38.

Index

Index